Animal Models in Medicine and Biology

Edited by Eva Tvrdá
and Sarat Chandra Yenisetti

Published in London, United Kingdom

IntechOpen

Supporting open minds since 2005

Animal Models in Medicine and Biology
http://dx.doi.org/10.5772/intechopen.80116
Edited by Eva Tvrdá and Sarat Chandra Yenisetti

Contributors
Enkhsaikhan Purevjav, Marcela Capcarova, Filip Benko, Mária Chomová, Oľga Uličná, Eva Tvrdá, Marie Noelle Giraud, Aurélien Frobert, Stéphane Cook, Guillaume Ajalbert, Jeremy Valentin, Milena Milojević, Sreten Mitrović, Živan Jokić, Silvester Poništ, Miloslav Zloh, Katarína Bauerová, Nicole Sommer, Annelie-Martina Weinberg, David Hahn, Mirian A.F. Hayashi, João Nani, Fabio Cruz, Benjamín Rodríguez, Georg Schmolzer, Megan O'Reilly, Po-Yin Cheung, Tze-Fun Lee, Pinky Kain, Juin-Hong Cherng, Shu-Jen Chang, Dewi Sartika, Gang-Yi Fan, Yi-Wen Wang, Saikat Samadder, Elena Golubkova, Anna Atsapkina, Anna K'ergaard, Ludmila Mamon

First published in London, United Kingdom, 2020 by IntechOpen
IntechOpen is the global imprint of INTECHOPEN LIMITED, registered in England and Wales, registration number: 11086078, 7th floor, 10 Lower Thames Street, London, EC3R 6AF, United Kingdom
Printed in Croatia

British Library Cataloguing-in-Publication Data
A catalogue record for this book is available from the British Library

Additional hard and PDF copies can be obtained from orders@intechopen.com

Animal Models in Medicine and Biology
Edited by Eva Tvrdá and Sarat Chandra Yenisetti
p. cm.
Print ISBN 978-1-83880-011-6
Online ISBN 978-1-83880-012-3
eBook (PDF) ISBN 978-1-78985-984-3

For EU product safety concerns:
IN TECH d.o.o., Prolaz Marije Krucifikse Kozulić 3, 51000 Rijeka, Croatia, info@intechopen.com or visit our website at intechopen.com.

Meet the editors

Eva Tvrdá obtained a MSc in Biotechnologies, followed by a PhD in Molecular Biology at the Slovak University of Agriculture in Nitra, Slovakia. She works as a senior researcher at the Department of Animal Physiology, Slovak University of Agriculture as well as the head of the Laboratory for Andrology and Molecular Toxicology at the AgroBioTech Research Center. Dr. Tvrdá has published more than 100 articles and eight book chapters on the topics of male reproduction in health and disease, oxidative balance and antioxidants. She has been collaborating with research groups from Spain, South Africa, Namibia, the United States and Poland. She is a former Erasmus, Sciex MNSch and Fulbright fellow, and member of the European Federation of Animal Science and the Association for Applied Animal Andrology.

Dr. Sarat Chandra Yenisetti is a professor and head of the Drosophila Neurobiology Laboratory (DNBL) in the Department of Zoology, Nagaland University (Central), Nagaland, India. He obtained a MSc from Bangaluru University, India, and a PhD from Kuvempu University, India. He completed post-doctoral training at the University of Regensburg, Germany, and in Neurogenetics at the National Institute of Neurological Disorders and Stroke (NINDS) of the National Institutes of Health (NIH), Bethesda, Maryland, USA. He follows *Drosophila* approaches to understand Parkinson's disease–associated neurodegeneration and to identify novel therapeutic targets that may help to reduce the burden of Parkinson's in humans. Dr. Yenisetti has visited the United States, Japan, Germany, Taiwan, South Korea, United Kingdom, Brazil, and Canada to participate in multiple academic assignments.

Contents

Preface **XI**

Chapter 1 **1**
Animal Models of Cardiomyopathies
by Enkhsaikhan Purevjav

Chapter 2 **23**
Animal Models in Orthopedic Research: The Proper Animal Model to Answer Fundamental Questions on Bone Healing Depending on Pathology and Implant Material
by Nicole Gabriele Sommer, David Hahn, Begüm Okutan, Romy Marek and Annelie-Martina Weinberg

Chapter 3 **43**
Animal Models of Burn Wound Management
by Shu-Jen Chang, Dewi Sartika, Gang-Yi Fan, Juin-Hong Cherng and Yi-Wen Wang

Chapter 4 **57**
Animal Models in Psychiatric Disorder Studies
by João Victor Nani, Benjamín Rodríguez, Fabio Cardoso Cruz and Mirian Akemi Furuie Hayashi

Chapter 5 **75**
Zucker Diabetic Fatty Rats for Research in Diabetes
by Marcela Capcarova and Anna Kalafova

Chapter 6 **93**
ZDF Rats: A Suitable Model to Study Male Reproductive Dysfunction in Diabetes Mellitus Type 2 Patients
by Filip Benko, Mária Chomová, Oľga Uličná and Eva Tvrdá

Chapter 7 **107**
Drosophila melanogaster: A Robust Tool to Study Candidate Drug against Epidemic and Pandemic Diseases
by Saikat Samadder

Chapter 8 **129**
Understanding Taste Using *Drosophila melanogaster*
by Shivam Kaushik and Pinky Kain

Chapter 9 151
Spermatogenesis in *Drosophila melanogaster*: Key Features and the Role of the NXF1 (Nuclear Export Factor) Protein
by Elena Golubkova, Anna Atsapkina, Anna K'ergaard and Ludmila Mamon

Chapter 10 167
A Porcine Model of Neonatal Hypoxia-Asphyxia to Study Resuscitation Techniques in Newborn Infants
by Megan O'Reilly, Po-Yin Cheung, Tze-Fun Lee and Georg M. Schmölzer

Chapter 11 183
High-Resolution Ultrasound Imaging System for the Evaluation of the Vascular Response to Stent or Balloon Injuries in the Rabbit Iliac Arteries
by Aurélien Frobert, Guillaume Ajalbert, Jérémy Valentin, Stéphane Cook and Marie-Noëlle Giraud

Chapter 12 195
Impact of Oxidative Stress on Inflammation in Rheumatoid and Adjuvant Arthritis: Damage to Lipids, Proteins, and Enzymatic Antioxidant Defense in Plasma and Different Tissues
by Silvester Ponist, Miloslav Zloh and Katarina Bauerova

Chapter 13 219
Effects of Morphometric Indicators on Incubation Values of Eggs and Sex of the Chicks of the Light Hen Hybrids
by Milena Milojević, Živan Jokić and Sreten Mitrović

Preface

Animals have been used in biological and medical research for thousands of years in order to advance the comprehension of living beings and to study the origin and basic principles of life. Furthermore, animal models have contributed significantly to a better understanding of general anatomy, physiology, pathology and pharmacology, as they can be utilized to create controlled situations and simulate specific physiological or pathological conditions.

Without a doubt, animal models have played essential roles in the most important advances in numerous medical and biological fields. From Louis Pasteur's demonstration of the germ theory of disease using anthrax in sheep and Thomas Hunt Morgan's work with the fruit fly *Drosophila melanogaster* identifying chromosomes as the vector of inheritance for genes, to Jonas Salk's studies on rhesus monkeys to isolate the polio virus, which led to the creation of a polio vaccine and modern medical advances such as organ transplant techniques and the heart–lung machine, animal models have significantly contributed to improvement in the quality of human life. New drugs, vaccines and treatment protocols for human and animal diseases have also been developed thanks to diverse animal models. The progress achieved thanks to the creation and use of animal models is indubitable, as the majority of Nobel Prize awardees in Physiology and Medicine used animal experiments in their breakthroughs.

This book is a collection of 13 state-of-the-art chapters written by renowned experts from all over the world. It represents a compilation of scientific and clinical data that emphasize findings in a wide variety of specialties, such as neurology, psychiatry, cardiology, and behavioral data including the eyes and ears, pediatrics, skin disorders, cardiovascular and musculoskeletal systems, reproduction, chronic diseases, epidemiology, and pain and inflammation management.

The book provides basic, up-to-date background on the topics as well as deals with clinical aspects, providing insightful information related to the use of animal models in the understanding of the pathophysiological mechanisms of action of various diseases.

A critical objective of the book is to provide real-time experimental approaches and data to scientists, clinicians, researchers and students for the identification of new therapeutic targets and biomarkers using animal models as well as investigate the pathogenesis and therapeutic strategies of human diseases.

We want to express our sincere gratitude to all of the authors who helped write the outstanding chapters in this book. Our special appreciation goes to Author Service Manager, Dajana Pemac, for her outstanding support from the beginning to the end. The editors are also thankful to the Slovak Research and Development Agency (15-0544) and the Department of Science and Technology-Science and Engineering

Research Board, India for their support in this important endeavor. We are truly honored to be able to bring this book to you.

Eva Tvrdá
Slovak University of Agriculture,
Slovakia

Sarat Chandra Yenisetti
Nagaland University,
India

Chapter 1

Animal Models of Cardiomyopathies

Enkhsaikhan Purevjav

Abstract

Cardiomyopathies are a heterogeneous group of disorders of heart muscle that ultimately result in congestive heart failure (CHF). Rapid progress in genetics as well as in molecular and cellular biology over the past three decades has greatly improved the understanding of pathogenic signaling pathways in inherited cardiomyopathies. This chapter will focus on animal models of different clinical forms of human cardiomyopathies with their summaries of triggered key molecules, and signaling pathways will be described.

Keywords: cardiomyopathy, heart failure, genetic mutation

1. From genetic abnormality to cardiomyopathy phenotype

It's widely accepted that inherited cardiomyopathies are a group of heterogeneous diseases of heart muscle resulting from genetic alterations in cardiac myocytes, the chief contractile cell type in the heart [1]. The genes encoding proteins that build muscle cytoskeleton and contractile apparatus are responsible for a cardiomyopathy phenotype with distinctive morpho-/histological cardiac remodeling [2]. Further, disruption of particular genetic and protein networks and pathways may intersect with other intracellular and intercellular pathways and disturbances in molecular signaling. Apoptosis, necrosis, autophagy, and metabolic and arrhythmogenic fluxes—which may present as the sole features or as overlapping signs of decompensated cardiac homeostasis—result in definitive forms of cardiac remodeling including fibrosis, cardiomyocyte hypertrophy, and atrophy. Typically, molecular signaling activates associated compensatory responses and cooperates with other modifiers such as genetic modifiers and environment, stress, or toxicity related that, in turn, may or may not influence the final cardiomyopathy phenotype. Alterations in cellular morphology and size, gene expression patterns, and metabolic shifts in cardiomyocytes initially compensate and maintain cardiac function in the subtle, preclinical stages of cardiomyopathy. Thus, inherited forms of cardiomyopathy, irrespective of the specific genetic or morpho-/clinical condition, may or may not present signs of a failing heart. Five types of inherited cardiomyopathies are distinguished based on clinical features: dilated cardiomyopathy (DCM), hypertrophic cardiomyopathy (HCM), restrictive cardiomyopathy (RCM), arrhythmogenic ventricular cardiomyopathies (ACM), and left ventricular noncompaction cardiomyopathy (LVNC) [3] as demonstrated in **Table 1**. DCM is characterized by left ventricular (LV) dilation and systolic dysfunction; HCM is characterized by LV hypertrophy with diastolic dysfunction; and RCM is accompanied by increased stiffness of the myocardium and dilated atria due to

Types	Cause / Inheritance /Prevalence	Clinical symptoms	Pathology	Features
DCM	Mutations in cytoskeletal genes, autosomal dominant, X-chromosome linked, 1:2500	Progressive CHF, arrhythmia, heart block, thromboembolism, sudden death	LV and atrial dilation, systolic dysfunction, and normal LV thickness, large, distorted myocytes	Most common cause of heart transplant
HCM	Mutations in sarcomeric genes, autosomal dominant, 1/500	Mitral regurgitation, dyspnea, syncope, myocardial ischemia, arrhythmia, sudden cardiac death	LV hypertrophy, fibrosis, myocardial disarray, mitochondrial abnormalities, thrombosis and obliteration of small vessels	Most common cause of sudden cardiac death in young and CHF disability
RCM	Mutations in sarcomeric and cytoskeletal genes, autosomal dominant, very rare.	Fatigue, dyspnea, ascites, JDV, peripheral edema, hepatomegally no cardiomegaly or systolic dysfucntion	Increased stiffness of myocardium, no dilation, no hypertrophy, dilated atria	The worst prognosis and lowest survival compared to other cardiomyopathies
ACM	Mutations in desmosomal genes, Autosomal dominant, 1/1000-1500	Ventricular tachyarrhthmia, syncope, or cardiac arrest, RV or LV chamber dilation	Myocyte loss, regional fatty of fibrofatty tissue replacement in RV and LV, or both ventricles	most common cause of sudden death in competitive athletes in Italy
LVNC	Autosomal dominant, X-chromosome linked recessive, 1/2000	Non-compacted "spongy" appearance of LV myocardium	Deep intertrabecular recesses (sinusoids) in ventricular muscle walls	Associated with thromboemboli, arrhythmia, CHF and sudden death

Table 1.
Clinical types of inherited cardiomyopathy and specific hallmarks of different types of cardiomyopathy.

diastolic dysfunction without significant hypertrophy [4]. Frequent and often life-threatening arrhythmias and associated sudden cardiac death and progressive heart failure are the main hallmarks of ACMs [5], while myocardial hypertrabeculation, intertrabecular recesses, and thin compact LV wall are the characteristics of LVNC [6]. Sustained maladaptive remodeling due to pathologic genetic insult results in the development of decompensated cardiomyopathy when the failing heart is unable to keep up with the hemodynamic demands at all levels, from the molecule to the whole organism. When compensatory mechanisms fail, additional neuroendocrine signaling and other pathways are activated on an organ and whole organism level, leading to CHF. Cellular and molecular level alterations of end-stage cardiomyopathy and CHF respond to irreversible cardiac remodeling with significant changes in membrane ion currents and intracellular Ca^{2+} metabolism, fibrosis, hypertrophic or atrophic remodeling, and cell death. Cardiac function is significantly depressed with depleted contractile force development and slowed relaxation [7].

2. Animal models of human cardiomyopathies

Translational comparative animal research is of considerable value in inherited cardiomyopathies, because animal models enable to explore and investigate the cellular and molecular pathology originating from the initial genetic assault but also may closely recapitulate the effects of cardiac remodeling culminating into a specific cardiomyopathy type seen in humans. Animal models carrying human gene mutations may not present clinical phenotypic signs of cardiomyopathy resembling the human disease until adulthood, supporting a temporal mechanism by which chronically altered cellular responses and cardiac remodeling lead to the clinically relevant phenotype.

2.1 Naturally occurring animal models of cardiomyopathy

Naturally occurring cardiomyopathy among small and large animals is commonly observed in canine and feline species [8, 9]. HCM is a common disease in pet cats, affecting 10–15% of the pet cat population [10], while DCM is more typical in dogs [11]. The similarity to human HCM or DCM, the rapid progression of disease, and the defined and readily determined endpoints of feline HCM or in canine DCM make them excellent natural models that are genotypically and phenotypically similar to human heart muscle disease [12]. The Maine Coon and Ragdoll cats are particularly valuable models of HCM associated with myosin binding protein C (*MyBP-C*) mutations and even higher disease incidence compared to the overall feline population [13, 14]. In canine, mutations in genes such as dystrophin (*DYST*) in German Shorthaired Pointers [15], desmin (*DES*) and α-actinin in the Doberman [16, 17], titin-cap (*TCAP*) in Irish Wolfhounds [18], and striatin in Boxers [19] were reported to be associated with DCM. In addition, many naturally occurring porcine HCM and DCM have been described offering the useful models for translational research [20–22].

2.2 Genetically engineered animal models of cardiomyopathy

Experimentally, numerous small and large animal models including fruit fly, fish, rodents, rabbit, canine, pig, and other species have been developed to discover pathogenetic mechanisms involved in cardiomyopathy in the research field [23–25]. Characterization of the mechanisms of cardiomyopathies using the study of animal models is challenging owing to the complexity of disease-causing mechanisms and modulators of pathology [25]. Moreover, animal models are successfully used for genome-wide screening, assessing of cardiac phenotypes and disease symptoms, genotype-phenotype association studies, and drug discovery and development assays. The accessibility of transgenic (TG), knockout (KO) and knock-in (KI) murine models has, however, been one of the most successful approaches for studying genetic cardiomyopathies [26]. With recent advances in CRISPR/Cas9 technology, researchers are able to achieve more effective and precise genome editing because of its simplicity, design, and efficiency over other traditional methods for genetic editing such as transgenesis and homologous recombination targeting techniques [27–29].

The lowest species that has typically been used for cardiomyopathy research is *Drosophila melanogaster* as a tool to study various developmental biological processes and mechanisms underlying congenital defects and inherited heart diseases [30, 31]. The *Drosophila* heart looks as a primitive linear tube similar to embryonic heart tube in vertebrates, and many heart development, function, and aging regulatory genes and networks such as NK-2, MEF2, GATA, Tbx, and Hand have been evolutionarily conserved. The conserved development of the heart in simple model organisms and vertebrates provides a unique ability to use many different animal models in cardiomyopathy research [32]. Important advantages of the use of animal models are the ability to manipulate gene expression and identify genes and mechanisms regulating heart development, cardiac pathology, and pathophysiology [33, 34]. Advanced systems to identify genes causing human cardiomyopathies such as UAS/GAL4 [35], techniques for accurate phenotyping of cardiac diseases such as optical coherence tomography [36], powerful electrophysiological, mechanical, and histological approaches to characterize heart development, cardiac tissue properties, and structure in the *Drosophila* heart have emerged as a pioneering model system in basic, genetic, and molecular studies of cardiac development, function, aging, and disease [37]. Numerous *Drosophila* models have been used to elucidate the pathophysiology of human HCM and DCM and other heart diseases, such as heart failure, cardiac tachycardia, atrial fibrillation, and congenital heart diseases [38–40].

The zebra fish (*Danio rerio*) model remains one of the most effective technologies for discovering and functional studying novel cardiomyopathy candidate genes, especially the ability to use morpholino knockdown techniques in fish models [26, 41, 42]. Compared with other vertebrate models, the zebra fish embryos are transparent allowing genetic engineering approaches to apply fluorescent reporter transgenes with genetic fate mapping strategies combined with high-resolution, high-throughput microscopy imaging in vivo of the heart [43, 44]. The transparency of the embryos allows to observe fluorescent proteins that are expressed in various cell types of the cardiovascular system, and these research advances have opened avenues to improve our knowledge of regulatory mechanisms of cardiomyocyte and other cardiac cells' differentiation [45, 46], regeneration [44], morphogenesis [47], drug effects and toxicity [48], and gene regulation [49]. The advancement in high-speed video imaging and automated image analysis techniques including light sheet planar illumination microscopy not only allows to precisely monitor morphologic and functional characteristics such as heart rate, arrhythmias, and ejection fraction in zebrafish but also progresses our current understanding of the different types of cardiomyopathy.

Rodent models are the most used model species for cardiomyopathy research, including genetics, pharmacology, and long-term survival considering that rodents have a short gestation time, have the ability to be genetically manipulated to generate transgenic or mutant strains, and are easy to handle and house with low maintenance costs [24, 50]. In addition, a fact that mice have short life span allows investigators to generate genetic models in a shorter time period and follow the natural history of genetic diseases at an accelerated pace, enabling to rapidly launch proof-of-principle experiments and potentially translating and exploiting the results into human studies. Significant advantages to rodents as the species of choice can limit the murine data's applicability to human cardiovascular function; there are significant differences between the mouse models and human disease presentation [25]. Rodents are phylogenetically farthest distant from humans compared to other mammals, and some pathophysiological features of cardiomyopathy phenotypes and their response to environmental stress and treatments may not be reliable for human diseases [23].

The rabbit and pig experimental models of cardiomyopathy offer significant advantages for cardiovascular research [50]. Compared with the mouse, the larger size and slower heart rate of the rabbit and pigs are advantageous for physiological analyses such as echocardiography and cardiac catheterization.

2.2.1 Hypertrophic cardiomyopathy animal models

Animal models of HCM mostly carry human mutations in sarcomeric protein-encoding genes such as a-MHC, a-tropomyosin, troponins, myosin binding protein C (MyBP-C), and other genes shown in **Table 1** [51–55]. Many models carry cardiac-specific (CS) expression or ablation of the proteins of interest. These models have demonstrated that HCM mutations enhance contractile properties with increased force generation, ATP hydrolysis, and actin-myosin sliding velocity, showing that the hypertrophy is not a compensatory response to diminished contractile function [56–58]. Models of HCM also show abnormal Ca^{2+} cycling in cardiomyocytes before overt histopathologic changes occurred in the myocardium and delayed myocardial relaxation that occurs before the onset of hypertrophy, suggesting that diastolic dysfunction is a direct consequence of HCM mutations [59, 60]. Hearts from models of HCM progressively accumulate myocardial fibrosis in the same manner as human patients, and fibrosis is considered to be a cellular substrate for cardiac arrhythmias and sudden cardiac death in humans [61–63].

2.2.2 Dilated cardiomyopathy animal models

Animal models of DCM mostly resemble human mutations in genes encoding cytoskeletal, sarcomeric, and Z-disk proteins and present with ventricular dilation and thinning of the ventricular walls correlated with loss of heart muscle mass. In addition, functional changes in non-myocytes induce fibrotic scars that

Gene	Human phenotype	Animal model	Animal phenotype	Pathogenesis
a-MHC	HCM	murine TG R403Q [51]	HCM	myocyte disarray, fibrosis, atrial dilation
Caveolin3	HCM	murine KO [52]	HCM, DCM, cardiac dysfunction	ERK1/2 activation, Src signaling
Caveolin3	HCM	murine TG P104L [53]	HCM, enhanced contractility, apoptosis	nNOS production, altered endoplasmic reticulum (ER) stress response
Caveolin3	HCM	zebrafish KO [54]	cardiac edema	myoblast fusion defects
Titin	DCM, HCM	zebrafish [55]	cardiac edema, poor contraction	blockage of sarcomere assembly
Tropomyosin	HCM	murine TG E180G [56]	HCM, fibrosis and atrial enlargement	increased myofilament sensitivity to Ca^{2+}.
Tropomyosin	HCM	murine TG D175N [57]	HCM, contractility and relaxation reduction	thin filament enhanced Ca^{2+} sensitivity
Troponin T	CM	murine TG MyHC [58]	HCM, reduced number of myocytes	multiple cellular mechanisms
Troponin T	CM	murine TG R92Q [60]	mitochondrial pathology, diastolic dysfunction	induction of ANP and bMHC
Troponin T		zebrafish KO [64]	sarcomere loss and myocyte disarray	dysregulation of thin filament protein expression
Troponin I	HCM	murine TG R145G [65]	HCM, diastolic dysfunction, death.	increased Ca^{2+} sensitivity and hypercontractility
Troponin I	HCM	rabbit TG R145G [66]	HCM and Cx43 disorganization	altered fractal pattern of the repolarization phase
Troponin I	HCM	murine KO [67]	acute HF, shortened sarcomeres	reduced Ca^{2+} sensitivity, elevated resting tension
MyBPC	HCM	murine TG [68]	sarcomere disorganization	stable truncated protein
MyBPC		murine KO [69]	HCM, reduced myofilament stiffness	abnormal sarcomere shortening velocity
MyBPC	HCM	cat TG [70]	sarcomeric disorganization	
Myopalladin	DCM HCM	murine TG Y20C [63]	HCM and heart failure	desmin, DPS, Cx43 and vinculin disruption
CARP	HCM DCM	murine TG αMHC [62]	HCM in response to pressure overload stress	reduced TGF-β, ERK1/2, MEK and Smad3
CARP	HCM DCM	murine KO [71]	No cardiac phenotype	
Talin	HCM	murine CS-KO [61]	HCM, hypercontraction to pressure overload	blunted ERK1/2, p38, Akt, and Gsk3 after stress
SGLT1		Tg CS-siRNA KD [72]	HCM, HF	
Meox1		Tg CS [73]	HCM	
ROCK		murine KO [74]	HCM	re-activation of fetal gene expression
cMyBPC	CM	murine KO [75]	HCM	dysregulation of Xirp2 and Zbtb16
β-MHC		rabbit TG R403Q [76]	HCM	reduced rates of force development and relaxation
CSRP3	HCM	murine KI C58G [77]	HCM	protein depletion via Bag3 and proteasomal overload
MYH7	HCM	pig KI R723G [78]	HCM, HF	myocyte disarray and malformed nuclei
TNNT2		murine R92Q; E163R [79]	HCM	altered myofilament Ca^{2+} sensitivity
αMHC	HCM	murine Arg403Gln [80]	HCM	altered repolarizing voltage-gated K^{+}(Kv) current
ERBB2		murine Tg [81]	HCM, diastolic dysfunction	ErbB2 signaling

Table 2.
Animal models of hypertrophic cardiomyopathy [51–58, 60–81].

stiffen the heart tissue and impede normal cardiomyocyte contractility. Novel DCM mechanisms such as impaired Z-disk assembly, sensitivity to apoptosis and abnormalities in myofibrillogenesis under metabolic stress, protein folding, inhibition of protein aggregation, and degradation of misfolded proteins have been explored (**Table 2**).

Gene	Human phenotype	Animal model	Animal phenotype	Pathogenesis
Sarcoglycan (delta)	DCM	murine KO [82]	focal necrosis, fibrosis after stress	destabilization of dystrophin glycoprotein complex (DGC), membrane permeability defect, Ca^{2+} imbalance
Sarcoglycan (delta)	DCM	murine KI S151A [83]	mild DCM	
Sarcospan	DMD	murine KO [84]	progressive DMD, extensive degeneration and regeneration.	
Laminin-a2		murine KO [85]	DCM	disruption of extracellular matrix (ECM) - cytoskeleton connection
Dystrophin	XL-DCM	murine [86]	dilated ventricles	destabilization of DGC, sarcolemma-actin connection, Ca^{2+} alteration
Dystrophin	XL-DCM	zebrafish [87]	mutants are less active	
Dystrophin	XL-DCM	canine [88]	DMD and DCM phenotype	
Tropomyosin	DCM	murine KO [89]	homozygous null mice are embryonic lethal (E8-E11.5)	
Tropomyosin	DCM	murine TG E54K [90]	DCM, impaired cardiac function	decrease in Ca^{2+} sensitivity and tension generation
Desmin	DCM	murine TG R173del179 [91]	DCM, intra-sarcoplasmic granular aggregates	blunted response to beta-agonist stimulation
Desmin	DCM	zebrafish [92]	disorganized muscles, small larvae	vulnerability during eccentric work
Desmin	DCM	murine KO [93]	DCM, mitochondrial abnormalities, necrosis	multisystem disruption of muscle architecture
MLP	DCM	murine KO [94]	DCM with hypertrophy and heart failure	altered mechano-sensation
Nebulette	DCM	murine TG [95]	DCM, mitochondrial abnormalities	stretch induced alteration of Z-disk assembly
Nexilin	DCM	zebrafish [96]	Z-disk damage, heart failure	stretch induced Z-disk destabilization
Telethonin	DCM	murine KO [97]	heart failure following biomechanical stress	modulation of nuclear p53 turnover after stress
Telethonin	DCM	zebrafish [98]	deformed muscle structure and impaired swimming ability	disruption of sarcomere-T-tubules ILK
Cypher/ ZASP	DCM	murine KO [99]	DCM, Z disk disruption, muscle weakness	a-actinin or other Z-line components disruption
Filamin C	DCM	medaka zacrofish K1680X [100]	DCM, myocardial wall rupture	Disrupted structure of cardiac and skeletal muscles
Lamin A	DCM	murine KO N195K [101]	nucleo-cytoplasmic shuttling of Mlk1	modulation of actin polymerization via Mlk1
Lamin A	DCM	murine KO [102]	Cardiomyocyte degeneration and mineralization	emerin dislocation
αMHC-cre		CS-Cre [103]	DCM	activated p38, JNK, p53, Bax
Dhcr24 x cTnT R141W		Double Tg Dhcr24 x cTnT [104]	DCM	activation of PI3K/Akt/HKII pathway
MST1 x Gal3		murine TG x KO [105]	DCM, HF	dysregulated transcriptional signaling
MGAT1		CS KO [106]	DCM	altered Ca^{2+} handling
RBM20	DCM	murine KI S637A [107]	DCM	disturbed nuclear localization of RBM20
MLP x MYBPC3	Varied CMs	Double KO [108]	DCM	increased Ca^{2+} sensitivity
FXR1		CS-KO [109]	DCM	altered levels of FCR1
GSK-3β x cTnT		KO, DKO [110]	DCM, HF	myocardial fibrosis, and cardiomyocyte apoptosis
NEXN	DCM	KO [111]	DCM, EFE	collagen and elastin deposits
BIN1		CS-KO [112]	DCM	mislocalization of the Cav1.2

Table 3.
Animal models of dilated cardiomyopathy [82–112].

2.2.3 Restrictive cardiomyopathy animal models

RCM is the least common but most lethal form of cardiomyopathy where impaired ventricular relaxation due to increased stiffness of the myocardium and pressure in the ventricles overcomes the changes in myofibrillar arrangement and cardiomyocyte gross abnormalities [113]. Animal models carrying human RCM-associated mutations have also been generated to mimic human RCM phenotype. These mutations are identified mainly in sarcomeric protein-encoding genes such as troponins, myosin and MYPN (summarized in **Table 3**).

Gene	Human phenotype	Animal model	Animal phenotype	Pathogenesis
cTnI	CMs	murine KO [67]	shortened sarcomeres and elevated resting tension	reduced myofilament Ca^{2+} sensitivity
cTnI	RCM	murine Tg R193H [114]	RCM	increased Ca^{2+} sensitivity
cTnI	RCM	murine Tg R145W [115]	diastolic dysfunction	prolonged force and intracellular Ca^{2+} transients
MYPN	RCM	murine KI Q529X [116]	disrupted intercalated discs, heart failure	desmin, DSP, connexin43 and vinculin disruption
myosin		E143K [117]	RCM	

Table 4.
Animal models of restrictive cardiomyopathy [67, 114–117].

Gene	Human phenotype	Animal model	Animal phenotype	Pathogenesis
Cx43	ARVC	murine KO, CS-KO [120, 121]	conduction abnormalities	intercellular channels abnormalities in SA node and ventricular conduction cells
Cx43	ARVC	murine TG aMHC [122]	cono-truncal abnormalities	
DSP	ACM	murine KO [123]	RV dilation, apoptosis, necrosis, fibro-fatty infiltration	Cell-cell contact disruption
DSP	ACM	murine TG [124]		
PKP2	ACM	murine KO [125, 126]	embryonic lethality	
PKP2		zebrafish [127]	disruption of heart development	
DSC2	ACM	zebrafish [128]	contractile dysfunction	loss of desmosomal plaque and midlines
DSG2	ACM	murine TG N271S [119]	biventricular dilatation, arrhythmias, death	Necrosis, cell-cell contact disruption
DSG2	ACM	murine TG Q558* [129]	ACM, fibrosis	miR-708-5p, miR-217-5p, miR-499-5p
DSG2	ACM	murine KO, CS-KO [119, 130]	embryonic lethality	
DSG2	ACM	murine TG [131]	LV dilation and arrhythmias	Wnt/b-Cat signaling
JUP	ACM	zebrafish KO [132]	bradicardia, cardiac edema	
JUP	ACM	murine KO [133]	VT and RV dilation and dysfunction	
SCN5a	ACM	murine KI delQKP 1510-1512 [134]	ACM	increased Na+ current and SR Ca^{2+} load
miR-130a		murine TG [135]	ACM	reduction in DSC2 3UTR
LKB1		murine KO [136]	AF, LV dysfunction	inflammation, fibrosis, apoptosis and necrosis
JUP	ACM	zebrafish TG 2057del2 [137]	ACM	redistribution of JUP, Cx43, and Nav1.5
TMEM43	ACM	murine KO, KI S358L [138]	No phenotype	
TMEM43	ACM	murine KI S358L [139]	ACM	NF-κB-TGFβ pathways

Table 5.
Animal models of arrhythmogenic ventricular cardiomyopathy [119–139].

Gene	Human phenotype	Animal model	Animal phenotype	Pathogenesis
alpha-dystrobrevin	DMD, LVNC	murine KO	Muscle dystrophy, cardiomyopathy	Alteration in cyclic GMP levels
NKX2-5	CHD	murine KI R52G [141]	LVNC, atrial septal anomalies	cardiomyocyte differentiation and heart development
NKX2-5	CHD	inducible Cx40-cre ERT2 [142]	hypertrabeculation, heart failure	cardiomyocyte differentiation and heart development
Fkbp1a		murine KO [143]	DCM, VSD and LVNC	immunoregulation and protein folding and trafficking
Jarid2		murine KO [144]	VSD, LVNC, Double outlet RV	dysregulated embryogenesis
Mest		murine KO [145]	thickness and less dense compact myocardium	dysregulated embryogenesis
Mib1		murine KO [146]	LVNC	dysregulated Notch signaling
BRAF		murine KI Q241R [147]	embryonic/neonatal lethality, LVNC	
CASZ1		murine KO [148]	hypoplasia of myocardium, VSD	abnormal genes expression
ANT2		murine KO [149]	embryonic lethality, LVNC	failure in cardiac developmental
Daam1		murine KO [150]	VSD, LVNC, Double outlet RV	Wnt/PCP signaling
S1PR1		murine KO [151]	LVNC, VSD	S1P signaling
NUMB / NUMBL		murine KO [152]	LVNC	ERBB2, YAP1 STAT5 signaling
RLF		murine KO [153]	LVNC	NOTCH pathway
LRP2		murine KO [154]	LVNC, aortic arch and coronary artery anomalies, VSD	
SLC39A8		murine KO [155]	LVNC, ECM accumulation	decrease MTF1 activity
DTNA		murine Tg N49S [156]	DCM, LVNC, cardiac dysfunction	
SRC-1/3		murine KO [157]	LVNC	up-regulate cyclin E2, cyclin B1 and myocardin
INO80		murine KO [158]	LVNC, defect in coronary vessels	upregulation of E2F-activated genes a
Tafazzin (TAZ)	LVNC	murine KD [159]	Neonatal death, LVNC, VSD	Fatty acid metabolism

Table 6.
Animal models of left ventricular noncompaction cardiomyopathy [141–159].

2.2.4 Arrhythmogenic ventricular cardiomyopathy models

Many models of ARVC with mutations in genes encoding desmosomal (DSP, PKP, DSC, DSG, and JUP) and non-desmosomal (RYR2, TMEM43, and ZASP) proteins have been developed [118]. Structural and functional alterations include progressive, diffuse, or segmental loss of cardiomyocytes, probably due to cardiomyocyte apoptosis or necrosis, and replacement with fibrotic and adipose tissue (**Table 4**). Fibro-fatty tissue primarily is seen in the right ventricle (RV), with common LV involvement in later stages of the disease [119] (**Table 5**).

2.2.5 Left ventricular noncompaction cardiomyopathy models

Animal models of LVNC typically demonstrate a spongiform ventricular myocardium and deep trabeculations, and many reports suggested that LV trabeculation and compaction processes are two distinct but tightly interconnected morphogenetic events resulting in the development of a functionally proficient ventricular chamber wall [140]. Animal models exhibiting LVNC phenotypes and potential pathogenetic mechanisms are summarized in **Table 6**.

3. Conclusion

Advances in molecular and genetic techniques have vastly improved the understanding of molecular mechanisms responsible for cardiomyopathies and cardiac dysfunction. The wide range of innovative technologies and techniques used in animal models in vivo has led to advances in our knowledge on the etiology, pathophysiology, and therapeutics of inherited cardiomyopathies. It is clear that mutant proteins in cardiomyocytes can perturb cardiac function whether the prime distress occurs in the contractile apparatus or neighboring cellular complexes, yet persistent cellular stress leads to tissue-, organ-, and organism-level pathology and pathophysiology. However, development and investigation of animal models are complex processes and the outcomes of which could be difficult to translate to humans due to differences in human and animal cardiovascular anatomy and physiology as well as differing pathophysiology of human cardiomyopathies and experimentally induced diseases in animals [160]. Therefore, the choice of appropriate animal model(s) for cardiomyopathy research should utterly rely on clinical knowledge of human cardiovascular diseases, proper research questions, sufficient number of study animals, and correct and relevant interpretation of results and outcomes in animals to human population. Although animal models of human cardiomyopathies often represent incomplete or inaccurate pathological and pathophysiological features seen in humans, the use of animal models not only has improved our knowledge on the etiology and mechanisms of cardiac muscle diseases and therapeutic interventions but also has greatly promoted an advancement in cardiac tissue engineering, induced pluripotent stem cells (iPSCs) technology, in silico and in vitro techniques, and preclinical assessment of drug discovery and development [161].

Author details

Enkhsaikhan Purevjav
Department of Pediatrics, The Heart Institute, University of Tennessee Health Science Center, Memphis, TN, USA

*Address all correspondence to: epurevja@uthsc.edu

References

[1] Harvey PA, Leinwand LA. The cell biology of disease: Cellular mechanisms of cardiomyopathy. The Journal of Cell Biology. 2011;**194**(3):355-365

[2] Bowles NE, Bowles KR, Towbin JA. The "final common pathway" hypothesis and inherited cardiovascular disease. The role of cytoskeletal proteins in dilated cardiomyopathy. Herz. 2000;**25**(3):168-175

[3] Maron BJ, Towbin JA, Thiene G, Antzelevitch C, Corrado D, Arnett D, et al. Contemporary definitions and classification of the cardiomyopathies: An American heart association scientific statement from the council on clinical cardiology, heart failure and transplantation committee; quality of care and outcomes research and functional genomics and translational biology interdisciplinary working groups; and council on epidemiology and prevention. Circulation. 2006;**113**(14):1807-1816

[4] Hershberger RE, Cowan J, Morales A, Siegfried JD. Progress with genetic cardiomyopathies: Screening, counseling, and testing in dilated, hypertrophic, and arrhythmogenic right ventricular dysplasia/cardiomyopathy. Circulation. Heart Failure. 2009;**2**(3):253-261

[5] Marcus FI, McKenna WJ, Sherrill D, Basso C, Bauce B, Bluemke DA, et al. Diagnosis of arrhythmogenic right ventricular cardiomyopathy/dysplasia: Proposed modification of the task force criteria. Circulation. 2010;**121**(13):1533-1541

[6] Towbin JA. Inherited cardiomyopathies. Circulation Journal. 2014;**78**(10):2347-2356

[7] Towbin JA, Bowles NE. The failing heart. Nature. 2002;**415**(6868):227-233

[8] Martin MW, Stafford Johnson MJ, Strehlau G, King JN. Canine dilated cardiomyopathy: A retrospective study of prognostic findings in 367 clinical cases. The Journal of Small Animal Practice. 2010;**51**(8):428-436

[9] Maron BJ, Fox PR. Hypertrophic cardiomyopathy in man and cats. Journal of Veterinary Cardiology. 2015;**17**(Suppl 1):S6-S9

[10] Kittleson MD, Meurs KM, Harris SP. The genetic basis of hypertrophic cardiomyopathy in cats and humans. Journal of Veterinary Cardiology. 2015;**17**(Suppl 1):S53-S73

[11] Simpson S, Edwards J, Ferguson-Mignan TF, Cobb M, Mongan NP, Rutland CS. Genetics of human and canine dilated cardiomyopathy. International Journal of Genomics. 2015;**2015**:204823

[12] Liu SK, Roberts WC, Maron BJ. Comparison of morphologic findings in spontaneously occurring hypertrophic cardiomyopathy in humans, cats and dogs. The American Journal of Cardiology. 1993;**72**(12):944-951

[13] Freeman LM, Rush JE, Stern JA, Huggins GS, Maron MS. Feline hypertrophic cardiomyopathy: A spontaneous large animal model of human HCM. Cardiology Research. 2017;**8**(4):139-142

[14] Longeri M, Ferrari P, Knafelz P, Mezzelani A, Marabotti A, Milanesi L, et al. Myosin-binding protein C DNA variants in domestic cats (A31P, A74T, R820W) and their association with hypertrophic cardiomyopathy. Journal of Veterinary Internal Medicine. 2013;**27**(2):275-285

[15] Nghiem PP, Hoffman EP, Mittal P, Brown KJ, Schatzberg SJ, Ghimbovschi S, et al. Sparing of the dystrophin-deficient cranial sartorius muscle is associated with classical

and novel hypertrophy pathways in GRMD dogs. The American Journal of Pathology. 2013;**183**(5):1411-1424

[16] Stabej P, Imholz S, Versteeg SA, Zijlstra C, Stokhof AA, Domanjko-Petric A, et al. Characterization of the canine desmin (DES) gene and evaluation as a candidate gene for dilated cardiomyopathy in the Dobermann. Gene. 2004;**340**(2):241-249

[17] O'Sullivan ML, O'Grady MR, Pyle WG, Dawson JF. Evaluation of 10 genes encoding cardiac proteins in Doberman pinschers with dilated cardiomyopathy. American Journal of Veterinary Research. 2011;**72**(7):932-939

[18] Philipp U, Vollmar A, Distl O. Evaluation of the titin-cap gene (TCAP) as candidate for dilated cardiomyopathy in Irish wolfhounds. Animal Biotechnology. 2008;**19**(4):231-236

[19] Cattanach BM, Dukes-McEwan J, Wotton PR, Stephenson HM, Hamilton RM. A pedigree-based genetic appraisal of boxer ARVC and the role of the Striatin mutation. The Veterinary Record. 2015;**176**(19):492

[20] Shyu JJ, Cheng CH, Erlandson RA, Lin JH, Liu SK. Ultrastructure of intramural coronary arteries in pigs with hypertrophic cardiomyopathy. Cardiovascular Pathology. 2002;**11**(2):104-111

[21] Collins DE, Eaton KA, Hoenerhoff MJ. Spontaneous dilated cardiomyopathy and right-sided heart failure as a differential diagnosis for hepatosis dietetica in a production pig. Comparative Medicine. 2015;**65**(4):327-332

[22] Lin JH, Huang SY, Lee WC, Liu SK, Chu RM. Echocardiographic features of pigs with spontaneous hypertrophic cardiomyopathy. Comparative Medicine. 2002;**52**(3):238-242

[23] Camacho P, Fan H, Liu Z, He JQ. Small mammalian animal models of heart disease. American Journal of Cardiovascular Disease. 2016;**6**(3):70-80

[24] Recchia FA, Lionetti V. Animal models of dilated cardiomyopathy for translational research. Veterinary Research Communications. 2007;**31**(Suppl 1):35-41

[25] Houser SR, Margulies KB, Murphy AM, Spinale FG, Francis GS, Prabhu SD, et al. Animal models of heart failure: A scientific statement from the American heart association. Circulation Research. 2012;**111**(1):131-150

[26] Duncker DJ, Bakkers J, Brundel BJ, Robbins J, Tardiff JC, Carrier L. Animal and in silico models for the study of sarcomeric cardiomyopathies. Cardiovascular Research. 2015;**105**(4):439-448

[27] Doudna JA, Charpentier E. Genome editing. The new frontier of genome engineering with CRISPR-Cas9. Science. 2014;**346**(6213):1258096

[28] Hsu PD, Lander ES, Zhang F. Development and applications of CRISPR-Cas9 for genome engineering. Cell. 2014;**157**(6):1262-1278

[29] Suzuki K, Tsunekawa Y, Hernandez-Benitez R, Wu J, Zhu J, Kim EJ, et al. In vivo genome editing via CRISPR/Cas9 mediated homology-independent targeted integration. Nature. 2016;**540**(7631):144-149

[30] Bier E, Bodmer R. Drosophila, an emerging model for cardiac disease. Gene. 2004;**342**(1):1-11

[31] Piazza N, Wessells RJ. Drosophila models of cardiac disease. Progress in Molecular Biology and Translational Science. 2011;**100**:155-210

[32] Tao Y, Schulz RA. Heart development in Drosophila. Seminars

in Cell & Developmental Biology. 2007;**18**(1):3-15

[33] Bryantsev AL, Cripps RM. Cardiac gene regulatory networks in Drosophila. Biochimica et Biophysica Acta. 2009;**1789**(4):343-353

[34] Bodmer R, Venkatesh TV. Heart development in Drosophila and vertebrates: Conservation of molecular mechanisms. Developmental Genetics. 1998;**22**(3):181-186

[35] Wolf MJ, Amrein H, Izatt JA, Choma MA, Reedy MC, Rockman HA. Drosophilaasamodelfortheidentification of genes causing adult human heart disease. Proceedings of the National Academy of Sciences of the United States of America. 2006;**103**(5):1394-1399

[36] Choma MA, Izatt SD, Wessells RJ, Bodmer R, Izatt JA. Images in cardiovascular medicine: In vivo imaging of the adult *Drosophila melanogaster* heart with real-time optical coherence tomography. Circulation. 2006;**114**(2):e35-e36

[37] Ocorr K, Vogler G, Bodmer R Methods to assess Drosophila heart development, function and aging. Methods. 2014;**68**(1):265-272

[38] Viswanathan MC, Kaushik G, Engler AJ, Lehman W, Cammarato A. A *Drosophila melanogaster* model of diastolic dysfunction and cardiomyopathy based on impaired troponin-T function. Circulation Research. 2014;**114**(2):e6-e17

[39] Vu Manh TP, Mokrane M, Georgenthum E, Flavigny J, Carrier L, Semeriva M, et al. Expression of cardiac myosin-binding protein-C (cMyBP-C) in Drosophila as a model for the study of human cardiomyopathies. Human Molecular Genetics. 2005;**14**(1):7-17

[40] Zhu JY, Fu Y, Nettleton M, Richman A, Han Z. High throughput in vivo functional validation of candidate congenital heart disease genes in Drosophila. eLife. 2017;**6**. pii: e22617

[41] Dvornikov AV, de Tombe PP, Xu X. Phenotyping cardiomyopathy in adult zebrafish. Progress in Biophysics and Molecular Biology. 2018;**138**:116-125

[42] Shi X, Chen R, Zhang Y, Yun J, Brand-Arzamendi K, Liu X, et al. Zebrafish heart failure models: Opportunities and challenges. Amino Acids. 2018;**50**(7):787-798

[43] Holtzman NG, Schoenebeck JJ, Tsai HJ, Yelon D. Endocardium is necessary for cardiomyocyte movement during heart tube assembly. Development. 2007;**134**(12):2379-2386

[44] Zhang R, Han P, Yang H, Ouyang K, Lee D, Lin YF, et al. In vivo cardiac reprogramming contributes to zebrafish heart regeneration. Nature. 2013;**498**(7455):497-501

[45] de Pater E, Clijsters L, Marques SR, Lin YF, Garavito-Aguilar ZV, Yelon D, et al. Distinct phases of cardiomyocyte differentiation regulate growth of the zebrafish heart. Development. 2009;**136**(10):1633-1641

[46] Staudt DW, Liu J, Thorn KS, Stuurman N, Liebling M, Stainier DY. High-resolution imaging of cardiomyocyte behavior reveals two distinct steps in ventricular trabeculation. Development. 2014;**141**(3):585-593

[47] Auman HJ, Coleman H, Riley HE, Olale F, Tsai HJ, Yelon D. Functional modulation of cardiac form through regionally confined cell shape changes. PLoS Biology. 2007;**5**(3):e53

[48] Lin KY, Chang WT, Lai YC, Liau I. Toward functional screening of cardioactive and cardiotoxic drugs with zebrafish in vivo using pseudodynamic three-dimensional imaging. Analytical Chemistry. 2014;**86**(4):2213-2220

[49] Olson EN. Gene regulatory networks in the evolution and development of the heart. Science. 2006;**313**(5795):1922-1927

[50] Milani-Nejad N, Janssen PM. Small and large animal models in cardiac contraction research: Advantages and disadvantages. Pharmacology & Therapeutics. 2014;**141**(3):235-249

[51] Geisterfer-Lowrance AA, Christe M, Conner DA, Ingwall JS, Schoen FJ, Seidman CE, et al. A mouse model of familial hypertrophic cardiomyopathy. Science. 1996;**272**(5262):731-734

[52] Woodman SE, Park DS, Cohen AW, Cheung MW, Chandra M, Shirani J, et al. Caveolin-3 knock-out mice develop a progressive cardiomyopathy and show hyperactivation of the p42/44 MAPK cascade. The Journal of Biological Chemistry. 2002;**277**(41):38988-38997

[53] Kuga A, Ohsawa Y, Okada T, Kanda F, Kanagawa M, Toda T, et al. Endoplasmic reticulum stress response in P104L mutant caveolin-3 transgenic mice. Human Molecular Genetics. 2011;**20**(15):2975-2983

[54] Nixon SJ, Wegner J, Ferguson C, Mery PF, Hancock JF, Currie PD, et al. Zebrafish as a model for caveolin-associated muscle disease; caveolin-3 is required for myofibril organization and muscle cell patterning. Human Molecular Genetics. 2005;**14**(13):1727-1743

[55] Xu X, Meiler SE, Zhong TP, Mohideen M, Crossley DA, Burggren WW, et al. Cardiomyopathy in zebrafish due to mutation in an alternatively spliced exon of titin. Nature Genetics. 2002;**30**(2):205-209

[56] Prabhakar R, Boivin GP, Grupp IL, Hoit B, Arteaga G, Solaro RJ, et al. A familial hypertrophic cardiomyopathy alpha-tropomyosin mutation causes severe cardiac hypertrophy and death in mice. Journal of Molecular and Cellular Cardiology. 2001;**33**(10):1815-1828

[57] Muthuchamy M, Pieples K, Rethinasamy P, Hoit B, Grupp IL, Boivin GP, et al. Mouse model of a familial hypertrophic cardiomyopathy mutation in alpha-tropomyosin manifests cardiac dysfunction. Circulation Research. 1999;**85**(1):47-56

[58] Tardiff JC, Factor SM, Tompkins BD, Hewett TE, Palmer BM, Moore RL, et al. A truncated cardiac troponin T molecule in transgenic mice suggests multiple cellular mechanisms for familial hypertrophic cardiomyopathy. The Journal of Clinical Investigation. 1998;**101**(12):2800-2811

[59] Wang L, Seidman JG, Seidman CE. Narrative review: Harnessing molecular genetics for the diagnosis and management of hypertrophic cardiomyopathy. Annals of Internal Medicine. 2010;**152**(8):513-520

[60] Tardiff JC, Hewett TE, Palmer BM, Olsson C, Factor SM, Moore RL, et al. Cardiac troponin T mutations result in allele-specific phenotypes in a mouse model for hypertrophic cardiomyopathy. The Journal of Clinical Investigation. 1999;**104**(4):469-481

[61] Manso AM, Li R, Monkley SJ, Cruz NM, Ong S, Lao DH, et al. Talin1 has unique expression versus Talin 2 in the heart and modifies the hypertrophic response to pressure overload. The Journal of Biological Chemistry. 2013;**288**(6):4252-4264

[62] Song Y, Xu J, Li Y, Jia C, Ma X, Zhang L, et al. Cardiac ankyrin repeat protein attenuates cardiac hypertrophy by inhibition of ERK1/2 and TGF-beta signaling pathways. PLoS ONE. 2012;7(12):e50436

[63] Purevjav E, Arimura T, Augustin S, Huby AC, Takagi K, Nunoda S, et al. Molecular basis for clinical heterogeneity

in inherited cardiomyopathies due to myopalladin mutations. Human Molecular Genetics. 2012;**21**(9):2039-2053

[64] Sehnert AJ, Huq A, Weinstein BM, Walker C, Fishman M, Stainier DY. Cardiac troponin T is essential in sarcomere assembly and cardiac contractility. Nature Genetics. 2002;**31**(1):106-110

[65] James J, Zhang Y, Osinska H, Sanbe A, Klevitsky R, Hewett TE, et al. Transgenic modeling of a cardiac troponin I mutation linked to familial hypertrophic cardiomyopathy. Circulation Research. 2000;**87**(9):805-811

[66] Sanbe A, James J, Tuzcu V, Nas S, Martin L, Gulick J, et al. Transgenic rabbit model for human troponin I-based hypertrophic cardiomyopathy. Circulation. 2005;**111**(18):2330-2338

[67] Huang X, Pi Y, Lee KJ, Henkel AS, Gregg RG, Powers PA, et al. Cardiac troponin I gene knockout: A mouse model of myocardial troponin I deficiency. Circulation Research. 1999;**84**(1):1-8

[68] Yang Q, Sanbe A, Osinska H, Hewett TE, Klevitsky R, Robbins J. A mouse model of myosin binding protein C human familial hypertrophic cardiomyopathy. The Journal of Clinical Investigation. 1998;**102**(7):1292-1300

[69] Palmer BM, McConnell BK, Li GH, Seidman CE, Seidman JG, Irving TC, et al. Reduced cross-bridge dependent stiffness of skinned myocardium from mice lacking cardiac myosin binding protein-C. Molecular and Cellular Biochemistry. 2004;**263**(1-2):73-80

[70] Meurs KM, Sanchez X, David RM, Bowles NE, Towbin JA, Reiser PJ, et al. A cardiac myosin binding protein C mutation in the Maine Coon cat with familial hypertrophic cardiomyopathy. Human Molecular Genetics. 2005;**14**(23):3587-3593

[71] Bang ML, Gu Y, Dalton ND, Peterson KL, Chien KR, Chen J. The muscle ankyrin repeat proteins CARP, Ankrd2, and DARP are not essential for normal cardiac development and function at basal conditions and in response to pressure overload. PLoS ONE. 2014;**9**(4):e93638

[72] Ramratnam M, Sharma RK, D'Auria S, Lee SJ, Wang D, Huang XY, et al. Transgenic knockdown of cardiac sodium/glucose cotransporter 1 (SGLT1) attenuates PRKAG2 cardiomyopathy, whereas transgenic overexpression of cardiac SGLT1 causes pathologic hypertrophy and dysfunction in mice. Journal of the American Heart Association. 2014;**3**(4). pii:e000899

[73] Lu D, Wang J, Li J, Guan F, Zhang X, Dong W, et al. Meox1 accelerates myocardial hypertrophic decompensation through Gata4. Cardiovascular Research. 2018;**114**(2):300-311

[74] Bailey KE, MacGowan GA, Tual-Chalot S, Phillips L, Mohun TJ, Henderson DJ, et al. Disruption of embryonic ROCK signaling reproduces the sarcomeric phenotype of hypertrophic cardiomyopathy. JCI Insight. 2019;**5**. pii:125172

[75] Farrell E, Armstrong AE, Grimes AC, Naya FJ, de Lange WJ, Ralphe JC. Transcriptome analysis of cardiac hypertrophic growth in MYBPC3-Null mice suggests early responders in hypertrophic remodeling. Frontiers in Physiology. 2018;**9**:1442

[76] Lowey S, Bretton V, Joel PB, Trybus KM, Gulick J, Robbins J, et al. Hypertrophic cardiomyopathy R403Q mutation in rabbit ¦Â-myosin reduces contractile function at the molecular and myofibrillar levels. Proceedings

of the National Academy of Sciences of the United States of America. 2018;**115**(44):11238-11243

[77] Ehsan M, Kelly M, Hooper C, Yavari A, Beglov J, Bellahcene M, et al. Mutant muscle LIM protein C58G causes cardiomyopathy through protein depletion. Journal of Molecular and Cellular Cardiology. 2018;**121**:287-296

[78] Montag J, Petersen B, Flögel AK, Becker E, Lucas-Hahn A, Cost GJ, et al. Successful knock-in of hypertrophic cardiomyopathy-mutation R723G into the MYH7 gene mimics HCM pathology in pigs. Scientific Reports. 2018;**8**(1):4786

[79] Ferrantini C, Coppini R, Pioner JM, Gentile F, Tosi B, Mazzoni L, et al. Pathogenesis of hypertrophic cardiomyopathy is mutation rather than disease specific: A comparison of the cardiac troponin T E163R and R92Q mouse models. Journal of the American Heart Association. 2017;**6**(7). pii:e005407

[80] Hueneke R, Adenwala A, Mellor RL, Seidman JG, Seidman CE, Nerbonne JM. Early remodeling of repolarizing K+ currents in the ¦ÁMHC403/+ mouse model of familial hypertrophic cardiomyopathy. Journal of Molecular and Cellular Cardiology. 2017;**103**:93-101

[81] Sr LL, Bedja D, Sysa-Shah P, Liu H, Maxwell A, Yi X, et al. Echocardiographic characterization of a murine model of hypertrophic obstructive cardiomyopathy induced by cardiac-specific overexpression of epidermal growth ractor receptor 2. Comparative Medicine. 2016;**66**(4):268-277

[82] Cordier L, Hack AA, Scott MO, Barton-Davis ER, Gao G, Wilson JM, et al. Rescue of skeletal muscles of gamma-sarcoglycan-deficient mice with adeno-associated virus-mediated gene transfer. Molecular Therapy. 2000;**1**(2):119-129

[83] Rutschow D, Bauer R, Gohringer C, Bekeredjian R, Schinkel S, Straub V, et al. S151A delta-sarcoglycan mutation causes a mild phenotype of cardiomyopathy in mice. European Journal of Human Genetics. 2014;**22**(1):119-125

[84] Araishi K, Sasaoka T, Imamura M, Noguchi S, Hama H, Wakabayashi E, et al. Loss of the sarcoglycan complex and sarcospan leads to muscular dystrophy in beta-sarcoglycan-deficient mice. Human Molecular Genetics. 1999;**8**(9):1589-1598

[85] Miyagoe-Suzuki Y, Nakagawa M, Takeda S. Merosin and congenital muscular dystrophy. Microscopy Research and Technique. 2000;**48**(3-4):181-191

[86] Sicinski P, Geng Y, Ryder-Cook AS, Barnard EA, Darlison MG, Barnard PJ. The molecular basis of muscular dystrophy in the mdx mouse: A point mutation. Science. 1989;**244**(4912):1578-1580

[87] Guyon JR, Mosley AN, Zhou Y, O'Brien KF, Sheng X, Chiang K, et al. The dystrophin associated protein complex in zebrafish. Human Molecular Genetics. 2003;**12**(6): 601-615

[88] Jones BR, Brennan S, Mooney CT, Callanan JJ, McAllister H, Guo LT, et al. Muscular dystrophy with truncated dystrophin in a family of Japanese spitz dogs. Journal of the Neurological Sciences. 2004;**217**(2):143-149

[89] Rethinasamy P, Muthuchamy M, Hewett T, Boivin G, Wolska BM, Evans C, et al. Molecular and physiological effects of alpha-tropomyosin ablation in the mouse. Circulation Research. 1998;**82**(1):116-123

[90] Rajan S, Ahmed RP, Jagatheesan G, Petrashevskaya N, Boivin GP, Urboniene

D, et al. Dilated cardiomyopathy mutant tropomyosin mice develop cardiac dysfunction with significantly decreased fractional shortening and myofilament calcium sensitivity. Circulation Research. 2007;**101**(2):205-214

[91] Wang X, Osinska H, Dorn GW 2nd, Nieman M, Lorenz JN, Gerdes AM, et al. Mouse model of desmin-related cardiomyopathy. Circulation. 2001;**103**(19):2402-2407

[92] Li M, Andersson-Lendahl M, Sejersen T, Arner A. Knockdown of desmin in zebrafish larvae affects interfilament spacing and mechanical properties of skeletal muscle. The Journal of General Physiology. 2013;**141**(3):335-345

[93] Milner DJ, Weitzer G, Tran D, Bradley A, Capetanaki Y. Disruption of muscle architecture and myocardial degeneration in mice lacking desmin. The Journal of Cell Biology. 1996;**134**(5):1255-1270

[94] Arber S, Hunter JJ, Ross J Jr, Hongo M, Sansig G, Borg J, et al. MLP-deficient mice exhibit a disruption of cardiac cytoarchitectural organization, dilated cardiomyopathy, and heart failure. Cell. 1997;**88**(3):393-403

[95] Purevjav E, Varela J, Morgado M, Kearney DL, Li H, Taylor MD, et al. Nebulette mutations are associated with dilated cardiomyopathy and endocardial fibroelastosis. Journal of the American College of Cardiology. 2010;**56**(18):1493-1502

[96] Hassel D, Dahme T, Erdmann J, Meder B, Huge A, Stoll M, et al. Nexilin mutations destabilize cardiac Z-disks and lead to dilated cardiomyopathy. Nature Medicine. 2009;**15**(11):1281-1288

[97] Knoll R, Linke WA, Zou P, Miocic S, Kostin S, Buyandelger B, et al. Telethonin deficiency is associated with maladaptation to biomechanical stress in the mammalian heart. Circulation Research. 2011;**109**(7):758-769

[98] Zhang R, Yang J, Zhu J, Xu X. Depletion of zebrafish Tcap leads to muscular dystrophy via disrupting sarcomere-membrane interaction, not sarcomere assembly. Human Molecular Genetics. 2009;**18**(21):4130-4140

[99] Zhou Q, Chu PH, Huang C, Cheng CF, Martone ME, Knoll G, et al. Ablation of Cypher, a PDZ-LIM domain Z-line protein, causes a severe form of congenital myopathy. The Journal of Cell Biology. 2001;**155**(4):605-612

[100] Fujita M, Mitsuhashi H, Isogai S, Nakata T, Kawakami A, Nonaka I, et al. Filamin C plays an essential role in the maintenance of the structural integrity of cardiac and skeletal muscles, revealed by the medaka mutant zacro. Developmental Biology. 2012;**361**(1):79-89

[101] Ho CY, Jaalouk DE, Vartiainen MK, Lammerding J, Lamin A. C and emerin regulate MKL1-SRF activity by modulating actin dynamics. Nature. 2013;**497**(7450):507-511

[102] Sullivan T, Escalante-Alcalde D, Bhatt H, Anver M, Bhat N, Nagashima K, et al. Loss of A-type lamin expression compromises nuclear envelope integrity leading to muscular dystrophy. The Journal of Cell Biology. 1999;**147**(5):913-920

[103] Rehmani T, Salih M, Tuana BS. Cardiac-specific cre induces age-dependent dilated cardiomyopathy (DCM) in mice. Molecules: A Journal of Synthetic Chemistry and Natural Product Chemistry. 2019;**24**(6):1189

[104] Dong W, Guan FF, Zhang X, Gao S, Liu N, Chen W, et al. Dhcr24 activates the PI3K/Akt/HKII pathway and protects against dilated cardiomyopathy in mice. Animal Models and Experimental Medicine. 2018;**1**(1):40-52

[105] Nguyen MN, Ziemann M, Kiriazis H, Su Y, Thomas Z, Lu Q, et al. Galectin-3 deficiency ameliorates fibrosis and remodeling in dilated cardiomyopathy mice with enhanced Mst1 signaling. American Journal of Physiology Heart and Circulatory Physiology. 2019;**316**(1):H45-H60

[106] Ednie AR, Deng W, Yip KP, Bennett ES. Reduced myocyte complex N-glycosylation causes dilated cardiomyopathy. FASEB Journal: Official Publication of the Federation of American Societies for Experimental Biology. 2019;**33**(1):1248-1261

[107] Murayama R, Kimura-Asami M, Togo-Ohno M, Yamasaki-Kato Y, Naruse TK, Yamamoto T, et al. Phosphorylation of the RSRSP stretch is critical for splicing regulation by RNA-binding motif protein 20 (RBM20) through nuclear localization. Scientific Reports. 2018;**8**(1):8970

[108] Li J, Gresham KS, Mamidi R, Doh CY, Wan X, Deschenes I, et al. Sarcomere-based genetic enhancement of systolic cardiac function in a murine model of dilated cardiomyopathy. International Journal of Cardiology. 2018;**273**:168-176

[109] Chu M, Novak SM, Cover C, Wang AA, Chinyere IR, Juneman EB, et al. Increased cardiac arrhythmogenesis associated with gap junction remodeling with upregulation of RNA-binding protein FXR1. Circulation. 2018;**137**(6):605-618

[110] Mohamed RM, Morimoto S, Ibrahim IA, Zhan DY, Du CK, Arioka M, et al. GSK-3¦Â heterozygous knockout is cardioprotective in a knockin mouse model of familial dilated cardiomyopathy. American Journal of Physiology Heart and Circulatory Physiology. 2016;**310**(11):H1808-H1815

[111] Aherrahrou Z, Schlossarek S, Stoelting S, Klinger M, Geertz B, Weinberger F, et al. Knock-out of nexilin in mice leads to dilated cardiomyopathy and endomyocardial fibroelastosis. Basic Research in Cardiology. 2016;**111**(1):6

[112] Laury-Kleintop LD, Mulgrew JR, Heletz I, Nedelcoviciu RA, Chang MY, Harris DM, et al. Cardiac-specific disruption of Bin1 in mice enables a model of stress- and age-associated dilated cardiomyopathy. Journal of Cellular Biochemistry. 2015;**116**(11):2541-2551

[113] Huang XP, Du JF. Troponin I, cardiac diastolic dysfunction and restrictive cardiomyopathy. Acta Pharmacologica Sinica. 2004;**25**(12):1569-1575

[114] Davis J, Wen H, Edwards T, Metzger JM. Thin filament disinhibition by restrictive cardiomyopathy mutant R193H troponin I induces Ca^{2+}-independent mechanical tone and acute myocyte remodeling. Circulation Research. 2007;**100**(10):1494-1502

[115] Wen Y, Xu Y, Wang Y, Pinto JR, Potter JD, Kerrick WG. Functional effects of a restrictive-cardiomyopathy-linked cardiac troponin I mutation (R145W) in transgenic mice. Journal of Molecular Biology. 2009;**392**(5):1158-1167

[116] Huby AC, Mendsaikhan U, Takagi K, Martherus R, Wansapura J, Gong N, et al. Disturbance in Z-disk mechanosensitive proteins induced by a persistent mutant myopalladin causes familial restrictive cardiomyopathy. Journal of the American College of Cardiology. 2014;**64**(25):2765-2776

[117] Yuan CC, Kazmierczak K, Liang J, Kanashiro-Takeuchi R, Irving TC, Gomes AV, et al. Hypercontractile mutant of ventricular myosin essential light chain leads to disruption of sarcomeric structure and function and results in restrictive cardiomyopathy

in mice. Cardiovascular Research. 2017;**113**(10):1124-1136

[118] Padron-Barthe L, Dominguez F, Garcia-Pavia P, Lara-Pezzi E. Animal models of arrhythmogenic right ventricular cardiomyopathy: What have we learned and where do we go? Insight for therapeutics. Basic Research in Cardiology. 2017;**112**(5):50

[119] Pilichou K, Remme CA, Basso C, Campian ME, Rizzo S, Barnett P, et al. Myocyte necrosis underlies progressive myocardial dystrophy in mouse dsg2-related arrhythmogenic right ventricular cardiomyopathy. The Journal of Experimental Medicine. 2009;**206**(8):1787-1802

[120] Thomas SA, Schuessler RB, Berul CI, Beardslee MA, Beyer EC, Mendelsohn ME, et al. Disparate effects of deficient expression of connexin43 on atrial and ventricular conduction: Evidence for chamber-specific molecular determinants of conduction. Circulation. 1998;**97**(7):686-691

[121] Lyon RC, Mezzano V, Wright AT, Pfeiffer E, Chuang J, Banares K, et al. Connexin defects underlie arrhythmogenic right ventricular cardiomyopathy in a novel mouse model. Human Molecular Genetics. 2014;**23**(5):1134-1150

[122] Ewart JL, Cohen MF, Meyer RA, Huang GY, Wessels A, Gourdie RG, et al. Heart and neural tube defects in transgenic mice overexpressing the Cx43 gap junction gene. Development. 1997;**124**(7):1281-1292

[123] Garcia-Gras E, Lombardi R, Giocondo MJ, Willerson JT, Schneider MD, Khoury DS, et al. Suppression of canonical Wnt/beta-catenin signaling by nuclear plakoglobin recapitulates phenotype of arrhythmogenic right ventricular cardiomyopathy. The Journal of Clinical Investigation. 2006;**116**(7):2012-2021

[124] Yang Z, Bowles NE, Scherer SE, Taylor MD, Kearney DL, Ge S, et al. Desmosomal dysfunction due to mutations in desmoplakin causes arrhythmogenic right ventricular dysplasia/cardiomyopathy. Circulation Research. 2006;**99**(6):646-655

[125] Rietscher K, Wolf A, Hause G, Rother A, Keil R, Magin TM, et al. Growth retardation, loss of desmosomal adhesion, and impaired tight junction function identify a unique role of plakophilin 1 in vivo. The Journal of Investigative Dermatology. 2016;**136**(7):1471-1478

[126] Dubash AD, Kam CY, Aguado BA, Patel DM, Delmar M, Shea LD, et al. Plakophilin-2 loss promotes TGF-beta1/p38 MAPK-dependent fibrotic gene expression in cardiomyocytes. The Journal of Cell Biology. 2016;**212**(4):425-438

[127] Moriarty MA, Ryan R, Lalor P, Dockery P, Byrnes L, Grealy M. Loss of plakophilin 2 disrupts heart development in zebrafish. The International Journal of Developmental Biology. 2012;**56**(9):711-718

[128] Heuser A, Plovie ER, Ellinor PT, Grossmann KS, Shin JT, Wichter T, et al. Mutant desmocollin-2 causes arrhythmogenic right ventricular cardiomyopathy. American Journal of Human Genetics. 2006;**79**(6):1081-1088

[129] Calore M, Lorenzon A, Vitiello L, Poloni G, Khan MAF, Beffagna G, et al. A novel murine model for arrhythmogenic cardiomyopathy points to a pathogenic role of Wnt signalling and miRNA dysregulation. Cardiovascular Research. 2019;**115**(4):739-751

[130] Kant S, Holthofer B, Magin TM, Krusche CA, Leube RE. Desmoglein 2-dependent arrhythmogenic cardiomyopathy is caused by a loss of adhesive function. Circulation.

Cardiovascular Genetics. 2015;**8**(4):553-563

[131] Krusche CA, Holthofer B, Hofe V, van de Sandt AM, Eshkind L, Bockamp E, et al. Desmoglein 2 mutant mice develop cardiac fibrosis and dilation. Basic Research in Cardiology. 2011;**106**(4):617-633

[132] Martin ED, Moriarty MA, Byrnes L, Grealy M. Plakoglobin has both structural and signalling roles in zebrafish development. Developmental Biology. 2009;**327**(1):83-96

[133] Kirchhof P, Fabritz L, Zwiener M, Witt H, Schafers M, Zellerhoff S, et al. Age- and training-dependent development of arrhythmogenic right ventricular cardiomyopathy in heterozygous plakoglobin-deficient mice. Circulation. 2006;**114**(17):1799-1806

[134] Montnach J, Chizelle FF, Belbachir N, Castro C, Li L, Loussouarn G, et al. Arrhythmias precede cardiomyopathy and remodeling of Ca^{2+} handling proteins in a novel model of long QT syndrome. Journal of Molecular and Cellular Cardiology. 2018;**123**:13-25

[135] Mazurek SR, Calway T, Harmon C, Farrell P, Kim GH. MicroRNA-130a regulation of desmocollin 2 in a novel model of arrhythmogenic cardiomyopathy. MicroRNA. 2017;**6**(2):143-150

[136] Ozcan C, Battaglia E, Young R, Suzuki G. LKB1 knockout mouse develops spontaneous atrial fibrillation and provides mechanistic insights into human disease process. Journal of the American Heart Association. 2015;**4**(3):e001733

[137] Asimaki A, Kapoor S, Plovie E, Karin Arndt A, Adams E, Liu Z, et al. Identification of a new modulator of the intercalated disc in a zebrafish model of arrhythmogenic cardiomyopathy. Science Translational Medicine. 2014;**6**(240):240ra274

[138] Stroud MJ, Fang X, Zhang J, Guimaraes-Camboa N, Veevers J, Dalton ND, et al. Luma is not essential for murine cardiac development and function. Cardiovascular Research. 2018;**114**(3):378-388

[139] Zheng G, Jiang C, Li Y, Yang D, Ma Y, Zhang B, et al. TMEM43-S358L mutation enhances NF-kappaB-TGFbeta signal cascade in arrhythmogenic right ventricular dysplasia/cardiomyopathy. Protein & Cell. 9 Feb 2018;**10**(2):104-119

[140] Liu Y, Chen H, Shou W. Potential common pathogenic pathways for the left ventricular noncompaction cardiomyopathy (LVNC). Pediatric Cardiology. 2018;**39**(6):1099-1106

[141] Ashraf H, Pradhan L, Chang EI, Terada R, Ryan NJ, Briggs LE, et al. A mouse model of human congenital heart disease: High incidence of diverse cardiac anomalies and ventricular noncompaction produced by heterozygous Nkx2-5 homeodomain missense mutation. Circulation. Cardiovascular Genetics. 2014;**7**(4):423-433

[142] Choquet C, Nguyen THM, Sicard P, Buttigieg E, Tran TT, Kober F, et al. Deletion of Nkx2-5 in trabecular myocardium reveals the developmental origins of pathological heterogeneity associated with ventricular non-compaction cardiomyopathy. PLoS Genetics. 2018;**14**(7):e1007502

[143] Shou W, Aghdasi B, Armstrong DL, Guo Q, Bao S, Charng MJ, et al. Cardiac defects and altered ryanodine receptor function in mice lacking FKBP12. Nature. 1998;**391**(6666):489-492

[144] Lee Y, Song AJ, Baker R, Micales B, Conway SJ, Lyons GE. Jumonji, a nuclear

protein that is necessary for normal heart development. Circulation Research. 2000;**86**(9):932-938

[145] King T, Bland Y, Webb S, Barton S, Brown NA. Expression of Peg1 (Mest) in the developing mouse heart: Involvement in trabeculation. Developmental Dynamics. 2002;**225**(2):212-215

[146] Luxan G, Casanova JC, Martinez-Poveda B, Prados B, D'Amato G, MacGrogan D, et al. Mutations in the NOTCH pathway regulator MIB1 cause left ventricular noncompaction cardiomyopathy. Nature Medicine. 2013;**19**(2):193-201

[147] Inoue S, Moriya M, Watanabe Y, Miyagawa-Tomita S, Niihori T, Oba D, et al. New BRAF knockin mice provide a pathogenetic mechanism of developmental defects and a therapeutic approach in cardio-facio-cutaneous syndrome. Human Molecular Genetics. 2014;**23**(24):6553-6566

[148] Liu Z, Li W, Ma X, Ding N, Spallotta F, Southon E, et al. Essential role of the zinc finger transcription factor Casz1 for mammalian cardiac morphogenesis and development. The Journal of Biological Chemistry. 2014;**289**(43):29801-29816

[149] Kokoszka JE, Waymire KG, Flierl A, Sweeney KM, Angelin A, MacGregor GR, et al. Deficiency in the mouse mitochondrial adenine nucleotide translocator isoform 2 gene is associated with cardiac noncompaction. Biochimica et Biophysica Acta. 2016;**1857**(8):1203-1212

[150] Li D, Hallett MA, Zhu W, Rubart M, Liu Y, Yang Z, et al. Dishevelled-associated activator of morphogenesis 1 (Daam1) is required for heart morphogenesis. Development. 2011;**138**(2):303-315

[151] Clay H, Wilsbacher LD, Wilson SJ, Duong DN, McDonald M, Lam I, et al. Sphingosine 1-phosphate receptor-1 in cardiomyocytes is required for normal cardiac development. Developmental Biology. 2016;**418**(1):157-165

[152] Hirai M, Arita Y, McGlade CJ, Lee KF, Chen J, Evans SM. Adaptor proteins NUMB and NUMBL promote cell cycle withdrawal by targeting ERBB2 for degradation. The Journal of Clinical Investigation. 2017;**127**(2):569-582

[153] Bourke LM, Del Monte-Nieto G, Outhwaite JE, Bharti V, Pollock PM, Simmons DG, et al. Loss of rearranged L-Myc fusion (RLF) results in defects in heart development in the mouse. Differentiation; Research in Biological Diversity. 2017;**94**:8-20

[154] Baardman ME, Zwier MV, Wisse LJ, Gittenberger-de Groot AC, Kerstjens-Frederikse WS, Hofstra RM, et al. Common arterial trunk and ventricular non-compaction in Lrp2 knockout mice indicate a crucial role of LRP2 in cardiac development. Disease Models & Mechanisms. 2016;**9**(4):413-425

[155] Lin W, Li D, Cheng L, Li L, Liu F, Hand NJ, et al. Zinc transporter Slc39a8 is essential for cardiac ventricular compaction. The Journal of Clinical Investigation. 2018;**128**(2):826-833

[156] Cao Q, Shen Y, Liu X, Yu X, Yuan P, Wan R, et al. Phenotype and functional analyses in a transgenic mouse model of left ventricular noncompaction caused by a DTNA mutation. International Heart Journal. 2017;**58**(6):939-947

[157] Chen X, Qin L, Liu Z, Liao L, Martin JF, Xu J. Knockout of SRC-1 and SRC-3 in mice decreases cardiomyocyte proliferation and causes a noncompaction cardiomyopathy phenotype. International Journal of Biological Sciences. 2015;**11**(9):1056-1072

[158] Rhee S, Chung JI, King DA, D'Amato G, Paik DT, Duan A, et al. Endothelial deletion of Ino80 disrupts coronary angiogenesis and causes congenital heart disease. Nature Communications. 2018;**9**(1):368

[159] Phoon CK, Acehan D, Schlame M, Stokes DL, Edelman-Novemsky I, Yu D, et al. Tafazzin knockdown in mice leads to a developmental cardiomyopathy with early diastolic dysfunction preceding myocardial noncompaction. Journal of the American Heart Association. 2012;**1**(2). pii:jah3-e000455

[160] Hearse DJ, Sutherland FJ. Experimental models for the study of cardiovascular function and disease. Pharmacological Research. 2000;**41**(6):597-603

[161] Savoji H, Mohammadi MH, Rafatian N, Toroghi MK, Wang EY, Zhao Y, et al. Cardiovascular disease models: A game changing paradigm in drug discovery and screening. Biomaterials. 2019;**198**:3-26

Chapter 2

Animal Models in Orthopedic Research: The Proper Animal Model to Answer Fundamental Questions on Bone Healing Depending on Pathology and Implant Material

Nicole Gabriele Sommer, David Hahn, Begüm Okutan, Romy Marek and Annelie-Martina Weinberg

Abstract

Different species vary in bone metabolism, especially in modeling and remodeling of the bone. Human-related diseases with severe outcomes on bone, such as osteoporosis or osteoarthritis, are often reflected in animal models, which cannot adequately mimic the human situation. The pre-clinical investigation of implant materials in vivo complicates the search for the ideal animal model, especially when combining pathologic bone diseases and implant material. For instance, while alterations in trabecular bone architecture are investigated in female osteoporotic rats, rodents commonly lack cortical bone remodeling or secondary osteon formation. Small ruminants are commonly used to study long bone defects or orthopedic materials, due to their comparability to humans regarding body weight, bone size, and fracture healing. Nevertheless, there are important differences between human and ruminant models: plexiform cortical bone, seasonal bone loss, and stronger trabecular bone appear in sheep compared to humans. This chapter will summarize fundamental differences in bone quality between different animal models used for orthopedic and implant material research. Thus, choosing the ideal animal model to answer the proposed research question remains the key to guarantee a solid and excellent scientific study.

Keywords: animal models, remodeling, orthopedics, pathological bone, biomaterials

1. Introduction

In vivo animal models are frequently used in orthopedic and trauma research. Due to ethical issues, researchers are focused to replace animal models by establishing novel in vitro systems. However, small tissue fragments are often used in in vitro systems thereby losing tissue's architecture at some times. Although it is possible to use in vitro organ and tissue cultures that allow preservation and differentiation, the control of dynamic cell properties and simulation of cell interactions remain difficult. In orthopedic and trauma research, the major disadvantages involve the placing of physiological loading and the cellular and molecular orchestration compared to the in vivo system. Hence, animal models are essential not only to evaluate pathological bone but also to study tissue response, biocompatibility and mechanical properties, especially when it comes to implant materials [1].

While small animals (mice and rats) are most commonly used due to lower costs (e.g., purchase, breeding, and housing), easy handling and the feasibility to enlarge the animal number, large animals (sheep and dogs) show several advantages including bone size, body weight, and bone quality when compared to humans. Murine animal models are commonly used to evaluate pathophysiology and novel treatment strategies [2]. For example, mice are highly adaptable to pathological conditions by experimental manipulation. Moreover, molecular tools, antibodies, and the well-characterized mouse strains (including knock-out or transgenic models) make the use of these animals more advantageous [3, 4]. There has been a long debate on whether rodents are appropriate to study osteophysiology due to the lack of true skeletal maturity (e.g., lack of Haversian remodeling and closure of epiphyseal growth plate) [2]. Larger animals, such as sheep and dogs, show several advantages over small models, including their life span and extended phases of skeletally matured bone, but seasonal bone loss and plexiform cortical bone especially occur in sheep. Thus, there is no animal model that entirely fulfills all requirements making it necessary to follow a particular research question and to confirm results obtained in research on small animal models in large animals before entering the clinics.

In this chapter, we will mainly focus on four animal models including mouse, rat, sheep, and dog. Since there are several bone pathologies, which need in vivo research models, we will particularly focus on osteoporosis and osteoarthritis as one of the major bone pathologies that are also associated with implant and scaffold research. Moreover, we will provide an overview about implant research in orthopedics and trauma surgery with specialization on bioresorbable implant technology.

2. Bone biology

Animal models are usually chosen by genetic background considerations that might influence bone phenotype, thereby assessing bone properties including bone mineral density, hardness, biomechanics, and elasticity [1]. Bone quality includes several variables such as geometry, architecture, composition (e.g., collagen and matrix components), cortical porosity, turnover, and damage and bone mineral density. However, bone quantity is classified as mineral mass or bone mineral content. In general, there are two major processes involved in bone development and maintenance.

2.1 Modeling

Bone modeling in general describes bone formation without prior osteoclastic resorption (uncoupled bone formation). This is the case during initial bone growth due to embryogenesis, as well as due to sequences in bone fracture healing and pathological bone situations, including inflammation or bone tumors. Bone modeling results in bone microstructures which are referred to as primary and woven bone. Histologically, primary bone can be separated into three types of structurally different bone tissues: primary lamellar bone, plexiform or laminar bone, and primary osteons. Depending on the vertebrate species, the state of development, but also the site of the skeleton, fulfills different kinds of functions. Primary bone is usually build up fast and gets remodeled to secondary bone during maturation. Woven bone, on the other hand, is a repair tissue, which builds the callus during fracture healing. There is no osseous or cartilage template (anlage) needed to build up woven bone. This kind of bone tissue shows a higher degree of mineralization and more porosity then secondary bone, but exhibits less mechanical qualities, as the embedded collagen-fibers are more or less disorganized. Typically, it also gets remodeled to secondary bone during bone maturation, with a few exceptions (e.g. alveolar bone and sutures of the cranium) [5, 6].

2.2 Remodeling

Bone is permanently rebuilt throughout the body to assure bone mineral homeostasis, to regenerate microfractures, or to adapt the bone to new load. Bone-degrading osteoclasts and bone-forming osteoblasts work together in a highly concerted procedure. The balance between bone resorption and formation is crucial for physiological bone metabolism. If the balance in between resorption and formation is disturbed, this can result in diverse disease patterns. In osteoporosis, for example, more bone is resorbed than is subsequently build; conversely, in osteopetrosis, more bone is formed than was previously degraded. Bone remodeling takes place within microscopical construction sites, the Basic Multicellular Units (BMUs). A BMU includes those osteoclasts, osteoblasts, and osteocytes involved in a particular remodeling event. BMUs in average are about 1–2 mm long, with a diameter between 0.2 and 0.4 mm. Cortical bone remodeling results in a secondary osteon (Havers' System), which in the center includes a neuro-vascular channel to provide the bone with nutrients and signals. Trabecular remodeling takes place in the spongy parts of the bone and results in so called avascular hemi-osteons. Trabecular bone is provided with nutrition by blood vessels from the medullary cavity [6–8].

3. How to choose the right animal model depending on bone pathology

In translational research, in vivo animal models are an important tool and have to be chosen carefully, when studying pathophysiology of diseases, implant materials or treatment options. To investigate diseases, there are several approaches including xenograft and genetically engineered models as well as inbred strains. However, results obtained from in vivo animal studies differ in their translatability to the clinical condition [9]. Generally, there are several other factors which have to be taken into account when choosing the animal model, for example, length of the experiment, costs for food and housing, experiment type and primary outcome measures.

Small animal models, especially mice and rats, exhibit several advantages including easy handling, lower costs, and quick experimentation, due to their short life span and enhanced metabolism. While small animals serve as ideal models to examine pathophysiology and pathogenesis as well as new treatment options, large animals, such as sheep and dogs are also often used to study long-term diseases processes and treatment options. Therefore, researchers suggest to additionally confirm treatment options' efficiency in large animal models before clinical use [10].

3.1 Rodents

Rodents are well-established in vivo models preferably used in translational research of different disciplines as well as in bioactivity and feasibility studies due to their well-defined genetics, biology, and immunology. Accordingly, the reproducibility is quite high. Due to their limited life span, rodents are favorably used for age-related bone metabolic and regenerative studies [11, 12]. Bone biology strongly depends on gender and age. However, there are also several differences within animal strains and after genetic manipulations. For example, fracture healing was strongly enhanced in C57BL/6 mice compared to C3H and DBA/2 [13]. Bone modeling (growth and reshaping) of the skeleton occurs throughout rodent's life cycle, and the epiphyseal growth plate still remains open throughout adulthood. Trabecular bone content is limited in rodents, and Haversian remodeling does not occur, whereas cancellous remodeling is established in rodents [14].

3.2 Large animals

Within processes which are related to body size or metabolic characteristics, like biomechanics (e.g., fracture fixation) or bone healing efficiency, respectively, the animal model (size and anatomy) should be as close to the human situation as possible [15, 16]. Martini et al. discussed the utilization of animal models in the field of orthopedic research from 1970 to 2001. Within the first decade (1990–2001), they reported a relative increase of sheep from ~6 to 8–9%, when compared with the two decades before. However, in parallel to the increase of sheep models, the relative amount of dogs used in orthopedic research decreased for about the same percentages (due to, for example, easier handling, ethical reasons) [17].

To the best of our knowledge, no deeper literature recherché was performed since then. Nevertheless, we expect a further increase of sheep being used as an animal model for orthopedics and traumatology.

3.2.1 Sheep

The cortical fraction of mature long bones in sheep is reported to exhibit a mixture of primary and secondary bone tissue. Plexiform bone appears close to the periosteum, while Haversian tissue occurs close to the endost, with a mixture of both in the mesosteal zone. Young animals up to 3–4 years in contrast exclusively show plexiform bone throughout whole sections of femora. For sheep, significantly higher bone densities have been observed compared to human bone. For example, the trabecular bone density of sheep femora is about 1.5–2-times higher than the density of that in humans. These values, however, are strongly related to the bone site where they have been measured and might not be predictive for the trabecular bone density of other bone locations, such as vertebrae [18]. Even though there are clear differences in bone microstructure, studies reported that sheep exhibit similar bone remodeling and turnover when compared to the human situation [19]. Sheep might be also an alternative model for studying osteoporosis. However, as there are differences in endocrinology and the gastrointestinal tract, it has been suggested to investigate the influence of these parameters on seasonal factors, hormones or low bone turnover during long days [20].

3.2.2 Dog

Bone composition, density, and quality were investigated in different species including chicken, cow, pig, dog, and sheep. On basis of the weight of ash, the content of hydroxyproline, extractable protein, and IGF-1, canine bone showed the greatest comparability to human bone. When it comes to bone density, dog and pig were suggested to closely mimic human bone. However, it was concluded that the canine model seems to represent the human situation the best [21]. Kimmel et al.

stated that there are similarities in trabecular bone; however, the bone turnover might be more difficult to match between human and the canine model, as even the same bone types from different sites of the same animal show high variability in turnover [22].

In comparison with the typical secondary osteonal microstructure in human cortical bone, Wang et al. reported that canine cortical bone rather consists of a secondary osteonal core, which is flanked to both sides (periosteal and endosteal) by plexiform bone. Plexiform or laminar bone is found predominantly in large and fast growing animals, but not in humans after embryogenesis [23].

3.3 Osteoporosis

3.3.1 Clinical significance

The advancing prevalence of post-menopausal osteoporosis is associated with increasing age of the population. Osteoporosis is characterized by weakening of the bone mass and density consecutively increasing the risk of bone fractures. In 2010, 3.5 million incident fragility fractures (fractures under osteoporotic conditions) were recorded in the European Union, which also increases the economic burden associated with high healthcare costs [24]. The strong increase in age is closely associated with the increase to suffer not only from a single fracture but also from multiple fractures at an advanced age. Worldwide, 1 in 3 women and 1 in 5 men over 50 will experience osteoporotic fractures [25, 26]. A quarter of those with hip fractures never walk again or even die [27].

Since the 1940s, when Fuller Albright demonstrated that estrogen can reverse negative calcium balance in post-menopausal women, there is a remarkable advance concerning osteoporotic drugs. However, concerns have been raised when it comes to anti-resorptive drugs, such as bisphosphonates, especially about rare side effects [28]. Therefore, researchers also focus on enhancing patient's acceptance and compliance with anti-resorptive drugs and in parallel evolving novel drugs without long-term side and prolonged anabolic effects.

3.3.2 Osteoporosis-related outcome on bone

Osteoporosis is a skeletal disorder that is generally subdivided into primary and secondary osteoporosis, latter describing osteoporosis as a secondary outcome to chronic diseases such as Cushing's syndrome. In contrast, primary osteoporosis involves type 1 post-menopausal and type 2 senile osteoporosis. Post-menopausal osteoporosis is a multifactorial disease characterized by weakening of the trabecular and cortical bone structure (**Figure 1**). During osteoporosis, loss of bone results in decreased total mineralization, leading to reduced tensile bone strength and increased risk of fracture. During bone fracture healing, mechanical and biological factors are negatively affected by osteoporosis [29]. Under healthy conditions, however, cellular and molecular events are carefully orchestrated, thereby producing a template for regeneration and remodeling of the fracture site, followed by bone function restoration, resulting in successful fracture healing [30]. Under osteoporotic conditions, reduced numbers and/or reduced activity of osteogenic cells including mesenchymal stem cells and osteoblasts, while osteoclast activity increases. An imbalance of anabolic and catabolic local factors has also been linked to osteoporosis [31]. Osteoporotic bone fractures are also associated with an impaired bone cell proliferation rate, reduced mechanical stress, and inhibited reactivity to local and systemic stimuli. Impaired vascularization has been observed under osteoporotic conditions [29]. However, spontaneously elevated

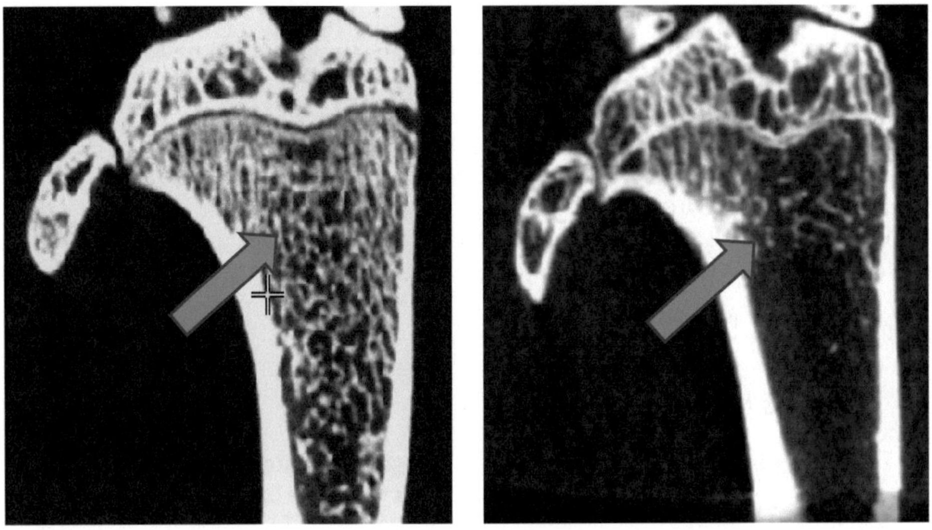

Figure 1.
Osteoporosis leads to reduced trabecular bone structure after ovariectomy in female Sprague Dawley rats. Micro-computed tomography pictures of the left proximal tibia are presented 4 weeks (left) and 8 weeks (right) after ovariectomy (unpublished data).

pro-inflammatory cytokine expression such as TNF-α, IL-6, and IL-1 and decreased bone forming factors (IGF-1 and TGF-β) are associated with osteoporosis [31].

3.3.3 Animal models for osteoporosis research

Depending on the research aspects of osteoporosis, animal models must be carefully chosen: on the one hand, animals are used to investigate anti-resorptive drugs (e.g., bisphosphonates), and on the other hand, bone fracture healing and novel treatment options (e.g., pharmaceutical, implants, etc.) for bone fractures are investigated in vivo [32, 33].

Before choosing the ideal animal model, one must consider different aspects in bone physiology. In general, there are different procedures to induce osteoporosis: on the one hand, surgical manipulation by ovariectomy, hypovasectomy, orchidectomy, and parathyroidectomy can be performed; on the other hand, diet modifications, drugs (e.g., steroids), and immobilization have been used to induce osteoporosis. Another possibility is to use aged animals or genetic modification to reflect senile osteoporosis. However, there have been several studies that demonstrated the relevance of rodent models to study post-menopausal (primary osteoporosis type 1) and senile osteoporosis (primary osteoporosis type 2). For example, the comparability of life time expectancy and closure of the epiphyseal growth plate is similar in mice and humans with about 20% in age ratio, and it markedly differs in rats with 30% as well as in sheep and dogs with 5–10% [2]. Moreover, the genetic uniformity in inbred rodents allows a smaller number of animals compared to outbred strains. Another important aspect has to be taken into account when conducting bone fracture studies: humans are mainly affected by metaphyseal fractures [34].

Mice: The average life span of laboratory mice is between 2 and 3 years, and after 8 months, BALB/c and C57BL/L mice show an age-depended decline in bone quality and mass (mice lack the Haversian remodeling), but aged animals show resorption cavities which are comparable to humans' Haversian canals [35]. The popular laboratory mouse strains, C57BL/L and BALB/c, develop senile osteoporosis-like bone phenotype with decreased bone mass and quality [36, 37]. For example, senescence-accelerated mouse (SAM) lines are reasonable models to study senile osteoporosis, because the aging phenotype is apparent even after 6–8 months [38].

Rats: In osteoporosis research, the rat model is most commonly used, especially for research on post-menopausal osteoporosis. Considering the bone physiology, the transition of modeling to remodeling occurs at 6–9 months of age in the proximal metaphysis of the tibia and at 12 months of age in the cortical bone in rat. In aged rats, Haversian canals are present, and at the age over 12 months, rats represent a good model for senile osteoporosis. However, the major issue with the rat model is that ovariectomy induces changes predominantly in the trabecular bone (**Figure 1**), and rats are preferably used to study late stages of bone fracture healing [39]. Another advantage when compared to mice is that this model is larger, which simplifies surgical procedures and investigation of mechanical properties.

Large animal models: Bone mass is only marginally reduced in dogs following ovariectomy and sheep exhibit plexiform bone arrangements in which age-related osteopenia does not occur. However, in general, sheep and mini-pig represent the most appropriate animal model for both post-menopausal and senile osteoporosis (>9 years of age). Nevertheless, extensive costs associated with housing and the variability of sheep regarding the aging process is a notable disadvantage for this large animal model.

A major disadvantage in aged large animals is that osteoporosis with low bone turnover develops only 24 months after hypothalamic-pituitary disconnection. Moreover, the typically ovariectomy-induced osteoclast recruitment has not been observed with this surgical method [40].

3.4 Osteoarthritis

3.4.1 Clinical significance

The definition of osteoarthritis (OA) depends on the way, how the disease was diagnosed including radiography, symptoms, self- or physician-diagnosed. Accordingly, the incident and prevalent numbers of OA dramatically vary and are also connected to OA with or without symptoms. OA is mainly characterized by deteriorated cartilage in joints, thereby resulting in rubbing of the bones leading to pain, stiffness, and impaired movement [10]. However, OA predominantly affects hands, feet, knees, and spine. OA is an age-depended disease, which is closely associated with several risk factors such as less physical activity, obesity, bone density, trauma, and gender [41, 42]. Especially, due to the age-related aspect of OA, it has been estimated that 15% (130 million) of people over 60 (20% of the population estimated by 2050) will exhibit OA-depending symptoms and one-third of those will be severely disabled (40 millions) [42]. Diagnostic tools for OA include magnetic resonance imaging (MRI), X-ray, and arthroscopy. However, the major problem associated with OA is non-modifiable risk factors such as age, gender, and genetics.

Hence, the disease must be properly understood to develop novel therapies to either stop or reverse the OA progression.

3.4.2 Osteoarthritis: pathogenesis and classification

This pathology leads to cartilage degradation, inflammation of joints, and abnormal bone formation [43]. Under healthy conditions, the meniscus, synovial membrane, subchondral bone, and articular cartilage support the joint: the meniscus is composed of type I collagen (also less amount of type II, III, V, and VI collagen), proteoglycans, and water and takes over several functions such as load bearing in the knee joint [44]; the synovial joints need the articular cartilage to move and latter one is composed of type II collagens and proteoglycans; the joint and the articular cartilage are nourished by synovial fluid, which is produced by the synovial membrane [45]; and the subchondral bone built up from mineralized type I cartilages serves to support the joint. The progression of OA can be stimulated by different factors; for example, mechanical abrasion tremendously degenerates type I and II collagen within the meniscus in the knee and further results in a pro-inflammatory situation with increased release of tumor-necrosis factor alpha (TNFα), IL-1, IL-4 or IL-13 and enzymes such as matrix-metalloproteinases (MMPs) might trigger the OA progress [46]. Due to MMP release, the collagen matrix is degraded, leading to articular cartilage degradation and in parallel, and the chondrocytes are not even more able to for new cartilage. Hence, abnormal remodeling of the subchondral bone, making the calcified cartilage and bone interface more acceptable to invades and leading to pain [10]. To date, novel treatment strategies are based on cytokines and the inflammatory situation, such as anti-rheumatic drugs [47]. Additionally, other treatment options such as scaffolds or lifestyle modifications might play a future role.

Similar to osteoporosis, OA was originally classified in primary and secondary OA: while primary OA was to be naturally occurring in either one (localized) or more (generalized) joints, secondary OA was associated with risk factors including diseases of bone or metabolism, trauma or others. However, there have been several debates on the classification of OA, which has been replaced based on recommendations and includes five phenotypes depending on aging, metabolism, genetics, trauma, and pain. On the basis of these phenotypes, the following ways to induce OA, including advantages and disadvantages have been proposed (**Table 1**):

3.4.3 Animal models to study osteoarthritis

In order to study the pathophysiology, pathogenesis, and therapeutic efficiency of novel treatment options for OA, there are several in vivo animal models [49]. The variability of this disease and the different outcomes for the patients make the choice of the ideal in vivo model much more difficult. While pathogenetic studies require naturally occurring OA models, molecular biological studies make use of genetic models. However, to test therapeutic strategies, surgical models are preferred (**Table 1**) [48]. Somebody has also to consider the morphology of the lesion and the pathogenesis-involved mediators, especially when testing pharmaceuticals [50].

Mice: Murine models are currently used to study primary OA, which is naturally occurring and is associated with the time consuming OA development [51]. The major disadvantage is huge husbandry costs due to the slow progression (**Table 1**), whereas the translatability to the human situation is given [48]. Genetic models, such as the prominent transgenetic model STR/ort with increased oxidative stress leading to the naturally development of OA, are particularly useful to investigate genes and their interaction with tissue components [52]. Transgenic mice are extensively used to both, induce and worsen OA progression, or to protect from the disease; to investigate molecular aspects underlying OA, inflammation and genetic contribution to OA. Surgical intervention in the knee of mice can be performed to induce OA: medial collateral ligament transection with partial medial meniscectomy [53] leads to moderate or severe medial cartilage degeneration with comparable lesion development in rats. Anterior cruciate ligament (ACL) has been described to result in severe lesions. However, the combination of a genetic

Model induction	Use	Advantage	Disadvantage	Animal model
Naturally occurring – no intervention needed	• Study pathogenesis of degenerative OA	• Variable disease like in humans	• Long time for disease development • Time consuming • High costs	Mouse, sheep, dog
Surgical intervention	• Test therapeutic efficacy of treatment options • Examine OA lesions and stages	• Rapid progression • Reproducibility • Severe lesions • Induction of traumatic OA	• Due to surgery, inappropriate for pathogenesis of degenerative OA	All
Chemical intervention	• Test therapeutic efficacy of treatment options • Examine OA lesions and stages	• Most rapid progression of OA • Less invasive • Easy implementation	• Not correlated to any type of human OA	All
Genetic intervention	• Test genetics of OA	• Genomic intervention	• Additional cartilage abnormalities or embryonic lethal deletions • High cost	Mouse

Table has been adapted from [48].

Table 1.
Methods to induce OA including the use, advantages, and disadvantages as well as most prominent animal models used according to the OA induction method.

model with a surgical intervention will be also beneficial to study detrimental factors or prophylactic effects of pharmaceuticals during different stages of OA [54]. However, chemical intervention using bacterial collagenase by intra-articular injection induces OA lesions which vary in severity [55]. Injections must be done carefully, otherwise damaging the cruciate ligaments thereby resulting in unwarranted OA lesions.

Rats: OA can only be induced surgically or chemically in rats, since there are only rare cases in which minimal focal areas of degenerative tibia can be seen [56]. However, OA in rats can be induced via medial meniscal tear or injection of iodoacetate, followed by ACL transection. After unilateral medial meniscal tear, OA-associated cartilage degeneration rapidly progresses [57] and large lesions can be observed. The major disadvantage of this model is the rapid degeneration of the cartilage thereby being difficult to observe protective effects. Importantly, toxicologic testing is the major advantage of the rat OA model, since efficacy of therapeutic interventions can be obtained easily and in a short duration and rats consistently respond to the surgery [50]. The intra-articular single-dose injection of iodoacetic acid (25–50 μl of 10 mg/ml) sufficiently kills chondrocytes by inhibition of aerobic glycolysis. The outcome on bone is remarkable and forms the basis for the development of cartilaginous lesions [58]. ACL transection in mature rats also leads to progressive changes, especially in the medial joint. In comparison to the meniscal tear model, OA progresses much slower after ACL transection and results obtained after ACL transection are comparable between rats and dogs. However, due to the slower progression, ACL transection is preferred when testing therapeutic interventions.

Disadvantages of this method are comparable to those in dogs, including variable severity of lesions and lesion locations [50].

Sheep: This large animal model is also used to study naturally occurring OA (primary OA) with similar advantages compared to mice. For example, sheep have been successfully used to study early changes of cartilage degeneration, meniscus changes, and treatment options [59].

Dogs: Dogs have been shown to be natural models to study therapeutic interventions: cranial cruciate ligament transection has been demonstrated to induce naturally occurring OA and serves as an interesting model to evaluate structural and functional benefits of treatment strategies that will give a better prediction for clinics [60]. Moreover, established canine OA models usually undergo ACL transection or partial medial meniscectomy. The major disadvantage is that dogs need large runs or plenty of exercise, otherwise resulting in mild, variable lesions. Additionally, surgical procedures must be carefully performed to avoid traumatic lesions. However, if surgical procedures are performed appropriately, OA lesions are consistent thereby allowing a relatively small number of animals per group (12–15 animals per treatment group). Another major advantage is a short screening and testing duration of 1 month [50].

Currently, there is no "gold-standardized OA model," and the most appropriate animal has to be chosen individually, depending on the research question. Moreover, extensive work is needed and advantages and disadvantages of the models must be clearly outlined in the future.

4. Orthopedic in vivo implant research

In vivo studies are essential to investigate novel implant materials and cannot be fully covered by in vitro testing. Preliminary safety tests with new implant materials using in vitro models give some information on acute toxicity and cytocompatibility. Nevertheless, some studies use the term of biocompatibility when testing implant material in vitro. However, biocompatibility tests need living organisms such as animals and humans; therefore, cytocompatibility needs to be correctly used when testing in vitro.

In order to test implant safety, adverse tissue reactions as well as corrosion and wear resistance need to be investigated to guarantee its long-term application in clinics. Hence, in vitro and in vivo tests are essential to evaluate new implant materials regarding cytocompatibility, biocompatibility, and mechanical stability.

The development of bioresorbable metal implants is one of the major goals in orthopedic and trauma surgery. Apparently, the advantages are the unnecessity to remove the implant due to material resorption and the associated avoiding of narcosis, mandatory for the second removal surgery. Since there is an increasing number of patients with metal sensitivity to permanent implants such as titanium (Ti), and long-term complications associated with currently available metal implants cannot be foreseen to date, there is a high demand to develop novel biocompatible and bioresorbable implants with good mechanical properties to stabilize bone fractures.

4.1 Implant design

To test bioresorbable orthopedic implants in animal models, the implant design and dimension is of utmost importance. Moreover, the implant number and size directly influences the number and species of animals used to test a research hypothesis. The most common implant designs used in small animal models such as mice and rats are cylindrical-shaped pins [61], whereas screws are the commonly used designs in large animal models such as sheep [62] (**Figure 2**). However, there are much more designs that are less commonly used such as plates and discs. More importantly, the size of the implant must be adjusted to the size of the animal which is comparable to common implant sizes in humans. In small animal models, cylindrical rods are used with a simple geometry that makes it easier to analyze implant degradation and behavior. However, these rods have to exactly fit ("press-fit"), otherwise the implants will become unstable and will be lost during investigation. Screws are more reliable when it comes to comparison with humans, since screws are commonly used to stabilize fractures or fix plates in humans [63]. However, the geometry is more complicated which makes the analysis more difficult.

Dimensions of implants differ according to the sizes of the animals. For example, the most appropriate dimension for cylindrical implants in rabbits is 6 mm in length and 2 mm in diameter, whereas the ideal dimension for large animals including goat, dog, and rabbit is 12 mm in length and 4 mm in diameter, according to ISO guidelines. Proper controls have to be chosen to investigate new implant material. According to ISO standards, it is recommended to use currently certified materials, which are already used in clinics, as a control [64]. In order to properly examine implant material, primary outcomes have to be specified: to test mechanical properties, bone tissue with implants are harvested and undergo pull-out/push-out tests (cylindrical implants) and torque removal tests (screws) (**Figure 2**). This test usually demonstrates proper integration of the implant in bone [1]. In case of resorbable biomaterials, degradation behavior and bone in-growth are the major primary outcomes besides mechanical properties. Real-time imaging techniques, such as in vivo micro-computed tomography (μCT) in small animals and clinical CT in large animals are used to observe material changes (degradation, bone in-growth, etc.) over the entire study duration. After reconstruction, 3-dimensional (3D) images can be reconstructed, and implant volume loss and bone formation can be calculated [61].

Other studies aim to investigate effects of implant surface modifications on bone formation and bone-implant interaction. To obtain accurate results, surface characteristics including chemical composition and surface topography must be determined. Therefore, visual observation (scanning electron microscopy, μCT) and numerical analysis (energy dispersive X-ray microscopy, profilometry) must be performed [65].

4.2 Implant material

Conventional alloys currently used in the treatment of fractures include Ti and stainless steel, which are more rigid with desirable advantages including biocompatibility, good resistance to material corrosion, and most importantly, these alloys do not show severe toxic effects on various immune cells and can bear weight soon

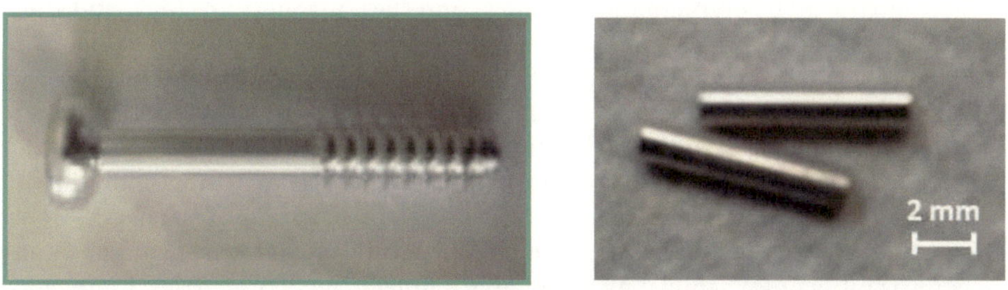

Figure 2.
Screws (left image) and cylindrical pins (right image) are often used in orthopedic and trauma research.

after implantation. However, Ti and stainless steel implants have a higher e-modulus (Ti: e-modulus 100–124 GPa; steel: e-modulus 200 GPa) than bone (e-modulus 6–24 GPa). Moreover, permanent implants can cause "stress-shielding" leading to loss of bone under the plate or between bone and implant, thereby increasing the risk of refractures, designated as peri-prosthetic or peri-implant fractures [29, 31]. Moreover, the FDA has already described possible metallic sensitivity or allergic reactions linked to Ti-based alloys, such as Ti-6Al-4V [66], and studies using vanadium have also shown adverse effects [67]. However, resorbable materials exhibit functional properties by supporting bone formation and in-growth on a molecular level. To date there are only few resorbable materials used in cardiac, dental and neuro-surgery and some not weight-bearing application in orthopedic surgery, but adequate materials for orthopedic and trauma surgeries need good mechanical properties. Therefore, the main focus in biomaterial research is to evolve materials and tools to develop the optimal implant for the respective bone condition under the necessity to bear weight.

4.2.1 Polymers

Poly-L-lactic acid (PLLA) and poly-lactic-co-glycolic acid (PLGA) are the most commonly used polymers considered for their use as osteosynthesis and bone grafts [68]. Disadvantages of polymers include poor mechanical properties (low strength and stiffness, high brittleness) and osteoconductivity. Degradation behavior depends on monomers and can be very slow thereby increasing the risk of adverse effects such as sclerotic areas in bone and fibrous encapsulation [69].

Alternatively, polyhydroxybutyrate (PHB) can serve as a polymeric implant material, which is produced by microorganisms, especially bacteria. However, PHB can induce toxicological effects. However, these effects have been reduced dramatically, raising the potential for its application in clinics. Nevertheless, functional properties (i.e., osteoinductivity or osteoconductivity) of polymeric implant material have not been discovered yet.

4.2.2 Ceramics

Ceramics are synthetic bone replacement materials with good biocompatibility and osteoconductivity, thereby showing good osseointegrative and non-immunogenic effects. Composed of hydroxyapatite (HA), or alpha (α)- and beta (β)-tricalcium phosphates (TCP), ceramics exhibit poor mechanical properties including low yield strength and high brittleness, which make them unattractive for their application in load-bearing regions.

4.2.3 Bioresorbable metals

In comparison to polymers and ceramics, iron (Fe), magnesium (Mg), and zinc (Zn) are more stable, tensile, and load-bearing, respectively. To process Fe-based alloys, the low melting point of Fe constitutes an interesting property. However, Hofstetter et al. demonstrated that limited access to oxygen was associated with slow degradation rates [70]. Metal implants based on Zn display several disadvantages including low rigidity and deformability, as well as corrosion inhibition. Therefore, Zn is likely more suited as an alloying element in combination with other materials. Finally, Mg-based alloys exhibit several advantages including good biocompatibility, resorbability, and favorable biomechanical properties. Moreover, some studies have demonstrated Mg's associated functional properties, especially its ability to support bone fracture healing [33]. For example, recent studies using

26 Mg isotope pins in a rat model demonstrated high Mg content in the bone-implant interface [71]. Good bone in-growth and a tight interface between bone and implant were observed. Additionally, drilled hole bone fractures showed full recovery after complete degradation of Mg implants. The serum concentration of Mg indicated a high tolerance of increased Mg levels which was controlled by urine excretion. Bone formation has been observed after implantation of XHP-Mg-0.45Zn-0.45Ca implants in young, growing small and large animal models [61].

5. Conclusion

Here, we summarized fundamental differences in small and large animal models concerning bone quality, composition as well as their individual advantages and disadvantages. Focusing on two major complications in orthopedics and traumatology, we wanted to underline the merits of an animal model by supporting with scientific results obtained from our intensive literature recherché. Implant research is a hot topic in orthopedics and trauma surgery. Based on our expertise, we wanted to give insights into implant technology, materials, and designs. Currently, permanent implants are the state-of-the-art material used to stabilize bone fractures in orthopedics and trauma surgery. However, to develop the ideal implant for a certain bone condition (e.g., osteoporosis and osteoarthritis), the underlying disease and the detrimental outcome on bone (e.g., bone mass, fracture risk, and bone density) have to be taken into account when choosing the implant material (e.g., Ti-, Mg-, Fe-based implants), design (e.g., pin, screw, plate, and scaffold), material properties (e.g., tensile strength, non- or bio-resorbable), and implantation site (e.g., knee, femur, and tibia).

Hence, it is of utmost importance to choose the most appropriate animal model according to the research question and warranted primary outcome measures.

Author details

Nicole Gabriele Sommer, David Hahn, Begüm Okutan, Romy Marek and Annelie-Martina Weinberg*
Department of Orthopedics and Traumatology, Medical University of Graz, Graz, Austria

*Address all correspondence to: annelie-martina.weinberg@medunigraz.at

References

[1] Pearce A, Richards R, Milz S, Schneider E, Pearce S. Animal models for implant biomaterial research in bone: A review. European Cells & Materials. 2007;**13**:1-10. DOI: 10.22203/eCM.v013a01

[2] Simpson AH, Murray IR. Osteoporotic fracture models. Current Osteoporosis Reports. 2015;**13**(1):9-15. DOI: 10.1007/s11914-014-0246-8

[3] Jacenko O, Olsen BR. Transgenic mouse models in studies of skeletal disorders. The Journal of Rheumatology. Supplement. 1995;**43**:39-41

[4] Houdebine L-M. Transgenic animal models in biomedical research. In: Sioud M, editor. Target Discovery and Validation Reviews and Protocols: Volume 1, Emerging Strategies for Targets and Biomarker Discovery. Totowa, NJ: Humana Press; 2007. pp. 163-202. [Methods in Molecular BiologyTM]. DOI: 10.1385/1-59745-165-7:163

[5] Hillier ML, Bell LS. Differentiating human bone from animal bone: A review of histological methods. Journal of Forensic Sciences. 2007;**52**(2):249-263. DOI: 10.1111/j.1556-4029.2006.00368.x

[6] Burr DB. Chapter 1 - bone morphology and organization. In: Burr DB, Allen MR, editors. Basic and Applied Bone Biology. Second ed. Amsterdam, The Netherlands: Academic Press; 2019. pp. 3-26. DOI: 10.1016/B978-0-12-813259-3.00001-4

[7] Hadjidakis DJ, Androulakis II. Bone remodeling. Annals of the New York Academy of Sciences. 2006;**1092**(1):385-396. DOI: 10.1196/annals.1365.035

[8] Eriksen EF. Cellular mechanisms of bone remodeling. Reviews in Endocrine & Metabolic Disorders. 2010;**11**(4):219-227. DOI: 10.1007/s11154-010-9153-1

[9] Wendler A, Wehling M. The translatability of animal models for clinical development: Biomarkers and disease models. Current Opinion in Pharmacology. 2010;**10**(5):601-606. DOI: 10.1016/j.coph.2010.05.009

[10] Kuyinu EL, Narayanan G, Nair LS, Laurencin CT. Animal models of osteoarthritis: Classification, update, and measurement of outcomes. Journal of Orthopaedic Surgery and Research. 2016;**11**(1):19. DOI: 10.1186/s13018-016-0346-5

[11] Santos LF, Singulani MP, Stringhetta-Garcia CT, Oliveira SHP, Chaves-Neto AH, Dornelles RCM. Oxytocin effects on osteoblastic differentiation of bone marrow mesenchymal stem cells from adult and aging female Wistar rats. Experimental Gerontology. 2018;**113**:58-63. DOI: 10.1016/j.exger.2018.09.023

[12] Hambright WS, Niedernhofer LJ, Huard J, Robbins PD. Murine models of accelerated aging and musculoskeletal disease. Bone. 2019;**125**:122-127. DOI: 10.1016/j.bone.2019.03.002

[13] Manigrasso MB, O'Connor JP. Comparison of fracture healing among different inbred mouse strains. Calcified Tissue International. 2008;**82**(6):465-474. DOI: 10.1007/s00223-008-9144-3

[14] Gomes PS, Fernandes MH. Rodent models in bone-related research: The relevance of calvarial defects in the assessment of bone regeneration strategies. Laboratory Animals. 2011;**45**(1):14-24. DOI: 10.1258/la.2010.010085

[15] den Boer FC, Patka P, Bakker FC, Wippermann BW, van LA, Vink GQM, et al. New segmental long bone defect model in sheep: Quantitative analysis

of healing with dual energy X-ray absorptiometry. Journal of Orthopaedic Research. 1999;**17**(5):654-660. DOI: 10.1002/jor.1100170506

[16] An YH, Freidman RJ. Animal Selections in Orthopaedic Research. Florida, US: CRC Press; 1998. pp. 39-57. 622 p

[17] Martini L, Fini M, Giavaresi G, Giardino R. Sheep Model in Orthopedic Research: A Literature Review [Internet]. 2001 [cited: 06 August 2019]. Available from: https://www.ingentaconnect.com/content/aalas/cm/2001/00000051/00000004/art00003

[18] Liebschner MAK. Biomechanical considerations of animal models used in tissue engineering of bone. Biomaterials. 2004;**25**(9):1697-1714. DOI: 10.1016/S0142-9612(03)00515-5

[19] Pastoureau P, Arlot M, Caulin F, Barlet J, Meunier P, Delmas P. Effects of oophorectomy on biochemical and histological indices of bone turnover in ewes. Journal of Bone and Mineral Research. 1989;**4**(1):58

[20] Aufdemorte TB, Boyan BD, Fox WC, Miller D. Diagnostic tools and biologic markers: Animal models in the study of osteoporosis and oral bone loss. Journal of Bone and Mineral Research. 1993;**8**(S2):S529-S534. DOI: 10.1002/jbmr.5650081320

[21] Aerssens J, Boonen S, Joly J, Dequeker J. Variations in trabecular bone composition with anatomical site and age: Potential implications for bone quality assessment. The Journal of Endocrinology. 1997;**155**(3):411-421. DOI: 10.1007/BF02500238

[22] Kimmel DB, Jee WSS. A quantitative histologic study of bone turnover in young adult beagles. The Anatomical Record. 1982;**203**(1):31-45. DOI: 10.1002/ar.1092030104

[23] Wang X, Mabrey JD, Agrawal CM. An interspecies comparison of bone fracture properties. Biomedical Materials and Engineering. 1998;**8**(1):1-9

[24] Svedbom A, Hernlund E, Ivergård M, Compston J, Cooper C, Stenmark J, et al. Osteoporosis in the European Union: A compendium of country-specific reports. Archives of Osteoporosis. 2013;**8**(1-2):218. Available from: https://www-1ncbi-1nlm-1nih-1gov-1pubmed.han.medunigraz.at/pmc/articles/PMC3880492/

[25] Kanis JA, Melton LJ, Christiansen C, Johnston CC, Khaltaev N. The diagnosis of osteoporosis. Journal of Bone and Mineral Research. 1994;**9**(8):1137-1141. DOI: 10.1002/jbmr.5650090802

[26] Melton LJ 3rd, Chrischilles EA, Cooper C, Lane AW, Riggs BL. How many women have osteoporosis? JBMR anniversary classic. JBMR, volume 7, number 9, 1992. Journal of Bone and Mineral Research. 2005;**20**(5):886-892. DOI: 10.1359/jbmr.2005.20.5.886

[27] Osteoporosis. International Osteoporosis Foundation [Internet]. [cited: 16 January 2018]. Available from: https://www.iofbonehealth.org/osteoporosis

[28] Khosla S, Hofbauer LC. Osteoporosis treatment: Recent developments and ongoing challenges. The Lancet Diabetes & Endocrinology. 2017;**5**(11):898-907. DOI: 10.1016/S2213-8587(17)30188-2

[29] Augat P, Simon U, Liedert A, Claes L. Mechanics and mechanobiology of fracture healing in normal and osteoporotic bone. Osteoporosis International. 2005;**16**(2):S36-S43. DOI: 10.1007/s00198-004-1728-9

[30] Mathavan N, Tägil M, Isaksson H. Do osteoporotic fractures constitute a greater recalcitrant

challenge for skeletal regeneration? Investigating the efficacy of BMP-7 and zoledronate treatment of diaphyseal fractures in an open fracture osteoporotic rat model. Osteoporosis International. 2017;**28**(2):697-707. DOI: 10.1007/s00198-016-3771-8

[31] Franchi M, Fini M, Giavaresi G, Ottani V. Peri-implant osteogenesis in health and osteoporosis. Micron. 2005;**36**(7-8):630-644. DOI: 10.1016/j.micron.2005.07.008

[32] Egermann M, Goldhahn J, Schneider E. Animal models for fracture treatment in osteoporosis. Osteoporosis International. 2005;**16**(2):S129-S138. DOI: 10.1007/s00198-005-1859-7

[33] Zhang Y, Xu J, Ruan YC, Yu MK, O'Laughlin M, Wise H, et al. Implant-derived magnesium induces local neuronal production of CGRP to improve bone-fracture healing in rats. Nature Medicine. 2016;**22**(10):1160. DOI: 10.1038/nm.4162

[34] Cheung WH, Miclau T, Chow SK, Yang FF, Alt V. Fracture healing in osteoporotic bone. Injury. 2016;**47**:S21-S26

[35] Nunamaker DM. Experimental models of fracture repair. Clinical Orthopaedics and Related Research. 1998;**355**:S56. DOI: 10.1097/00003086-199810001-00007

[36] Perkins SL, Gibbons R, Kling S, Kahn AJ. Age-related bone loss in mice is associated with an increased osteoclast progenitor pool. Bone. 1994;**15**(1):65-72. DOI: 10.1016/8756-3282(94)90893-1

[37] Ferguson VL, Ayers RA, Bateman TA, Simske SJ. Bone development and age-related bone loss in male C57BL/6J mice. Bone. 2003;**33**(3):387-398. DOI: 10.1016/S8756-3282(03)00199-6

[38] Okamoto Y, Takahashi K, Toriyama K, Takeda N, Kitagawa K, Hosokawa M, et al. Femoral peak bone mass and osteoclast number in an animal model of age-related spontaneous osteopenia. The Anatomical Record. 1995;**242**(1):21-28. DOI: 10.1002/ar.1092420104

[39] Kubo T, Shiga T, Hashimoto J, Yoshioka M, Honjo H, Urabe M, et al. Osteoporosis influences the late period of fracture healing in a rat model prepared by ovariectomy and low calcium diet. The Journal of Steroid Biochemistry and Molecular Biology. 1999;**68**(5):197-202. DOI: 10.1016/S0960-0760(99)00032-1

[40] Oheim R, Beil FT, Köhne T, Wehner T, Barvencik F, Ignatius A, et al. Sheep model for osteoporosis: Sustainability and biomechanical relevance of low turnover osteoporosis induced by hypothalamic–pituitary disconnection. Journal of Orthopaedic Research. 2013;**31**(7):1067-1074. DOI: 10.1002/jor.22327

[41] Laupattarakasem W, Laopaiboon M, Laupattarakasem P, Sumananont C. Arthroscopic debridement for knee osteoarthritis. Cochrane Database of Systematic Reviews. 2008;(1). DOI: 10.1002/14651858.CD005118.pub2/abstract

[42] WHO. Chronic Rheumatic Conditions [Internet]. WHO. [cited: 01 August 2019]. Available from: http://www.who.int/chp/topics/rheumatic/en/

[43] Kraus VB, Blanco FJ, Englund M, Karsdal MA, Lohmander LS. Call for standardized definitions of osteoarthritis and risk stratification for clinical trials and clinical use. Osteoarthritis and Cartilage. 2015;**23**(8):1233-1241. DOI: 10.1016/j.joca.2015.03.036

[44] Fox AJS, Bedi A, Rodeo SA. The basic science of human knee menisci: Structure, composition, and function.

Sports Health. 2012;**4**(4):340-351. DOI: 10.1177/1941738111429419

[45] de Sousa EB, Casado PL, Neto VM, Duarte MEL, Aguiar DP. Synovial fluid and synovial membrane mesenchymal stem cells: Latest discoveries and therapeutic perspectives. Stem Cell Research & Therapy. 2014;**5**(5):112. DOI: 10.1186/scrt501

[46] Man G, Mologhianu G. Osteoarthritis pathogenesis – A complex process that involves the entire joint. Journal of Medicine and Life. 2014;**7**(1):37-41

[47] Orlowsky EW, Kraus VB. The role of innate immunity in osteoarthritis: When our first line of defense goes on the offensive. The Journal of Rheumatology. 2015;**42**(3):363-371. DOI: 10.3899/jrheum.140382

[48] Lampropoulou-Adamidou K, Lelovas P, Karadimas EV, Liakou C, Triantafillopoulos IK, Dontas I, et al. Useful animal models for the research of osteoarthritis. European Journal of Orthopaedic Surgery and Traumatology. 2014;**24**(3):263-271. DOI: 10.1007/s00590-013-1205-2

[49] McCoy AM. Animal models of osteoarthritis: Comparisons and key considerations. Veterinary Pathology. 2015;**52**(5):803-818. DOI: 10.1177/0300985815588611

[50] Bronner F, Farach-Carson MC, editors. Bone and Osteoarthritis. London: Springer-Verlag; 2007. (Topics in Bone Biology). Available from: https://www.springer.com/de/book/9781846285134

[51] Little CB, Hunter DJ. Post-traumatic osteoarthritis: From mouse models to clinical trials. Nature Reviews Rheumatology. 2013;**9**(8):485-497. DOI: 10.1038/nrrheum.2013.72

[52] Kyostio-Moore S, Nambiar B, Hutto E, Ewing PJ, Piraino S, Berthelette P, et al. STR/ort Mice, a Model for Spontaneous Osteoarthritis, Exhibit Elevated Levels of Both Local and Systemic Inflammatory Markers [Internet]. 2011 [cited: 02 August 2019]. Available from: https://www.ingentaconnect.com/content/aalas/cm/2011/00000061/00000004/art00007

[53] Clements KM, Price JS, Chambers MG, Visco DM, Poole AR, Mason RM. Gene deletion of either interleukin-1β, interleukin-1β–converting enzyme, inducible nitric oxide synthase, or stromelysin 1 accelerates the development of knee osteoarthritis in mice after surgical transection of the medial collateral ligament and partial medial meniscectomy. Arthritis and Rheumatism. 2003;**48**(12):3452-3463. DOI: 10.1002/art.11355

[54] Kamekura S, Hoshi K, Shimoaka T, Chung U, Chikuda H, Yamada T, et al. Osteoarthritis development in novel experimental mouse models induced by knee joint instability. Osteoarthritis and Cartilage. 2005;**13**(7):632-641. DOI: 10.1016/j.joca.2005.03.004

[55] van der Kraan PM, Vitters EL, van de Putte LB, van den Berg WB. Development of osteoarthritic lesions in mice by "metabolic" and "mechanical" alterations in the knee joints. The American Journal of Pathology. 1989;**135**(6):1001-1014

[56] Smale G, Bendele A, Horton W Jr. Comparison of age-associated degeneration of articular cartilage in Wistar and fisher 344 rats. Laboratory Animal Science. 1995;**45**(2):5

[57] Janusz MJ, Bendele AM, Brown KK, Taiwo YO, Hsieh L, Heitmeyer SA. Induction of osteoarthritis in the rat by surgical tear of the meniscus: Inhibition of joint damage by a matrix metalloproteinase

inhibitor. Osteoarthritis and Cartilage. 2002;**10**(10):785-791. DOI: 10.1053/joca.2002.0823

[58] Janusz MJ, Little CB, King LE, Hookfin EB, Brown KK, Heitmeyer SA, et al. Detection of aggrecanase- and MMP-generated catabolic neoepitopes in the rat iodoacetate model of cartilage degeneration. Osteoarthritis and Cartilage. 2004;**12**(9):720-728. DOI: 10.1016/j.joca.2004.06.004

[59] Vandeweerd J-M, Hontoir F, Kirschvink N, Clegg P, Nisolle J-F, Antoine N, et al. Prevalence of naturally occurring cartilage defects in the ovine knee. Osteoarthritis and Cartilage. 2013;**21**(8):1125-1131. DOI: 10.1016/j.joca.2013.05.006

[60] Moreau M, Pelletier J-P, Lussier B, d'Anjou M-A, Blond L, Pelletier J-M, et al. A posteriori comparison of natural and surgical destabilization models of canine osteoarthritis. BioMed Research International. 2013;**2013**(1-12):12. Available from: https://www.hindawi.com/journals/bmri/2013/180453/abs/

[61] Grün NG, Holweg P, Tangl S, Eichler J, Berger L, van den Beucken JJJP, et al. Comparison of a resorbable magnesium implant in small and large growing-animal models. Acta Biomaterialia. 2018;**78**:378-386. DOI: 10.1016/j.actbio.2018.07.044

[62] Shen X, Qu F, Li C, Qi W, Lu X, Li H, et al. Comparison between a novel human cortical bone screw and bioabsorbable interference screw for graft fixation of ACL reconstruction. European Review for Medical and Pharmacological Sciences. 2018;**22**:111-118

[63] Köse A, Topal M, Engin MÇ, Şencan A, Dinçer R, Baran T. Comparison of low-profile plate-screw and Kirschner-wire osteosynthesis outcomes in extra-articular unstable proximal phalangeal fractures. European Journal of Orthopaedic Surgery and Traumatology. 2019;**29**(3):597-604. DOI: 10.1007/s00590-018-2342-4

[64] ISO 10993-6:1994 [Internet]. ISO. [cited: 05 August 2019]. Available from: http://www.iso.org/cms/render/live/en/sites/isoorg/contents/data/standard/01/89/18966.html

[65] Myrissa A, Nezha Ahmad A, Lu Y, Martinelli E, Eichler J, Szakacs G, et al. In vitro and in vivo comparison of binary Mg alloys and pure Mg. Materials Science and Engineering: C. 2016;**61**:865-874. DOI: 10.1016/j.msec.2015.12.064

[66] Khadija G, Saleem A, Akhtar Z, Naqvi Z, Gull M, Masood M, et al. Short term exposure to titanium, aluminum and vanadium (Ti_6Al_4V) alloy powder drastically affects behavior and antioxidant metabolites in vital organs of male albino mice. Toxicology Reports. 2018;**5**:765-770. DOI: 10.1016/j.toxrep.2018.06.006

[67] Mravcová A, Jírová D, Jančí H, Lener J. Effects of orally administered vanadium on the immune system and bone metabolism in experimental animals. Science of the Total Environment. 1993;**134**:663-669. DOI: 10.1016/S0048-9697(05)80069-5

[68] Bizenjima T, Takeuchi T, Seshima F, Saito A. Effect of poly (lactide-co-glycolide) (PLGA)-coated beta-tricalcium phosphate on the healing of rat calvarial bone defects: A comparative study with pure-phase beta-tricalcium phosphate. Clinical Oral Implants Research. 2016;**27**(11):1360-1367. DOI: 10.1111/clr.12744

[69] Gentile P, Chiono V, Carmagnola I, Hatton PV. An overview of poly(lactic-co-glycolic) acid (PLGA)-based biomaterials for bone tissue engineering. International Journal of Molecular Sciences. 2014;**15**(3):3640-3659. DOI: 10.3390/ijms15033640

[70] Hofstetter J, Martinelli E, Weinberg AM, Becker M, Mingler B, Uggowitzer PJ, et al. Assessing the degradation performance of ultrahigh-purity magnesium in vitro and in vivo. Corrosion Science. 2015;**91**(Supplement C):29-36. DOI: 10.1016/j.corsci.2014.09.008

[71] Draxler J, Martinelli E, Weinberg AM, Zitek A, Irrgeher J, et al. The potential of isotopically enriched magnesium to study bone implant degradation in vivo. Acta Biomaterialia. 2017;**51**:526-536. DOI: 10.1016/j.actbio.2017.01.054

Chapter 3

Animal Models of Burn Wound Management

Shu-Jen Chang, Dewi Sartika, Gang-Yi Fan, Juin-Hong Cherng and Yi-Wen Wang

Abstract

Burn injury is known as the most traumatic wound. In the clinical, most patients with burn injury suffer from extreme pain during wound management; hence, the effective treatment that involved advanced medication is needed. In the evaluation of burn wound care devices, the use of animal model is considered suitable as valuable tools to investigate the burn pathophysiology as well as the efficacy of treatment strategies due to the complexity and heterogeneous nature of the burn. This chapter aimed to review the preclinical small and large animal models of burn injury for translational applications and to highlight their benefits and limitations for the burn treatment design that are clinically applicable to humans.

Keywords: burn wound, treatment, animal model

1. Introduction

The skin is the largest organ of the body, and its destruction, especially caused by burn injuries, is sufficient to be life threatening. Burns are responsible for many pathophysiological changes, resulting in a severe form of trauma that initiates several complications such as an escalation in infection and mortality rates as well as prolonged hospitalization and time of inactivity [1, 2]. For affected large surface area, burns may turn into a systemic problem affecting a various range of organs [3]. Consequently, there will be an intense inflammatory process and prolonged hypermetabolism, coordinated by hormones, cytokines, and acute phase proteins, which are associated with delayed wound healing process, enormous catabolism, multi-organ failure, and death [4]. Further, burn patients also will associate with anxiety, sleep disturbances, social avoidance, depression, and a disruption in activities of daily living after physical rehabilitation [5].

In decades, many important advances have been made for the improvements of the burn care; however, more still needs to be undertaken. The comprehensive study of burn pathophysiology is vital for further improvements of the current treatment strategies. Numerous experimental models have been established and can be applied to address the systemic, cellular, or molecular responses that occur in burn injuries, particularly the development of animal models. The use of these burn animal models is crucial for burn research especially for investigating the properties of new medicines, as it is known that novel treatment strategies should be initially tested at the experimental level before the clinical use [6]. For accurately investigating any therapeutic approaches and relevantly translating to the clinical, the utilization of animal models has to be reproducible and as close as possible to burn lesions occurring in humans. Nevertheless, each animal model has advantages and limitations that determine its translational significance for burn treatments. In addition, the selection of the model should consider the anatomical and physiological characteristics of interspecies that reflect the differences in how different types of wounds heal and analytical techniques be applied. This chapter will further discuss the common animal models of burn injury as well as provide researchers with a better understanding of their benefits and limitations for the burn treatment design that is proposed to be clinically applicable to humans.

2. Burn wound management

Burn injuries differ in their cause types and severity; hence, its treatment can be challenging to be managed. The first and second degrees of burn injuries usually are treated with the moisturizer, the topical agents, and/or an antimicrobial creams advised by the doctor [7]. This condition will typically heal within 2 weeks. On the other hand, because third degree of burn injuries destroys all of the skin layers, the majority of wound will tend to severely long-term consequences and cannot be managed by the primary healing process, so the additional surgical procedures, including skin grafting, skin substitutes, and the application of advanced wound dressing, are required [8, 9]. They act as filler to increase the dermal component of wound, improve the re-epithelization, and reduce the inhibitory factors and the inflammatory responses of wound healing, and therefore subsequent scarring [9, 10]. Numerous options for skin substitutes, dermal analogs, and advanced dressings existed, which can be broadly divided and utilized depending on the severity of burn injuries [11]. However, removing the eschar and covering the wound as early as possible are crucial since the main challenge in treating third degree of burn injuries is avoiding infection from any contaminations. In addition, appropriate deep burn care providing protection from physical damage and supporting the circulation of gas and moisture as well as a comfort to enhance the functional recovery should also be the priorities in severe burn wound care.

Advanced burn care has been associated with a deeper insight of the pathophysiology of burn wound healing as it demands the collaboration of many different tissues and cells that contribute to each phase of wound healing [12]. In severe burn, the phases of wound healing including inflammation, proliferation, matrix synthesis, and contraction, are dynamic and complex and tend to overlap [13]. Therefore, a better understanding of these phases is a key concept to continuously develop an advanced severe burn wound management.

Experimental model is essential when studying on the burns and its underlying mechanisms. Many animal models of burn injuries using mice, rats, rabbits, dogs, and pigs are reported. They have been widely used to examine the burn wound pathology, the effect of systemic drug application, local therapy, and the effect of burn trauma on the entire organism [14–16]. The use of animal models is considered suitable as valuable tools to examine the burn pathophysiology instead of *in vitro* experiment due to the heterogeneous nature of the burns and its similarity to the characteristics of the human skin. The accurate animal model that closely mimics the overlapping phases of severe burn wound would enable the researchers to investigate the potential of novel treatments and study each phase more precisely. However, each animal model of burn has its own advantages and limitations, so the evaluation of several models of burn wound in animals is important and will be further described below.

3. Animal models in burn wound studies

The use of animals as experimental models in various biological researches for transposition into human physiology was initially provoked by Bernard in 1865 [17]. Over time, the notable similarities of anatomy and physiology between humans and animals have further encouraged many researchers to investigate a large range of mechanisms and therapies in the animal models before translating their findings to humans. In burn studies, there are some common techniques for producing wound burns in the animal model including hot water, hot metal tools, electricity, and heated paraffin [18–20]. In these methods, the back of the animal is shaved, and a heated material is executed to the skin to induce the desired burn surface area. The specific parameters such as raised temperatures and duration of exposure are required in each different burn models [21–23]. Furthermore, the integral planning for the burn animal model experiment is also crucial to be estimated. The most significant difference in the skin histology between human and animal is the density of hair. The rapidity of reepithelialization and the morphology of hair follicle are extremely influenced by the hair cycle; it would affect the planimetry area of wound and the microscope data of observable skin biopsy [24–26]. For instance, the hair cycle of rodents is short (approximately around 23–28 days). In order to avoid their hair cycle effects for the evaluation of the wound, rodents with a similar birth date should be used. Because different animals possess different hair cycles, the specific time consideration of each animal model is necessary to be highlighted. In addition, the hair might reduce the heat transfer, and some serious infections source could be hiding in the hair; thus the animal hair needs to be thoroughly depilated. Shaving by hair clipper and then applying with hair removal cream can remove the hair entirely. However, the hair removal cream might induce contact dermatitis so its administration time should be carefully controlled. Last but not least, appropriate post-operation care is needed to be considered too in order to elevate the survival rate of animal. The rational use of antibiotics can prevent wound infections, and the proper administration of analgesics can improve the appetite and self-harm of the animal [27, 28]. Moreover, large areas of burns can also cause severe loss of body fluids; therefore, intensive monitoring and handling for the dehydration of animals are necessary.

The right choice of method of burn induction and its maintenance in animal models are important as this impacts the burn outcome and determines how the wounds are treated. There is diversity among the species in the structure and anatomy of the skin along with their pros and cons as an experimental burn injury model. In this section, several animal models of burn in literature will be evaluated.

3.1 Mouse

As a research model, mouse contains the major layers of the human skin (e.g., epidermis, dermis) and provides the main insights of the signaling pathways associated in the healing process due to the variety of mouse-specific reagents and transgenic feasibility in mouse. Mouse also shares several physiological and pathological features with human, including cardiovascular, musculoskeletal, and other internal organ systems [29]. Additionally, the morbidity of mouse in research is relatively low owing to an extensively reduced healing time and superior immune system [30, 31].

In burn, mouse animal models are usually used to understand the burn wound healing process and have a reproducible model. Recently, Lateef et al. demonstrated a highly reproducible partial-thickness injury in mouse that mimics the key aspects of the inflammatory and hypertrophic scarring responses observed in humans [32]. Further, Calum et al. have established a 6% third-degree burn injury mouse model with a hot air blower [33]. This model resembles the clinical situation and provides an opportunity to examine or develop new strategies such as new antibiotics and immune therapy for handling burn wound. Moreover, a 25% third-degree burn injury was demonstrated by exposure to boiling water for examining the efficacy of new formula-based traditional medicine [34]. Although burn mouse model has its specific advantages, evidently this model fails to completely mimic the wound healing process of humans. Mouse wound healing occurs mainly through wound contraction and the presence of enriched progenitor cells from their dense skin's hair, which facilitates rapid skin healing and keratinization [30, 35]. In order to alleviate the wound contraction issue, the splinting strategy (performing mechanical fixation of the skin by using devices or splints) has been developed [36]. This method could maintain the wound volume to remain relatively constant, so it allows the histomorphometric or biomolecular quantification of the cellular response under well-controlled, experimental conditions. Another issue is the differences of chemokines and chemokine receptor system between human and mouse including chemokines IL-8, neutrophil-activating peptide-2, inducible T cell chemoattractant, and monocyte chemoattractant, which is critical for wound repair as they contribute to the inflammatory events and reparative processes [31, 37]. Because management strategies for burn injuries are advancing, it becomes essential to consider the potential limitations when assessing the translational accuracy from mouse to humans.

3.2 Rat

Rat is one of the most widely used animal models in burn studies and mainly shares similar features with mouse burn model. Both of them have the cheapest cost in terms of housing, maintenance, and reproduction. Compared with the mouse, rat possesses a larger body size and also is easier to handle as well as less easily stressed by human contact. Despite their popularity, the rapid wound healing mechanics in rats are opposed to the wound healing process seen in humans. This limitation is because rodents (rat, mouse) own a subcutaneous panniculus carnosus muscle that facilitates skin healing by both wound contraction and collagen formation [30, 38]. However, this rapid wound contraction enables the researchers to quickly study the comprehensive mechanics of wound healing to develop advanced treatment strategies.

Motamed et al. have demonstrated third-degree burn rat animal model to investigate the efficacy of amniotic membrane combined with adipose-derived stem cell treatment. The burn wound was fabricated using a hot bar (boiled in water) suppression on the dorsal site for 30 seconds [39]. In our previous study, we have developed a similar model using the implementation of 190°C brass block onto the rats' backs parallel to the midline for 20 seconds [40]. This model was used to evaluate the medical dressing's treatment on severe burn wound as well as its inflammatory responses and healing mechanisms. Recently, a rat model of polytrauma (the combination of severe burns, bone fracture, and blunt force trauma) was established to investigate the abnormal immune response leading to inadequate healing and resolution [41]. This model is proposed to create a useful model of battlefield injuries or severe traumatic injuries in a civilian population for evaluating the interventional strategies to enhance wound healing outcomes. Nevertheless, while the rat burn model is relatively simple, it loses significance when it purposes to learning the complex post-burn etiology of hypermetabolism. In the early post-burn phase with high total body surface area in humans, hyperglycemia will occur and initiate an overall increase of glucose and lactate [42]. As the burn wound of greater than 60% of total body surface area in rats results in reduced survivability and is not maintainable for the experimentation [14], therefore, it needs to be considered to have a burn injury model with high total body surface area to recapture the hypermetabolism observed in human burns.

3.3 Pig

It is well known that the pig's skin characteristics such as structure, function, and cellular components most closely resemble that of humans. The epidermis and dermis of the pig are thick just like the case in humans, and their epidermis ranges from 30 to 140 μm and from 50 to 120 μm, respectively [16, 43]. Physiologically, the pig's skin responds as the human skin does to various growth factors and cytokines and displays the reepithelialization rather than contraction during the wound healing process, similar to that observed in humans. In addition, they also share important similarities such as epidermal enzyme forms, epidermal tissue turnover time, the keratinous proteins, and the composition of the lipid film of the skin surface [16]. Based on those aforementioned great anatomical and physiological similarities between pig and human, pig then has been extensively used as the experimental burn models than nearly every other animal model.

Severe burn injuries cause hypertrophic scarring that generates the painful permanent hard, red, and raised scars. With great similar skin characteristics to human, pig appears to produce scarring most identical to human hypertrophic scarring. Cuttle et al. have demonstrated a pig model of hypertrophic scarring after burns using a glass bottle containing water at 92°C to the skin of a large white pig for 14 seconds to create the partial-thickness burn wound [44]. This model of hypertrophic scarring after deep dermal partial-thickness burn injury can be used to further understand the pathophysiology of burn wound healing and scar formation as well as for the testing of various agents which could potentially improve the outcome of the burn wound. Another report demonstrated the reproducible burn hypertrophic scar model using the Bama miniature pig by applying a homemade heating device for 35 seconds followed by debridement surgery [45]. This model has displayed a similar macroscopic, histologic, and biologic criteria of burn wound compared to the human hypertrophic scars. As some burn characteristics in human can be practically well mimicked, hence, the examination of various treatment strategies for severe burn injuries can be specifically applied to gain a comprehensive understanding of the mechanisms of burn healing.

Several studies developed the severe burn pig model in order to evaluate the advanced strategy for the reconstruction of burn injuries. Our laboratory has demonstrated a severe pig burn model using a minimally invasive surgical technique with an easy-to-learn, cost-effective, and reproducible method [46]. This model provides crucial tools for the evaluation of any clinical dressings and uncovers the pathophysiology of burn wound healing. Recently, full-thickness burn wounds in pig model were utilized to evaluate the effect of fractional CO_2 laser therapy on objectively measured scar outcomes including scar area, pigmentation, erythema, roughness, histology, and biomechanics [47]. This model offers a powerful platform to examine the efficacy of laser therapy as a function of many treatment parameters such as the timing of therapy initiation, energy, and laser density. The use of pig as a large animal model provides the standardized location of burn injury and the therapy investigation in greater depth of wound via noninvasive and invasive analyses. Further, Singer et al. established a partial-thickness burn in pig model to investigate the efficacy of topical nitric oxide application to the burn wound [48]. They found that topical application of a nitric oxide-releasing agent accelerated wound reepithelialization and angiogenesis in this model. As there are similarities in skin anatomy and physiology between pig and human, therefore, this treatment can be considered as alternative burn care in patients. However, future studies should discover other approaches to deliver nitric oxide to burn wounds and improve long-term outcomes.

Besides those advantages to capture most pig burn model can be quite challenging to implement due to its risk of infection and high expense of housing with the greatest care.

3.4 Rabbit

Severe burn injuries are known to induce analogous hypermetabolic and pathological systemic alterations in rabbits and humans [49]. Hence, due to their remarkable similarity in metabolic characteristics, rabbit was considered as a promising animal model for burn research. Rabbit is also a cost-effective choice as burn animal model compared to the use of pig.

Rabbit model provides facilities to conduct the systemic effects of burn injury such as dynamic changes in whole-body amino acid and substrate metabolism [49]. It has also been revealed that rabbits present a high level of resting energy expenditure after a thermal injury that indicates the same evidence in burn patients [49]. Moreover, rabbit model has proven to demonstrate the involvement of leucine as an important amino acid in muscle anabolism that shows the similar kinetics and pattern of change post-thermal injury in human patients [50]. Recently, Friedrich et al. have demonstrated a quantifiable deep partial-thickness burn model in the rabbit ear using a dry-heated brass rod for 10 and 20 seconds at 90°C, resulting in a measurable burn progression and minimization of burn healing by contraction [51]. This animal model could be an important new tool to guide the treatment strategies of burn hypertrophic scarring.

3.5 Dog

Instead of several animal models that have been developed in early research, dog can be performed as a mature model for burn-blast combined injury studies. Hu et al. have established the Beagle dogs in the development of a stabilized, controllable, and repeatable animal model that can mimic the actual site of the burn-blast combined injury using explosion and napalm burns [52]. The hemodynamic changes in the early shock stage of burn were successfully investigated in this model, and it also can be used as a good research platform on the mechanisms of fluid resuscitation during burn-blast combined injury shock. Another dog burn-blast combined injury model was established including blast injury caused by explosion immediately followed by seawater immersion that is known to induce the hemodynamic changes and metabolic acidosis [53]. This model supports the investigation of the early symptoms and unique pathophysiology of the blast-burn combined injury that will be valuable in defining the suitable management of such patients. However, the use of dog burn animal model for examining the comprehensive of wound healing process needs to be more considered due to the ethical regulations, limited standardized reagents, and its looser skin over the body/trunk which results in a wound that heals primarily by contraction. Rapid contraction is a common feature of animals with loose skin, while in the tight-skinned species (human, porcine), the wound closure occurs principally as the result of reepithelialization.

4. Clinical advantages of animal models in burn research

In clinical purposes, animal research models should be determined by maximizing their translational relevance to humans. Besides that each animal model has the unique strengths and limitations (summarized in **Table 1**), its most important value is the capability to represent the nature of disease and accurately evaluate the outcomes. There are several reasons the treatment strategies are considerably tested on animal models: (1) animals offer a degree of environmental and genetic manipulations that are rarely feasible in humans as well as unique insights into the pathophysiology and etiology of disease that frequently reveal novel targets for directed treatments; (2) if preliminary testing on animals shows their not clinically useful results, it may not be essential to test on humans; and (3) the authorities concerned with public protection have to ensure the toxicity and safety of the treatment strategies through animal testing [54].

Progress has been made in the area of assessment and measurement, either the comprehensive evaluation of burn pathological mechanisms or novel therapeutic approaches, by involving the animal models of burn. As we have discussed before, there are numerous animal models of burn established to disclose these issues. The ultimate goal of these animal studies is to examine a safe and effective test condition

Species	Advantages	Disadvantages	References
Mouse	• Shares several physiological and pathological features with human (e.g., the skin, cardiovascular, musculoskeletal, other internal organ systems) • Superior immune system • Provides various mouse-specific reagents and transgenic feasibility • Low morbidity • Cost-effective • Easy handling	• Rapid healing along with wound contraction issue • Different chemokines and chemokine receptors system • Looser skin with dense hair structure	[29–31, 35, 37]
Rat	• Similar to mouse but possesses a larger body size and is less easily stressed by human contact	• Similar to mouse	
Pig	• Possesses great anatomical and physiological similarities with human	• Risks of infection and morbidity • High expense of housing and care	[16, 43]
Rabbit	• Shares remarkable similarity in metabolic and pathological alterations of burn with human • Lower cost than pig	• Risks of infection and morbidity	[49]
Dog	• Similar environment to human • Can mimic the actual site of the burn-blast combined injury so it can be used as a good research platform on the mechanisms of fluid resuscitation during burn-blast combined injury shock as well as its early symptoms and unique pathophysiology	• Ethical regulations • Limited standardized reagents • Cost hurdles • Looser skin over the body/trunk	[52, 53]

Table 1.
Comparison of the advantages and disadvantages of burn animal model.

for clinical trials in humans with burn injuries. Generally, the choice of animals for burn model is mainly based on cost and ethics and further is based on which species will give the best correlation to human trials.

Several substantial advancements have been made in burn patient care such as controlling wound healing, developing novel graft and coverage preferences, optimizing dietary needs, and testing unique pharmacological interventions, resulting in an improved patient's survival and decreased hospitalization period [11]. For example, Wang et al. established a clinical scar in a pig burn model that is found to greatly correlate with scar histology, wound size, and reepithelialization data [55]. This clinical scar scale demonstrated a reliable and independent tool for assessing the burn wound healing outcomes without using other healing and scar measuring systems. Clinically, scar appearance and function are the major concerns to both burn victims and their carers, so its minimization is one of the ultimate goals of burn care, which relies on the appropriate evaluation of the scars.

5. Conclusion

Burn injury is one of the most severe forms of trauma that is related with significant pain and various physical, psychological, and social diminishments; therefore, the exploration of advanced treatment strategies in order to obviously heal and reduce the lifelong burn wound recovery phases is demanding. Burn animal models have been proposed as valuable tools that provide considerable insights into the burn pathophysiology as well as for investigating the properties of new medicines before the clinical use. Accordingly, the standardization of animal models is crucial for all scientific research, and it can merely be achieved with a comprehensive description of the experimental techniques along with their advantages and limitations. With a better understanding of burn underlying phenomenon as animal models paved the road to its mechanisms, progressive research is expected to continuously identify novel treatment strategies to improve the quality of life for burn patients.

Acknowledgements

The authors would like to thank Prof. Jiang-Chuan Liu and Prof. Nien-Hsien Liou (Department and Graduate Institute of Biology and Anatomy, National Defense Medical Center, Taipei, Taiwan (ROC)) for the knowledge and moral support and also Dr. Chih-Hsin Wang (Department of Plastic and Reconstructive Surgery, Tri-Service General Hospital, National Defense Medical Center, Taipei, Taiwan (ROC)) for his helpful discussion during this chapter writing.

Conflict of interest

The authors declared no potential conflicts of interest with respect to the research, authorship, and/or publication of this article.

Author details

Shu-Jen Chang[1,2], Dewi Sartika[2], Gang-Yi Fan[1,2], Juin-Hong Cherng[2,3] and Yi-Wen Wang[3]*

1 Independent Research Fellow, Tri-Service General Hospital, National Defense Medical Center, Taipei, Taiwan, ROC

2 Laboratory of Stem Cell and Tissue Regeneration, National Defense Medical Center, Taipei, Taiwan, ROC

3 Department and Graduate Institute of Biology and Anatomy, National Defense Medical Center, Taipei, Taiwan, ROC

*Address all correspondence to: christmas1035@hotmail.com

References

[1] Ashburn MA. Burn pain: The management of procedure-related pain. The Journal of Burn Care & Rehabilitation. 1995;**16**(3 Pt 2):365-371

[2] Summer GJ, Puntillo KA, Miaskowski C, Green PG, Levine JD. Burn injury pain: The continuing challenge. The Journal of Pain. 2007;**8**(7):533-548. DOI: 10.1016/j.jpain.2007.02.426

[3] Horton JW. Left ventricular contractile dysfunction as a complication of thermal injury. Shock. 2004;**22**(6):495-507. DOI: 10.1097/01.shk.0000145205.51682.c3

[4] Jeschke MG, Gauglitz GG, Kulp GA. Long-term persistance of the pathophysiologic response to severe burn injury. PLoS One. 2011;**6**(7):e21245. DOI: 10.1371/journal.pone.0021245

[5] Rumsey N, Clarke A, White P. Exploring the psychosocial concerns of outpatients with disfiguring conditions. Journal of Wound Care. 2003;**12**:247-252. DOI: 10.12968/jowc.2003.12.7.26515

[6] Asko-Seljavaara S. Burn research--animal experiments. Acta Physiologica Scandinavica. Supplementum. 1986;**554**:209-213

[7] Palmieri TL, Greenhalgh DG. Topical treatment of pediatric patients with burns. American Journal of Clinical Dermatology. 2002;**3**(8):529-534. DOI: 10.2165/00128071-200203080-00003

[8] Pereira DST, Lima-Ribeiro MH, Pontes-Filho NT, et al. Development of animal model for studying deep second-degree thermal burns. Journal of Biomedicine & Biotechnology. 2012;**2012**:460841. DOI: 10.1155/2012/460841

[9] Wang Y, Beekman J, Hewa J, et al. Burn injury: Challenges and advances in burn wound healing, infection, pain and scarring. Advanced Drug Delivery Reviews. 2018;**123**:3-17. DOI: 10.1016/j.addr.2017.09.018

[10] Shores JT, Gabriel A, Gupta S. Skin substitutes and alternatives: A review. Advances in Skin & Wound Care. 2007;**20**(9 Pt 1):493-508. DOI: 10.1097/01.ASW.0000288217.83128.f3

[11] Rowan MP, Cancio LC, Elster EA, et al. Burn wound healing and treatment: Review and advancements. Critical Care. 2015;**19**:243. DOI: 10.1186/s13054-015-0961-2

[12] Gurtner GC, Werner S, Barrandon Y, Longaker MT. Wound repair and regeneration. Nature. 2008;**453**:314-321. DOI: 10.1038/nature07039

[13] Martin P. Wound healing--aiming for perfect skin regeneration. Science. 1997;**276**:75-81. DOI: 10.1126/science.276.5309.75

[14] Abdullahi A, Amini-Nik S, Jeschke M. Animal models in burn research. Cellular and Molecular Life Sciences. 2014;**71**(17):3241-3255. DOI: 10.1007/s00018-014-1612-5

[15] Andrews CJ, Kempf M, Kimble R, et al. Development of a consistent and reproducible porcine scald burn model. PLoS One. 2016;**11**(9):e0162888. DOI: 10.1371/journal.pone.0162888

[16] Dahiya P. Burns as a model of SIRS. Frontiers in Bioscience. 2009;**14**:4962-4967. DOI: 10.2741/3580

[17] Bernard C. An Introduction to the Study of Experimental Medicine. New York: Dover Publications Inc; 1957. p. 272

[18] Venter NG, Monte-Alto-Costa A, Marques RG. A new model for the

standardization of experimental burn wounds. Burns. 2015;**41**(3):542-547. DOI: 10.1016/j.burns.2014.08.002

[19] Singer AJ, Taira BR, Anderson R, et al. Does pressure matter in creating burns in a porcine model? Journal of Burn Care & Research. 2010;**31**(4):646-651. DOI: 10.1097/BCR.0b013e3181e4ca73

[20] Pfurtscheller K, Petnehazy T, Goessler W, et al. Innovative scald burn model and long-term dressing protector for studies in rats. Journal of Trauma and Acute Care Surgery. 2013;**74**(3):932-935. DOI: 10.1097/TA.0b013e31827d0fc3

[21] Hoekstra MJ, Hupkens P, Dutrieux RP, et al. A comparative burn wound model in the New York shire pig for the histopathological evaluation of local therapeutic regimens: Silver sulfadiazine cream as a standard. British Journal of Plastic Surgery. 1993;**46**(7):585-589. DOI: 10.1016/0007-1226(93)90111-N

[22] Campelo APBS, Campelo MWS, de Castro Britto GA, et al. An optimized animal model for partial and total skin thickness burns studies. Acta Cirúrgica Brasileira. 2011;**26**(1):38-42. DOI: 10.1590/S0102-86502011000700008

[23] Gurfinkel R, Singer AJ, Cagnano E, et al. Development of a novel animal burn model using radiant heat in rats and swine. Academic Emergency Medicine. 2010;**17**(5):514-520. DOI: 10.1111/j.1553-2712.2010.00736.x

[24] Ansell DM, Kloepper JE, Thomason HA, et al. Exploring the "hair growth-wound healing connection": Anagen phase promotes wound re-epithelialization. The Journal of Investigative Dermatology. 2011;**131**(2):518-528. DOI: 10.1038/jid.2010.291

[25] Essayem S, Kovacic-Milivojevic B, Baumbusch C, et al. Hair cycle and wound healing in mice with a keratinocyte-restricted deletion of FAK. Oncogene. 2006;**25**(7):1081-1089. DOI: 10.1038/sj.onc.1209130

[26] Lin KK, Chudova D, Hatfield GW, et al. Identification of hair cycle-associated genes from time-course gene expression profile data by using replicate variance. Proceedings of the National Academy of Sciences of the United States of America. 2004;**101**(45):15955-15960. DOI: 10.1073/pnas.0407114101

[27] Nguyen HB, Rivers EP, Abrahamian FM, et al. Severe sepsis and septic shock: Review of the literature and emergency department management guidelines. Annals of Emergency Medicine. 2006;**48**(1):28-54. DOI: 10.1016/j.annemergmed.2006.02.015

[28] Peterson NC, Nunamaker EA, Turner PV. To treat or not to treat: The effects of pain on experimental parameters. Comparative Medicine. 2017;**67**(6):469-482

[29] Rosenthal N, Brown S. The mouse ascending: Perspectives for human-disease models. Nature Cell Biology. 2007;**9**(9):993-999. DOI: 10.1038/ncb437

[30] Wong VW, Sorkin M, Glotzbach JP, et al. Surgical approaches to create murine models of human wound healing. Journal of Biomedicine & Biotechnology 2011;2011:969-618. DOI: 10.1155/2011/969618

[31] Mestas J, Hughes CC. Of mice and not men: Differences between mouse and human immunology. Journal of Immunology. 2004;**172**(5):2731-2738. DOI: 10.4049/jimmunol.172.5.2731

[32] Lateef Z, Stuart G, Jones N, et al. The cutaneous inflammatory response to thermal burn injury in a murine model. International Journal

of Molecular Sciences. 2019;**20**(3):538. DOI: 10.3390/ijms20030538

[33] Calum H, Høiby N, Moser C. Burn mouse models. In: Filloux A, Ramos JL, editors. Pseudomonas Methods and Protocols. Methods in Molecular Biology (Methods and Protocols). Vol 1149. New York: Humana Press; 2013. pp. 793-802. DOI: 10.1007/978-1-4939-0473-0_60

[34] Mehrabani M, Seyyedkazemi SM, Nematollahi MH, et al. Accelerated burn wound closure in mice with a new formula based on traditional medicine. Iranian Red Crescent Medical Journal. 2016;**18**(11):e26613. DOI: 10.5812/ircmj.26613

[35] Ito M, Liu Y, Yang Z, et al. Stem cells in the hair follicle bulge contribute to wound repair but not to homeostasis of the epidermis. Nature Medicine. 2005;**11**(12):1351-1354. DOI: 10.1038/nm1328

[36] Davidson JM, Yu F, Opalenik SR. Splinting strategies to overcome confounding wound contraction in experimental animal models. Advances in Wound Care. 2013;**2**(4):142-148. DOI: 10.1089/wound.2012.0424

[37] Werner S, Grose R. Regulation of wound healing by growth factors and cytokines. Physiological Reviews. 2003;**83**(3):835-870. DOI: 10.1152/physrev.2003.83.3.835

[38] Dorsett-Martin WA. Rat models of skin wound healing: A review. Wound Repair and Regeneration. 2004;**12**(6):591-599. DOI: 10.1111/j.1067-1927.2004.12601.x

[39] Motamed S, Taghiabadi E, Molaei H, et al. Cell-based skin substitutes accelerate regeneration of extensive burn wounds in rats. American Journal of Surgery. 2017;**214**(4):762-769. DOI: 10.1016/j.amjsurg.2017.04.010

[40] Wang CH, Chang SJ, Tzeng YS, et al. Enhanced wound-healing performance of a phyto-polysaccharide-enriched dressing—A preclinical small and large animal study. International Wound Journal. 2017;**14**:1359-1369. DOI: 10.1111/iwj.12813

[41] Mangum LH, Avila JJ, Hurtgen BJ, et al. Burn and thoracic trauma alters fracture healing, systemic inflammation, and leukocyte kinetics in a rat model of polytrauma. Journal of Orthopaedic Surgery and Research. 2019;**14**:58. DOI: 10.1186/s13018-019-1082-4

[42] Kulp GA, Tilton RG, Herndon DN, et al. Hyperglycemia exacerbates burn-induced liver inflammation via noncanonical nuclear factor-kappaB pathway activation. Molecular Medicine. 2012;**18**:948-956. DOI: 10.2119/molmed.2011.00357

[43] Sullivan TP, Eaglstein WH, Davis SC, et al. The pig as a model for human wound healing. Wound Repair and Regeneration. 2001;**9**(2):66-76. DOI: 10.1046/j.1524-475x.2001.00066.x

[44] Cuttle L, Kempf M, Phillips GE, et al. A porcine deep dermal partial thickness burn model with hypertrophic scarring. Journal of the International Society for Burn Injuries. 2006;**32**:806-820. DOI: 10.1016/j.burns.2006.02.023

[45] Deng X, Chen Q, Qiang L, et al. Development of a porcine full-thickness burn hypertrophic scar model and investigation of the effects of shikonin on hypertrophic scar remediation. Frontiers in Pharmacology. 2018;**9**:590. DOI: 10.3389/fphar.2018.00590

[46] Fan GY, Cherng JH, Chang SJ, et al. Severe burn injury in a swine model for clinical dressing assessment. Journal of Visualized Experiments. 2018;**141**:e57942. DOI: 10.3791/57942

[47] Baumanna ME, Clairmontea IA, DeBrulerb DM, et al. FXCO2 laser therapy of existing burn scars does not significantly improve outcomes in a porcine model. Burns Open. 2019;**3**(3):89-95. DOI: 10.1016/j.burnso.2019.04.004

[48] Singer AJ, Choi Y, Rashe M, et al. The effects of topical nitric oxide on healing of partial thickness porcine burns. Burns. 2018;**44**(2):423-428. DOI: 10.1016/j.burns.2017.07.017

[49] Hu RH, Yu YM, Costa D, et al. A rabbit model for metabolic studies after burn injury. The Journal of Surgical Research. 1998;**75**(2):153-160. DOI: 10.1006/jsre.1998.5274

[50] Zhang XJ, Chinkes DL, Wolfe RR. Leucine supplementation has an anabolic effect on proteins in rabbit skin wound and muscle. The Journal of Nutrition. 2004;**134**(12):3313-3318. DOI: 10.1093/jn/134.12.3313

[51] Friedrich EE, Niknam-Bienia S, Xie P, et al. Thermal injury model in the rabbit ear with quantifiable burn progression and hypertrophic scar. Wound Repair and Regeneration. 2017;**25**(2):327-337. DOI: 10.1111/wrr.12518

[52] Hu Q, Chai J, Hu S, et al. Development of an animal model for burn-blast combined injury and cardiopulmonary system changes in the early shock stage. The Indian Journal of Surgery. 2015;**77**(3):977-984. DOI: 10.1007/s12262-014-1095-5

[53] Hu Y, Mao Q, Ye S, et al. Blast-burn combined injury followed by immediate seawater immersion induces hemodynamic changes and metabolic acidosis: An experimental study in a canine model. Clinical Laboratory. 2016;**62**(7):1193-1199. DOI: 10.7754/Clin.Lab.2015.150929

[54] Hackam DG. Translating animal research into clinical benefit. BMJ. 2007;**334**(7586):163-164. DOI: 10.1136/bmj.39104.362951.80

[55] Wang XQ, Kravchuk O, Liu PY, et al. The evaluation of a clinical scar scale for porcine burn scars. Burns. 2009;**35**(4):538-546. DOI: 10.1016/j.burns.2008.10.005

Chapter 4

Animal Models in Psychiatric Disorder Studies

João Victor Nani, Benjamín Rodríguez, Fabio Cardoso Cruz and Mirian Akemi Furuie Hayashi

Abstract

Among the diseases affecting the brain, special attention has been paid to psychiatric disorders (PDs) due to high prevalence and significant debilitating clinical features. Many difficulties need to be overcome to find good animal models for PDs, due to their multifactorial origins, high heterogeneity and symptoms, as for instance the hallucinations and delusions, which usually cannot be easily assessed employing ordinary experimental animal models. The use of animal models reproducing at least some specific traits that can be studied individually, known as endophenotypes, is often reported. However, since altered biological pathways are common to many of these disorders, each of these behaviors may also reflect different PDs. In this context, it is possible to perform several approaches, to elicit changes in the endophenotypes of interest, not only in vertebrate models like rodents, but also in invertebrate models which have important advantages due to high conservation of essential pathways, lower complexity, and shorter life cycle compared to mammals. Therefore, animal models are also helpful for elucidating the etiology underlying PDs, by allowing easier access to biological samples that are usually not accessible in clinical studies, as for instance, fresh brain samples, from embryos to adults.

Keywords: animal model, psychiatric disorders, neurodevelopment, biomarkers, CNS, endophenotypes

1. Introduction

According to the World Health Organization (WHO), psychiatric disorders (PDs) comprise a broad range of dysfunctions, with several and some common symptoms. PDs are generally characterized by the combination of symptoms as abnormal thoughts, emotions, behavior, and social interaction. The most common PDs include schizophrenia (SCZ), bipolar disorder (BD), major depression disorder (MDD), attention deficit hyperactivity disorder (ADHD), intellectual disabilities, drug abuse disorders, among others [1].

1.1 The need and the value of animal models for PD studies

There are several reasons to use animal models in the studies of disorders affecting the brain. The poor understanding of the etiopathogenesis and pathophysiology of PDs is clearly reflected by the unmet clinical need for better pharmacological treatments. Therefore, good models are clearly needed to clarify the neurobiology involved in PDs, as well as for the identification of biomarkers useful to assist diagnosis and/or for the development of novel therapies. It is also implausible to move forward in clinical trials with a novel drug tested only in a cell model, without any evidence about its efficacy in animal experiments. The value of animal models to drug development has been demonstrated empirically. For example, the first and the most efficacious drugs available for complex PDs such as SCZ (e.g., chlorpromazine and clozapine) was discovered observing the alterations in behaviors of experimental animals in response to each drug administration. In fact, in the last decades, most of the CNS drugs approved were discovered employing a phenotypic screening approach in animal models [2, 3].

1.2 Challenges to model PDs in animals

A reliable animal model must share several similarities with the studied target to allow a successful translation from the basic to the clinical research. However, several limitations need to be overcome. First, the heterogeneous behavioral symptom characteristics of PDs are in some grade uniquely expressed in humans, and they are certainly impossible to be reproduced authentically in animals as rodents, fishes or worms [4]. Second, there is a lack of an objective measure to unequivocally diagnose mental illness [5], which adds complexity to the modeling any mental disorder in experimental animals. Third, in order to develop meaningful animal models for PDs with potential translational power, the disease phenotypes must be represented in the experimental animals. The selection and update of these phenotypes, in agreement with the recent findings in clinical psychiatry and neuroscience, represents a challenge, as evidenced by the recognized gap between the clinical and basic scientific research [6]. In addition, a rising question is what are the specific traits or phenotypes that an animal model should express to be translatable to specific disorder? (**Figure 1**).

1.3 How to develop an animal model for PD studies

The traditional approach to establish an animal model in PDs is based on three classic constructs proposed by Willner in 1984: face validity, which determines how much a phenotype presented by a patient is represented by the animal model (corresponds to similarity between the model and the PDs assessed, that includes symptoms, signs, and pharmacological features); construct validity, which demonstrates whether it is possible to reproduce the pathological condition based on processes that are already known to be altered (correspondence between the physiological dysfunctions in the human population and in the animal model); predictive validity, which tries to evaluate if a pharmacological or non-pharmacological intervention is capable to reverse the pathological condition (in other words, if the treatment that is effective in reversing PDs in humans would reverse the changes seen in animals) [7–10]. However, in practice, no animal models fully meet these three criteria of validity.

Many authors have proposed that instead of these three proposed criteria defining an external validation, in addition, the validity of an animal model should not be simply organisms that resemble human dysfunction, but they would also reproduce the processes by which animals and humans enter this state, and therefore, this could be better exploited by adding a new validation criteria [9]. For instance, the validity by homology, which proposes, for instance, an invertebrate model, such as *Drosophila*, may not be the ideal animal model for studying complex changes in a brain circuitry, but in turn, it may represent a great choice to study the genetic control of early embryonic development [11]. In fact, the nematode *Caenorhabditis elegans* is a reliable model with conserved neurobiological systems

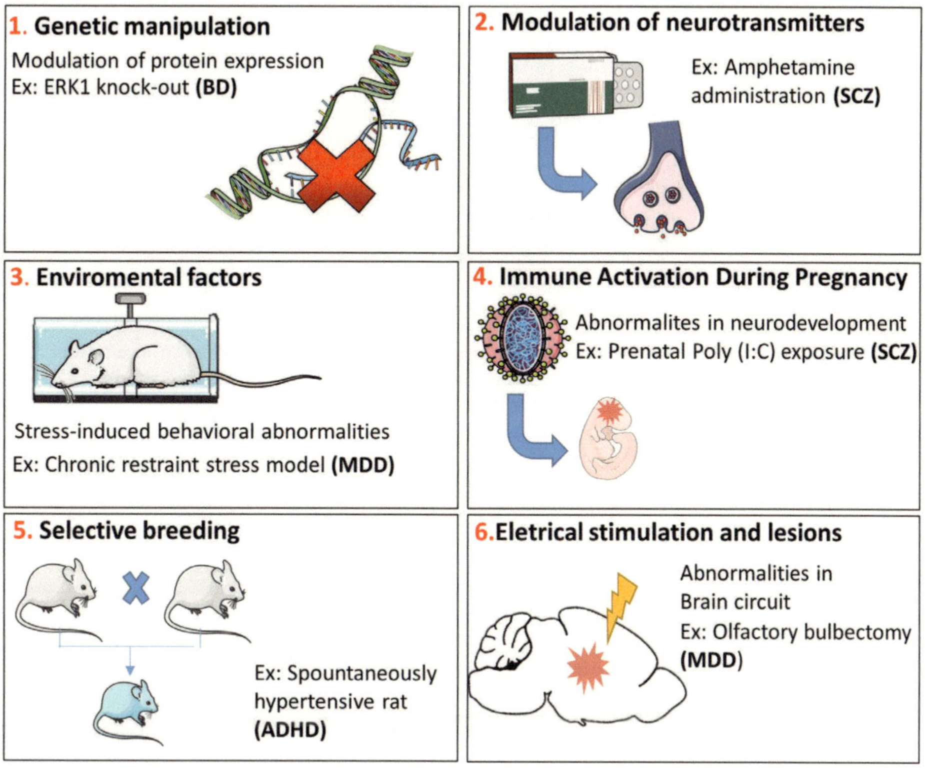

Figure 1.
Different approaches to construct animal models for neuropsychiatric disorders studies.

that has been helpful in the discovery of molecular mechanisms that underlie learning and memory, and, in addition, this animal model has a fully sequenced genome and other several molecular and genetic tools available for researchers [12, 13].

1.4 Symptoms versus endophenotypes in experimental model animals

There is a consensus about the low reliability of the diagnostic construct provided for the employment of Diagnostic and Statistical Manual of Mental Disorders or DMS (which is a manual that determines the criteria for the clinical diagnosis of PDs). The heterogeneity implicit in this classification system and the imprecise quantification of the symptoms make it impossible to deconstruct PDs within model organisms. In fact, an etiology-based nosology system has been advocate for psychiatry, and it has been proposed to identify the endophenotypes that occur in both healthy individuals and subjects with different psychopathologies [14]. Endophenotypes are basically quantitative trait-like deficits that are possible to assess by laboratory-based methods rather than by clinical observation. An endophenotype should be state-independent, heritable, occurring at a high rate in affected families, and in addition, it should be associated to genetic variants of the disorder, as it should be involved the same brain circuits associated with the symptoms of the illness in patients (**Table 1**).

The Research Domains Criteria (RDoC) framework was introduced as an alternative categorization system for psychopathological states [15–17]. This system provides a platform to improve the translatability of studies from animals to humans, since it supports the endophenotype-based comparison of animals and humans on an objective neurobiological basis across all behavioral domains. In fact, the endophenotypes have been reverse-translated into animal models successfully and allows the evaluation

Endophenotype	Description	What can be evaluated
Locomotor activity	Distance travelled, time spent, and frequency of the movements measured during or after a habituation period or after some stimuli (i.e. drug administration)	Behavioral sensitization (BD; ADHD; SCZ); Depressive-like behaviors (MDD); etc...
Latent inhibition	Latent inhibition is the ability of a pre-exposed nonreinforced stimulus to inhibit later stimulus-response learning	Cognitive impairments (SCZ); etc...
Pre-pulse inhibition (PPI)	Decrease of the startle reflex after exposure to a pre-pulse before the pulse	Cognitive impairments (SCZ); etc...
Working memory and learning	Describes short-term memory, in a olfactory domain and spatial domain	Cognitive impairments (PDs in general); etc...
Social interaction	Evaluation of time spent on exploring a social stimulus.	Anxiety-like behaviors; Depressive-like behaviors; etc...
Rearing	Measure of activity, investigation and exploratory behavior induced by a drug or/and novelty	Anxiety-like behaviors; etc...
Grooming	A maintenance behavior evaluated by the cleaning of the fur; is displayed as reaction to unexpected stimuli and in conflict situations	Anxiety-like behaviors; Depressive-like behaviors; etc...
Aggressiveness	Evaluation of attack and defensive behavior as reaction to a stimuli or other animal	Anxiety-like behaviors; Depressive-like behaviors; etc...
Food intake	Amount of food ingested by the animal	Anxiety-like behaviors; Depressive-like behaviors; etc...
Sucrose preference test	Assesses the sensitivity to reward based on the rodent's natural preference for sweets. This test measures the amount of a sweet-tasting solution that the animal ingests	Depressive-like behaviors; etc...
Fear conditioning	Classical conditioning paradigm, in which an aversive stimulus is paired with some neutral stimuli. Used to assess associative fear learning and memory in rodents.	Cognitive impairments (PDs in general); etc...
Forced swim test	Measures the scoring of swimming and climbing (active behavior), and immobility (passive behavior) when animals are placed in an inescapable cylinder filled with water	Depressive-like behaviors; etc...

Table 1.
Most common endophenotypes used to evaluate behaviors associated with psychiatric disorders (PDs).

of the neural neurobiological substrates and their circuit dysfunctions [18]. Thus, it has been demonstrated that the modeling of neurobiological and behavioral endophenotypes to reproduce PDs in experimental animals is possible.

The ideal animal model should be derived from risk factors or the causative agent of the human disease. One of the strategies used during the construction of a model is focused on a specific factor that can reproduce the condition as a whole or an aspect of the disease [19]. The choice for the methodology used in establishing a model is fundamentally important to guide which aspect of the disease should be explored, and it is an essential component in the validation of a model known as construct validity.

2. PDs and animal models

In the following sections, selected examples of animal models used in the context of investigating PDs will be demonstrated, indicating which changes are observed in behavioral and molecular levels.

2.1 Animal models in schizophrenia (SCZ)

Schizophrenia (SCZ) is a severe brain disorder, characterized by a set of positive and negative symptoms and cognitive disorders, which are the basis for the clinical diagnosis of individuals who needs to present at least two or more of those symptoms, according to the DSM. SCZ is one of the most debilitating mental disorders, affecting about 21 million people worldwide. The antipsychotics used to treat SCZ patients can soften the development of the disorder, and this pharmacological treatment was the basis for the most accepted theory to explain the neurobiology of SCZ, as noticed by the alterations in the dopamine transmission. In addition, several other theories have been suggested soon after, as for instance, the serotoninergic, glutamatergic, GABAergic, and the neurodevelopmental susceptibility hypothesis, among others [20]. However, none of these theories had allowed the characterization of the etiology or the identification of strong biomarker for the diagnosis of SCZ. Many efforts are being made to characterize a model for SCZ, but there is a great difficulty in reproduce endophenotypes that frame all the groups of symptoms related to this disease, or which allow associating all risk factors that are already known. Below, we exemplify some of these models, and for a more detailed review of SCZ models can be found elsewhere [21].

Most of the models are based on the theory of neurotransmitter imbalance, and they are induced by the disruption of these pathways, other models explore changes in the levels of expression of candidate genes involved in the processes of SCZ susceptibility. It should be considered that SCZ is a multifactorial disorder, and thus, the genetic component should be evaluated in addition to changes in the environment, as in contrast to the models based on genetic alterations, there are those taking into account the environmental changes, such as the prenatal insults, which impose changes in the neurodevelopment processes. Some of these models are exemplified in **Table 2**.

All of these models show behavioral and molecular changes that can be associated with SCZ.

2.2 Animal models in major depressive disorder (MDD)

Major depressive disorder (MDD) is a common, complex, and heterogeneous mental disorder, characterized by persistent sadness and loss of interest in general activities, affecting about 10% of the population worldwide, and which is caused by multifactorial mechanisms not fully understood yet, characterizing MDD as a disorder with many variations in clinical features among the patients, imposing a consequent high variability in the diagnosis, time course of response and remission [45], which is one of the main reasons justifying the intensive search for animal models and biomarkers, aiming for advances in MDD diagnosis [46]. In addition, these advances could be helpful for a better classification for depressive spectrum, and thereby for improving the treatment [47]. The animal models of depression have been developed based on acute or chronic stress exposure, exogenous administration of glucocorticoids, injuries in brain regions and/or genetic manipulations [48–50]. There is a great variation in the number of protocols that

Model	Endophenotype	Molecular alterations	References
Drug-induced models			
Amphetamine model of SCZ	**Acute:** ↓ Latent inhibition; ↑ locomotion **Chronic:** Same as acute but with ↓ PPI	↑ Mesolimbic dopamine response; ↑ Acetilcholine in PFc	[22–26]
Glutamatergic manipulation (Phencyclidine; MK-801; Ketamine)	↑ Locomotion; ↓ working memory; ↓ Reversal learning performance; ↓ Social interaction; ↓ PPI	↓ PV-immunoreactive neurons in PFc and hippocampus	[27–29]
Genetic manipulation			
DISC-1 mutations			
Missense mutations models	↓ PPI; ↓ latent inhibition; ↑ Depressive-like phenotype	↓ Brain volume; ↓ PDE4B activity and binding to DISC1; ↓ PV-immunoreactive; ↓ Dendritic density	[30–32]
Dominant-negative isoforms of DISC1	↑ Amphetamine sensibility; ↓ working memory	↓ Dopamine, DOPAC; ↓ PV-immunoreactive	[33, 34]
Knockdown	↑ Amphetamine sensibility; ↓ PPI; ↓ working memory	↓ Dopamine; ↓ PV-immunoreactive	[35]
Overexpression	↑ Amphetamine sensibility; ↑ rearing behavior; ↑ locomotion; ↓ learning in rotarod task	↑ Increase in high-affinity D2R; ↑ Translocation of dopamine transporter; ↑ Dopamine inflow	[36]
Neuregulin1, ErbB4, and dysbindin			
Knock-out	↑ Amphetamine sensibility; ↑ locomotion; ↓ PPI; ↓ Working memory; ↓ social interaction	**Neuregulin1; ErbB4:** ↓ Hippocampal spine density; ↑ Lateral ventricles; **Dysbindin:** ↑ HVA/DA ratio; ↑ Excitability of PFc pyramidal neurones	[37–39]
Developmental models			
Neonatal excitotoxic hippocampal lesion	↓ PPI; ↓ Working memory; ↓ Social interaction; ↑ Amphetamine sensibility; ↑ MK-801/PCP sensibility; ↑ locomotion	↑ Mesolimbic dopamine response; ↑ Acetilcholine in PFc	[40, 41]
Methylazomethanol (MAM) and polyinosinic-polycytidylic acid (poly I:C)	↑ Locomotion; ↑ Amphetamine sensibility; ↑ MK-801/PCP sensibility; ↓ Social interaction; ↓ PPI; ↓ Working memory	↓ PV-immunoreactive neurons in PFc and hippocampus	[42–44]

All of these models show behavioral and molecular changes that can be associated with SCZ.
PPI = prepulse inhibition; PFc = prefrontal cortex; PV = parvalbumin; PDE4B = cAMP-specific 3",5"-cyclic phosphodiesterase 4B; DISC1 = disrupted-in-schizophrenia 1; DOPAC = dihydroxyphenylacetic acid; HVA = homovanillic acid; DA = dopamine; poly I:C = Polyinosinic:polycytidylic acid.

Table 2.
Some examples of SCZ models induced by drugs, genetic manipulation, and prenatal insults.

can be used to induce these changes, in which the stressor, time of exposure to the stimulus, and other parameters may vary. For more detailed review of MDD models, see also [51] (**Table 3**).

Model	Endophenotype	Molecular alterations	References
Stress-induced models			
Learned Helplessness	↓ Locomotion; ↑ aggressiveness ↓ Grooming; ↓ response to rewards ↑ Sleep disturbance	↓ Norepinephrine ; ↑ BDNF; aberrant miRNA brain- region specific expression	[52–57]
Unpredictable chronic mild stress	↓ Food intake; ↓ growth rate; ↓ Locomotion; ↑ aggressiveness; ↓ Response to rewards	↑ Corticosterone; ↓ glucocorticoid receptor expression; ↓ endogenous ATP	[58–61]
Chronic restraint stress model	↑ Aggressiveness; ↑ fear conditioning; ↓ locomotion; ↓ food intake	↑ CA3 dendritic atrophy and damage; ↓ neurogenesis in dentate gyrus; ↑ apoptotic cell death; ↑ corticosteroid	[62–64]
Social defeat	↓ Locomotion; ↓ exploratory activity; ↓ Aggression; ↓ sexual behavior; ↑ Anhedonia; ↑ sleep disturbance ; ↓ Growth rate	↓ Volume and cell proliferation in hippocampus and PFc; ↑ corticosteroid; ↓ serotonin; ↓ BDNF	[65–67]
Early life stress model	↑ Anxiety-like behavior; ↑ Depression-like behavior; ↑ Novelty responsivity	↑ BDNF expression PFC and hippocampus	[68, 69]
Brain lesion model			
Olfactory bulbectomy	↑ Locomotion; ↓ working memory; ↓ response to rewards; ↓ food intake; ↑ sleep disturbance; ↑ responsivity to stressors	Dysfucntion in HPA and neuro-immune axis; ↓ neurotransmitters; ↑ neuronal degeneration; ↑ BDNF; ↓ neuropeptides	[70, 71]
Selective inbreeding			
Wistar-Kyoto	↓ Locomotion; ↑ immobility in forced swim test; ↑ social avoidance; ↑ freezing to context	↑ Adrenal glands; ↑ corticosterone	[72, 73]
Flinders Sensitive Line rat	↓ Activity in enclosed arena; ↑ immobility in forced swim test; ↓ sucrose intake under stress	↓ Serotonin synthesis; dysfunction in dopaminergic and noradrenergic systems	[74–77]

BDNF = brain-derived neurotrophic factor; miRNA = microRNA; ATP = adenosine triphosphate; PFc = prefrontal cortex; HPA = hypothalamic–pituitary–adrenal axis.

Table 3.
Examples of models for MDD induced by stressors, injuries in brain regions, and by selective inbreeding.

2.3 Animal models in bipolar disorder (BD)

Bipolar disorder (BD) is a chronic mood disorder, characterized by fluctuations between mania and depressive episodes, which affects approximately 1% of the global population irrespective of nationality, ethnic origin, or socioeconomic status [78]. Due to the complex mood alterations, misdiagnosis in BD is very common, as other mental illnesses as depression and SCZ share several common symptoms, in addition to the specific and common endophenotypes and brain structural changes [79, 80]. The search for advances in diagnosis is important for these disorders, since early diagnosis would be essential to foster earlier suited pharmacological treatment in BD, which was proved to be beneficial to prevent the cognitive deficits and disabilities in these BD patients [81], as also demonstrated for SCZ patients [82]. The major limitation in evaluating a model for BD is the difficulty in reproducing the phases of mania and depression observed in the clinic. Many of these models present only one of these parameters, and they are often developed by genetic alterations in genes known to be involved in this disorder or stressors, mainly involved in the circadian cycle as also demonstrated for other PDs. Another interesting approach used for the development of animal models for BD is the one induced by psychostimulant sensitization (which causes mania-like behavior), as withdrawal from psychostimulants is accompanied by depressive-like behavior, which together leads to changes and compulsory behaviors. Some of these models are exemplified in **Table 4**. A more detailed review of BD models can be found elsewhere [93].

BDNF = brain-derived neurotrophic factor; ERK1 = Extracellular signal-regulated kinase 1; DAT = dopamine transporter.

Model	Endophenotype	Molecular alterations	References
Genetic manipulation			
BDNF haploinsufficient	↑ Locomotion; ↑ agressive behavior; ↑ food intake	↓ Brain volume; ↓ BDNF; ↓ dopamine	[83, 84]
ERK1 Knock-out	↑ Amphetamine sensibility; ↓ learing in fear conditioning; ↑ locomotion; ↓ immobility in forced swim	↓ Phospho-RSK1/3 in PFC and striatum; shift of activity rhythm	[85–86]
DAT Knock-down	↑ Locomotion; ↓ anxiety; ↑ rearing	↑ Dopamine	[87–89]
Environmental stress			
Sleep deprivation	↑ Locomotion; ↑ agressive behavior; ↑ exploratory behavior	—	[90, 91]
Photoperiod lenghts	↑ Anxiety; ↑ helplessness	Switch in dopamine neurotransmission to somatostatin	[92]
Sensitization model			
Chronic amphetamine administration followed by withdrawal	↑ Locomotion; ↑ anxiety; ↑ anhedonia; ↓ motivation; ↓ working memory	↓ Dopamine responsiveness ↑ serotonin sensitivty	[94–96]

BDNF = brain-derived neurotrophic factor; ERK1 = Extracellular signal-regulated kinase 1; DAT = dopamine transporter.

Table 4.
Examples of models for BD induced by genetic manipulation, environmental stressors, and induced by sensitization, which lead to some aspects of molecular and behavioral changes related to BD.

2.4 Animal models in attention-deficit/hyperactivity disorder (ADHD)

Attention-deficit/hyperactivity disorder (ADHD) is a neurodevelopmental disorder, affecting approximately 2.2–2.8% of worldwide, with multifactorial inducement, as reflected by the heterogeneity found in this disorder, and as indicated by the diversity in its psychiatric comorbidities [97]. This disorder is defined by inappropriate levels of attention deficits and/or hyperactivity behavior, which directly interfere with the normal life and functioning of an individual [98]. While there is no cure for ADHD, currently available treatments can help reducing the symptoms and improving the general functioning, although with a peculiar wide variability due to the clinically and scientifically difficulties to exactly determine the specificity and the origin of the symptoms [99]. As for other PDs, due to the high heritability, animal models for ADHD are mostly derived from genetic alterations or breeding selection or from neonatal insults that can lead to neurodevelopmental changes. Models related to dopaminergic neurotransmission are also important to evaluate ADHD, as also listed in **Table 2**, and which includes the administration of psychostimulants as amphetamine. A more detailed review on ADHD animal models can be found elsewhere [100] (**Table 5**).

Model	Endophenotype	Molecular alterations	References
Genetic manipulation			
Spontaneously hypertensive rats	↓ Attention; ↑ motor impulsiveness ↑ Locomotion; ↑ exploratory behavior	↑ Dopamine ↓ Dopamine transporter 1 expression ↓ Brain volume	[101–104]
Coloboma mouse mutant	↑ Locomotion; ↑ exploratory behavior; ↑ amphetamine sensibility	↑ Noraedrenergic function ↓ Dopamine ↓ DOPAC and HVA	[105–108]
Neonatal insults			
6-hydroxydopamine	↓ Working memory; ↑ locomotion; ↑ Exploratory behavior	↓ Dopamine ↑ Dopamine receptor 4 ↓ Serotonin transporter binding in striatum	[109–111]
Neonatal anoxia	↑ Locomotion; ↑ exploratory behavior; ↓ spatial memory	Transient changes in neurotransmitters ↑ Dopamine turnover ↓ Noraepinephrine and 5-HIAA ↓ CA1 cell density	[112–114]

DOPAC = 3,4-Dihydroxyphenylacetic acid; HVA = Homovanillic acid; 5-HIIA = 5-Hydroxyindoleacetic acid.

Table 5.
Examples of models for ADHD induced by genetic manipulation in susceptibility genes and selective inbreeding and by prenatal insults.

3. Conclusion

There is a consensus about the critical role of animal models for the advance and understanding the functioning of brain and brain disorders, as well as for the development of new treatments. However, it is important to use them judiciously and avoid the over interpretations derived for the findings, as it is noticeable that the results obtained on experimental animals are not necessarily confirmed in clinical studies.

As it has been shown, there are several approaches to obtain an animal model for studies in psychiatry, but there is still a limitation in reproducing all the conditions involved in the pathophysiology of the disorder, and it is extremely crucial to recognize this limitation. An alternative that has proved to be efficient is to direct the study to a specific symptom domain that can answer at least in part, the significance of these findings to concretely improve the knowledge in PDs, and thereby bring advances in treatment. The crisis of the classification system is evidenced in the diagnostic inflation in psychiatry, which adds complexity to the preclinical research and complicates the modeling of PDs within the available experimental laboratory animals. The recent and alternative approaches as the RDoC to study the brain and behavior are in a relative infancy, but promises bringing new perspectives in how models that can be improved to become indeed helpful to benefit the quality of life of patients with PDs.

Acknowledgements

This work was supported by the São Paulo Research Foundation (Fundação de Amparo à Pesquisa do Estado de São Paulo) (FAPESP No. 2013/13392-4 and 2017/02413-1 for M.A.F.H) and the National Council of Technological and Scientific Development (Conselho Nacional de Desenvolvimento Científico e Tecnológico - CNPq) (477760/2010-4, 557753/2010-4; 508113/2010-5; 311815/2012-0; 475739/2013-2; 311815/2012-0 and 309337/2016-0 for M.A.F.H). Both João V. Nani and Benjamín Rodríguez receive fellowship from CAPES. This study was financed in part by the Coordenação de Aperfeiçoamento de Pessoal de Nível Superior (CAPES), Brazil - Finance Code 001.

Author details

João Victor Nani[1,2], Benjamín Rodríguez[1], Fabio Cardoso Cruz[1] and Mirian Akemi Furuie Hayashi[1,2]*

1 Department of Pharmacology, Escola Paulista de Medicina (EPM), Universidade Federal de São Paulo (UNIFESP), São Paulo, SP, Brazil

2 National Institute for Translational Medicine (INCT-TM, CNPq/FAPESP/CAPES), Brazil

*Address all correspondence to: mhayashi@unifesp.br; mirianhayashi@yahoo.com

References

[1] Sansone RA, Sansone LA. Psychiatric disorders: A global look at facts and figures. Psychiatry. 2010;**7**(12):16-19

[2] Swinney DC, Anthony J. How were new medicines discovered? Nature Reviews Drug Discovery. 2011;**10**(7):507-519

[3] Alexandrov V, Brunner D, Hanania T, Leahy E. High-throughput analysis of behavior for drug discovery. European Journal of Pharmacology. 2015;**750**(5):82-89

[4] Salgado JV, Sandner G. A critical overview of animal models of psychiatric disorders: Challenges and perspectives. Revista Brasileira de Psiquiatria. 2013;**35**(2):77-81

[5] Lema YY, Gamo NJ, Yang K, Ishizuka K. Trait and state biomarkers for psychiatric disorders: Importance of infrastructure to bridge the gap between basic and clinical research and industry. Psychiatry and Clinical Neurosciences. 2018;**72**(7):482-489

[6] Kesby JP, Eyles DW, McGrath JJ, Scott JG. Dopamine, psychosis and schizophrenia: The widening gap between basic and clinical neuroscience. Translational Psychiatry. 2018;**8**:30

[7] McKinney WT, Bunney WE. Animal model of depression. I. Review of evidence: Implications for research. Archives of General Psychiatry. 1969;**21**(2):240-248

[8] Willner P. The validity of animal models of depression. Psychopharmacology. 1984;**83**(1):1-16

[9] Belzung C, Lemoine M. Criteria of validity for animal models of psychiatric disorders: Focus on anxiety disorders and depression. Biology of Mood & Anxiety Disorders. 2011;**1**(1):9

[10] Vervliet B, Raes F. Criteria of validity in experimental psychopathology: Application to models of anxiety and depression. Psychological Medicine. 2012;**43**(11):2241-2244

[11] Lewis E. A gene complex controlling segmentation in drosophila. Nature. 1978;**276**(5688):565-570

[12] Kandel ER. The molecular biology of memory storage: A dialogue between genes and synapses. Science. 2001;**294**(5544):1030-1038

[13] Hulme SE, Whitesides GM. Chemistry and the worm: Caenorhabditis elegans as a platform for integrating chemical and biological research. Angewandte Chemie International Edition in English. 2011;**50**(21):4774-4807

[14] Surís A, Holliday R, North CS. The evolution of the classification of psychiatric disorders. Behavioral Science. 2016;**6**(1):5

[15] Insel T, Cuthbert B, Garvey M, Heinssen R, Pine DS, Quinn K, et al. Research domain criteria (RDoC): Toward a new classification framework for research on mental disorders. American Journal of Psychiatry. 2010;**167**(7):748-751

[16] Cuthbert BN, Insel TR. Toward new approaches to psychotic disorders: The NIMH research domain criteria project. Schizophrenia Bulletin. 2010;**36**(6):1061-1062

[17] Cuthbert BN. The RDoC framework: Facilitating transition from ICD/DSM to dimensional approaches that integrate neuroscience and psychopathology. World Psychiatry. 2014;**3**(1):28-35

[18] Greenwood TA, Shutes-David A, Tsuang DW. Endophenotypes in schizophrenia: Digging deeper to identify genetic mechanisms. Journal

of Psychiatry and Brain Science. 2019;**4**(2):e190005

[19] Nestler EJ, Hyman SE. Animal models of neuropsychiatric disorders. Nature Neuroscience. 2010;**13**(10):1161-1169

[20] Owen M, Sawa A, Mortensen P. Schizophrenia. The Lancet. 2016; **388**(10039):86-97

[21] Jones CA, Watson DJ, Fone KC. Animal models of schizophrenia. British Journal of Pharmacology. 2011;**164**(4):1162-1194

[22] Murphy C, Fend M, Russig H, Feldon J. Latent inhibition, but not prepulse inhibition, is reduced during withdrawal from an escalating dosage schedule of amphetamine. Behavioral Neuroscience. 2001;**115**(6):1247-1256

[23] Peleg-Raibstein D, Sydekum E, Russig H, Feldon J. Withdrawal from repeated amphetamine administration leads to disruption of prepulse inhibition but not to disruption of latent inhibition. Journal of Neural Transmission. 2005;**113**(9):1323-1336

[24] Martinez V, Parikh V, Sarter M. Sensitized Attentional performance and Fos-Immunoreactive cholinergic neurons in the basal forebrain of amphetamine-pretreated rats. Biological Psychiatry. 2005;**57**(10):1138-1146

[25] Tenn C, Fletcher P, Kapur S. Amphetamine-sensitized animals show a sensorimotor gating and neurochemical abnormality similar to that of schizophrenia. Schizophrenia Research. 2003;**64**(2-3):103-114

[26] Turner K, Burne T. Improvement of attention with amphetamine in low- and high-performing rats. Psychopharmacology. 2016;**233**(18):3383-3394

[27] Castañé A, Santana N, Artigas F. PCP-based mice models of schizophrenia: Differential behavioral, neurochemical and cellular effects of acute and subchronic treatments. Psychopharmacology. 2015;**232**(21-22):4085-4097

[28] Mouri A, Noda Y, Enomoto T, Nabeshima T. Phencyclidine animal models of schizophrenia: Approaches from abnormality of glutamatergic neurotransmission and neurodevelopment. Neurochemistry International. 2007;**51**(2-4):173-184

[29] Neill J, Barnes S, Cook S, Grayson B, Idris N, McLean S, et al. Animal models of cognitive dysfunction and negative symptoms of schizophrenia: Focus on NMDA receptor antagonism. Pharmacology & Therapeutics. 2010;**128**(3):419-432

[30] Jaaro-Peled H. Gene models of schizophrenia: DISC1 mouse models. Progress in Brain Research. 2009;**179**:75-86

[31] Clapcote S, Lipina T, Millar J, Mackie S, Christie S, Ogawa F, et al. Behavioral phenotypes of disc1 missense mutations in mice. Neuron. 2007;**54**(3):387-402

[32] Lee F, Fadel M, Preston-Maher K, Cordes S, Clapcote S, Price D, et al. Disc1 point mutations in mice affect development of the cerebral cortex. Journal of Neuroscience. 2011;**31**(9):3197-3206

[33] Pletnikov M, Ayhan Y, Nikolskaia O, Xu Y, Ovanesov M, Huang H, et al. Inducible expression of mutant human DISC1 in mice is associated with brain and behavioral abnormalities reminiscent of schizophrenia. Molecular Psychiatry. 2008;**13**(2):173-186

[34] Hikida T, Jaaro-Peled H, Seshadri S, Oishi K, Hookway C, Kong S, et al. Dominant-negative DISC1 transgenic mice display schizophrenia-associated phenotypes detected by measures

translatable to humans. Proceedings of the National Academy of Sciences. 2007;**104**(36):14501-14506

[35] Niwa M, Kamiya A, Murai R, Kubo K, Gruber A, Tomita K, et al. Knockdown of DISC1 by in utero gene transfer disturbs postnatal dopaminergic maturation in the frontal cortex and leads to adult behavioral deficits. Neuron. 2010;**65**(4):480-489

[36] Trossbach S, Bader V, Hecher L, Pum M, Masoud S, Prikulis I, et al. Misassembly of full-length disrupted-in-schizophrenia 1 protein is linked to altered dopamine homeostasis and behavioral deficits. Molecular Psychiatry. 2016;**21**(11):1561-1572

[37] Mei L, Xiong W. Neuregulin 1 in neural development, synaptic plasticity and schizophrenia. Nature Reviews Neuroscience. 2008;**9**(6):437-452

[38] Papaleo F, Yang F, Garcia S, Chen J, Lu B, Crawley J, et al. Dysbindin-1 modulates prefrontal cortical activity and schizophrenia-like behaviors via dopamine/D2 pathways. Molecular Psychiatry. 2010;**17**(1):85-98

[39] Karlsgodt K, Robleto K, Trantham-Davidson H, Jairl C, Cannon T, Lavin A, et al. Reduced dysbindin expression mediates n-methyl-d-aspartate receptor hypofunction and impaired working memory performance. Biological Psychiatry. 2011;**69**(1):28-34

[40] Sams-Dodd F, Lipska B, Weinberger D. Neonatal lesions of the rat ventral hippocampus result in hyperlocomotion and deficits in social behaviour in adulthood. Psychopharmacology. 1997;**132**(3):303-310

[41] Lipska B. Using animal models to test a neurodevelopmental hypothesis of schizophrenia. Journal of Psychiatry & Neuroscience. 2004;**29**(4):282-286

[42] Moore H, Jentsch J, Ghajarnia M, Geyer M, Grace A. A neurobehavioral systems analysis of adult rats exposed to methylazoxymethanol acetate on E17: Implications for the neuropathology of schizophrenia. Biological Psychiatry. 2006;**60**(3):253-264

[43] Meyer U. Prenatal poly(I:C) exposure and other developmental immune activation models in rodent systems. Biological Psychiatry. 2014;**75**(4):307-315

[44] Winship I, Dursun S, Baker G, Balista P, Kandratavicius L, Maia-de-Oliveira J, et al. An overview of animal models related to schizophrenia. The Canadian Journal of Psychiatry. 2018;**64**(1):5-17

[45] Belmaker RH, Agam G. Major Depressive Disorder. The New England Journal of Medicine. 2008;**358**:55-68

[46] Redei EE, Mehta NS. The promise of biomarkers in diagnosing major depression in primary care: The present and future. Current Psychiatry Reports. 2015;**17**(8):601

[47] Woods AG, Iosifescu DV, Darie CC. Biomarkers in major depressive disorder: The role of mass spectrometry. Advances in Experimental Medicine and Biology. 2014;**806**:545-560

[48] Caspi A, Moffitt TE. Gene-environment interactions in psychiatry: Joining forces with neuroscience. Nature Reviews Neuroscience. 2006;7(7):583-590

[49] McGonagle KA, Kessler RC. Chronic stress, acute stress, and depressive symptoms. American Journal of Community Psychology. 1990;**18**(5):681-706

[50] Uher R, McGuffin P. The moderation by the serotonin transporter gene of environmental adversity in the etiology of depression: 2009 update. Molecular Psychiatry. 2010;**15**(1):18-22

[51] Wang Q, Timberlake MA 2nd, Prall K, Dwivedi Y. The recent progress in animal models of depression. Progress in Neuro-Psychopharmacology & Biological Psychiatry. 2017;77:99-109

[52] Weiss J, Simson PG, Ambrose M, Webster A, Hoffman L. Neurochemical basis of behavioral depression. Advances in Behavioral Medicine. 1985;**1**:233-275

[53] Weiss J, Bailey WH, Pohorecky LA, Korzeniowski D, Grillione G. Stress-induced depression of motor activity correlates with regional changes in brain norepinephrine but not in dopamine. Neurochemical Research. 1980;**5**(1):9-22

[54] Zacharko R, Bowers W, Kokkinidis L, Anisma H. Region-specific reductions of intracranial self-stimulation after uncontrollable stress: Possible effects on reward processes. Behavioural Brain Research. 1983;**9**(2):129-141

[55] Corum R, Thurmond J. Effects of acute exposure to stress on subsequent aggression and locomotion performance. Psychosomatic Medicine. 1977;**39**(6):436-443

[56] Dwivedi Y, Mondal A, Shukla P, Rizavi H, Lyons J. Altered protein kinase a in brain of learned helpless rats: Effects of acute and repeated stress. Biological Psychiatry. 2004;**56**(1):30-40

[57] Aznar S, Klein A, Santini M, Knudsen G, Henn F, Gass P, et al. Aging and depression vulnerability interaction results in decreased serotonin innervation associated with reduced BDNF levels in hippocampus of rats bred for learned helplessness. Synapse. 2010;**64**(7):561-565

[58] Katz R, Roth K, Carroll B. Acute and chronic stress effects on open field activity in the rat: Implications for a model of depression. Neuroscience & Biobehavioral Reviews. 1981;**5**(2):247-251

[59] Willner P, Muscat R, Papp M. Chronic mild stress-induced anhedonia: A realistic animal model of depression. Neuroscience & Biobehavioral Reviews. 1992;**16**(4):525-534

[60] Boyle M, Brewer J, Funatsu M, Wozniak D, Tsien J, Izumi Y, et al. Acquired deficit of forebrain glucocorticoid receptor produces depression-like changes in adrenal axis regulation and behavior. Proceedings of the National Academy of Sciences. 2005;**102**(2):473-478

[61] Crema L, Schlabitz M, Tagliari B, Cunha A, Simão F, Krolow R, et al. Na+, K+ ATPase activity is reduced in amygdala of rats with chronic stress-induced anxiety-like behavior. Neurochemical Research. 2010;**35**(11):1787-1795

[62] Conrad C, Magariños A, LeDoux J, McEwen B. Repeated restraint stress facilitates fear conditioning independently of causing hippocampal CA3 dendritic atrophy. Behavioral Neuroscience. 1999;**113**(5):902-913

[63] Wood G, Young L, Reagan L, McEwen B. Acute and chronic restraint stress alter the incidence of social conflict in male rats. Hormones and Behavior. 2003;**43**(1):205-213

[64] Zhang L, Luo J, Zhang M, Yao W, Ma X, Yu S. Effects of curcumin on chronic, unpredictable, mild, stress-induced depressive-like behaviour and structural plasticity in the lateral amygdala of rats. The International Journal of Neuropsychopharmacology. 2014;**17**(05):793-806

[65] Koolhaas J, Meerlo P, De Boer S, Strubbe J, Bohus B. The temporal dynamics of the stress response. Neuroscience & Biobehavioral Reviews. 1997;**21**(6):775-782

[66] Crawford L, Rahman S, Beck S. Social stress alters inhibitory synaptic

input to distinct subpopulations of raphe serotonin neurons. ACS Chemical Neuroscience. 2013;**4**(1):200-209

[67] Hollis F, Kabbaj M. Social defeat as an animal model for depression. ILAR Journal. 2014;**55**(2):221-232

[68] Liu H, Atrooz F, Salvi A, Salim S. Behavioral and cognitive impact of early life stress: Insights from an animal model. Progress in Neuro-Psychopharmacology & Biological Psychiatry. 2017;**78**:88-95

[69] Boulle F, Pawluski JL, Homberg JR, Machiels B, Kroeze Y, Kumar N, et al. Prenatal stress and early-life exposure to fluoxetine have enduring effects on anxiety and hippocampal BDNF gene expression in adult male offspring. Developmental Psychobiology. 2016;**58**(4):427-438

[70] Harkin A, Kelly J, Leonard B. A review of the relevance and validity of olfactory bulbectomy as a model of depression. Clinical Neuroscience Research. 2003;**3**(4-5):253-262

[71] Hellweg R, Zueger M, Fink K, Hörtnagl H, Gass P. Olfactory bulbectomy in mice leads to increased BDNF levels and decreased serotonin turnover in depression-related brain areas. Neurobiology of Disease. 2007;**25**(1):1-7

[72] Nam H, Clinton S, Jackson N, Kerman I. Learned helplessness and social avoidance in the Wistar-Kyoto rat. Frontiers in Behavioral Neuroscience. 2014;**8**:109

[73] Will C, Aird F, Redei E. Selectively bred Wistar–Kyoto rats: An animal model of depression and hyper-responsiveness to antidepressants. Molecular Psychiatry. 2003;**8**(11):925-932

[74] Zangen A, Overstreet D, Yadid G. High serotonin and 5-hydroxyindoleacetic acid levels in limbic brain regions in a rat model of depression; normalization by chronic antidepressant treatment. Journal of Neurochemistry. 2002;**69**(6):2477-2483

[75] Overstreet D, Russell R. Selective breeding for diisopropyl fluorophosphate-sensitivity: Behavioural effects of cholinergic agonists and antagonists. Psychopharmacology. 1982;**78**(2):150-155

[76] Overstreet D, Friedman E, Mathé A, Yadid G. The Flinders sensitive line rat: A selectively bred putative animal model of depression. Neuroscience & Biobehavioral Reviews. 2005;**29**(4-5):739-759

[77] Nishi K, Kanemaru K, Hasegawa S, Watanabe A, Diksic M. Both acute and chronic buspirone treatments have different effects on regional 5-HT synthesis in Flinders sensitive line rats (a rat model of depression) than in control rats. Neurochemistry International. 2009;**54**(3-4):205-214

[78] Merikangas KR, Jin R, He JP, Kessler RC, Lee S, Sampson NA, et al. Prevalence and correlates of bipolar spectrum disorder in the world mental health survey initiative. Archives of General Psychiatry. 2011;**68**(3):241-251

[79] Grande I, Berk M, Birmaher B, Vieta E. Bipolar disorder. Lancet. 2016;**387**(10027):1561-1572

[80] Johansson V, Hultman CM, Kizling I, Martinsson L, Borg J, Hedman A, et al. The schizophrenia and bipolar twin study in Sweden (STAR). Schizophrenia Research. 2019;**204**:183-192

[81] Sanchez-Moreno J, Martinez-Aran A, Vieta E. Treatment of functional impairment in patients with bipolar disorder. Current Psychiatry Reports. 2017;**19**(1):3

[82] Lewandowski KE, Whitton AE, Pizzagalli DA, Norris LA, Ongur D,

Hall MH. Reward learning, neurocognition, social cognition, and symptomatology in psychosis. Frontiers in Psychiatry. 2016;7:100

[83] Beyer D, Freund N. Animal models for bipolar disorder: From bedside to the cage. International Journal of Bipolar Disorders. 2017;**5**(1):35

[84] Magariños A, Li C, Gal Toth J, Bath K, Jing D, Lee F, et al. Effect of brain-derived neurotrophic factor haploinsufficiency on stress-induced remodeling of hippocampal neurons. Hippocampus. 2011;**21**(3):253-264

[85] Kernie S. BDNF regulates eating behavior and locomotor activity in mice. The EMBO Journal. 2000;**19**(6):1290-1300

[86] Engel S, Creson T, Hao Y, Shen Y, Maeng S, Nekrasova T, et al. The extracellular signal-regulated kinase pathway contributes to the control of behavioral excitement. Molecular Psychiatry. 2008;**14**(4):448-461

[87] Young JW, Cope ZA, Romoli B, Schurs E, Joosen A, Enkhuzien J, et al. Mice with reduced DAT levels recreate seasonal-induced switching between states in bipolar disorder. Neuropsychopharmacology. 2018;**43**(8):1732-1731

[88] van Enkhuizen J, Henry B, Minassian A, Perry W, Milienne-Petiot M, Higa K, et al. Reduced dopamine transporter functioning induces high-reward risk-preference consistent with bipolar disorder. Neuropsychopharmacology. 2014;**39**(13):3112-3122

[89] Giros B, Jaber M, Jones S, Wightman R, Caron M. Hyperlocomotion and indifference to cocaine and amphetamine in mice lacking the dopamine transporter. Nature. 1996;**379**(6566):606-612

[90] Benedetti F, Fresi F, Maccioni P, Smeraldi E. Behavioural sensitization to repeated sleep deprivation in a mice model of mania. Behavioural Brain Research. 2008;**187**(2):221-227

[91] Gessa G, Pani L, Fadda P, Fratta W. Sleep deprivation in the rat: An animal model of mania. European Neuropsychopharmacology. 1995;**5**:89-93

[92] Dulcis D, Jamshidi P, Leutgeb S, Spitzer N. Neurotransmitter switching in the adult brain regulates behavior. Science. 2013;**340**(6131):449-453

[93] Paulson P, Camp D, Robinson T. Time course of transient behavioral depression and persistent behavioral sensitization in relation to regional brain monoamine concentrations during amphetamine withdrawal in rats. Psychopharmacology. 1991;**103**(4):480-492

[94] Barr A, Fiorino D, Phillips A. Effects of withdrawal from an escalating dose schedule of d-amphetamine on sexual behavior in the male rat. Pharmacology Biochemistry and Behavior. 1999;**64**(3):597-604

[95] Barr A, Phillips A. Increased successive negative contrast in rats withdrawn from an escalating-dose schedule of d-amphetamine. Pharmacology Biochemistry and Behavior. 2002;**71**(1-2):293-299

[96] Marszalek-Grabska M, Gibula-Bruzda E, Jenda M, Gawel K, Kotlinska J. Memantine improves memory impairment and depressive-like behavior induced by amphetamine withdrawal in rats. Brain Research. 1642;**2016**:389-396

[97] Luo Y, Weibman D, Halperin JM, Li X. A review of heterogeneity in attention deficit/hyperactivity disorder (ADHD). Frontiers in Human Neuroscience. 2019;**13**:42

[98] Franke B, Michelini G, Asherson P, Banaschewski T, Bilbow A, Buitelaar JK, et al. Live fast, die young? A review on the developmental trajectories of ADHD across the lifespan. European Neuropsychopharmacology. 2018;**28**(10):1059-1088

[99] Nigg JT. Attention-deficit/hyperactivity disorder and adverse health outcomes. Clinical Psychology Review. 2013;**33**(2):215-228

[100] Russell VA, Sagvolden T, Johansen E. Animal models of attention-deficit hyperactivity disorder. Behavioral and Brain Functions. 2005;**1**:9

[101] Sagvolden T, Russell V, Aase H, Johansen E, Farshbaf M. Rodent models of attention-deficit/hyperactivity disorder. Biological Psychiatry. 2005;**57**(11):1239-1247

[102] Sagvolden T. Behavioral validation of the spontaneously hypertensive rat (SHR) as an animal model of attention-deficit/hyperactivity disorder (AD/HD). Neuroscience & Biobehavioral Reviews. 2000;**24**(1):31-39

[103] Carboni E, Silvagni A, Valentini V, Di Chiara G. Effect of amphetamine, cocaine and depolarization by high potassium on extracellular dopamine in the nucleus accumbens shell of SHR rats. An *in vivo* microdyalisis study. Neuroscience & Biobehavioral Reviews. 2003;**27**(7):653-659

[104] Linthorst A, van Giersbergen P, Gras M, Versteeg D, de Jong W. The nigrostriatal dopamine system: Role in the development of hypertension in spontaneously hypertensive rats. Brain Research. 1994;**639**(2):261-268

[105] Wilson M. Coloboma mouse mutant as an animal model of hyperkinesis and attention deficit hyperactivity disorder. Neuroscience & Biobehavioral Reviews. 2000;**24**(1):51-57

[106] Jones M, Williams M, Hess E. Expression of catecholaminergic mRNAs in the hyperactive mouse mutant coloboma. Molecular Brain Research. 2001;**96**(1-2):114-121

[107] Raber J, Mehta P, Kreifeldt M, Parsons L, Weiss F, Bloom F, et al. Coloboma hyperactive mutant mice exhibit regional and transmitter-specific deficits in neurotransmission. Journal of Neurochemistry. 2002;**68**(1):176-186

[108] Jones M, Hess E. Norepinephrine regulates locomotor hyperactivity in the mouse mutant coloboma. Pharmacology Biochemistry and Behavior. 2003;**75**(1):209-216

[109] Luthman J, Fredriksson A, Lewander T, Jonsson G, Archer T. Effects ofd-amphetamine and methylphenidate on hyperactivity produced by neonatal 6-hydroxydopamine treatment. Psychopharmacology. 1989;**99**(4):550-557

[110] Zhang K. Role of dopamine D4 receptors in motor hyperactivity induced by neonatal 6-hydroxydopamine lesions in rats. Neuropsychopharmacology. 2001;**25**(5):624-632

[111] Zhang K, Davids E, Tarazi F, Baldessarini R. Serotonin transporter binding increases in caudate-putamen and nucleus accumbens after neonatal 6-hydroxydopamine lesions in rats: Implications for motor hyperactivity. Developmental Brain Research. 2002;**137**(2):135-138

[112] Dell'Anna M. Neonatal anoxia induces transitory hyperactivity, permanent spatial memory deficits and CA1 cell density reduction in developing rats. Behavioural Brain Research. 1999;**45**:125-134

[113] Dell'Anna M, Calzolari S, Molinari M, Iuvone L, Calimici R. Neonatal anoxia induces transitory hyperactivity, permanent spatial memory

deficits and CA1 cell density reduction in developing rats. Behavioural Brain Research. 1991;**45**(2):125-134

[114] Iuvone L, Geloso M, Dell'Anna E. Changes in open field behavior, spatial memory, and hippocampal parvalbumin immunoreactivity following enrichment in rats exposed to neonatal anoxia. Experimental Neurology. 1996;**139**(1):25-33

Chapter 5

Zucker Diabetic Fatty Rats for Research in Diabetes

Marcela Capcarova and Anna Kalafova

Abstract

The rising incidence of diabetes mellitus (DM) worldwide presents a global public health problem. DM is classified into two main groups: type 1 (T1DM) and type 2 (T2DM). T1DM requires insulin treatment. T2DM is complex, heterogeneous, polygenic disease defined primarily by insulin resistance, ongoing hyperglycemia, and β cells' dysfunction. For research in diabetes, an appropriate experimental model reflecting symptoms and complications of human T2DM is required for understanding the pathogenesis, molecular nature, and the possibilities of the treatment. Among the many animal models, rodent models that develop DM spontaneously are frequently used in the studies due to their similarity to the humans and economic effectiveness. This work gives a detailed overview of the literature, covering the characteristic of DM, its symptoms and complications, the description of Zucker diabetic fatty (ZDF) rats as an appropriate model for research in T2DM, and the possibility of the treatment.

Keywords: diabetes, animal model, pancreatic β cells, Zucker diabetic fatty rats, treatment

1. Introduction

1.1 Diabetes mellitus

Diabetes mellitus (DM) is often incident endocrine disorder in many countries [1]. The International Diabetes Federation reported that 6 million people die directly from diabetes every year, and additional 318 million people are suffering with DM. This number is predicted to reach 642 million by 2040 [2] and 693 million by 2045 [3]. DM is a heterogeneous group of chronic disease characterized by a relative or absolute lack of insulin resulting in hyperglycemia [4]. It causes a variety of complications as cardiovascular disease, renal failure, neuropathy, and retinopathy [5]. Chronic hyperglycemia mostly deteriorates the vascular tree and promotes the development of micro- and macrovascular disease [6]. It was reported that hyperglycemia accelerates the development of DM complications through some mechanisms such as increased aldose reductase-related polyol pathway flux, formation of advanced glycation end products (AGEs), increased hexosamine pathway flux, activation of protein kinase C isoforms, and rising generation of reactive oxygen species [7]. Metabolic imbalance in the peripheral nervous system that is activated in the diabetic milieu of hyperglycemia, impaired insulin signaling, and dyslipidemia are the key parameters in the development of diabetic neuropathy [8]. The determining points involve multiple mechanisms of glucose toxicity including polyol pathway activity, hexosamine pathway, nonenzymatic glycations of proteins, and altered protein kinase C activity [9]. Activation of these pathways can eventually flow into inflammatory and oxidative stress in neurons and adjacent microvascular system [10].

DM is divided into two main forms—type 1 and type 2 [4]. Type 1 diabetes mellitus (T1DM) or insulin-dependent diabetes mellitus (IDDM) is an autoimmune disease and is a result of β cells' death, because a foreign protein is incorporated into islet β cells. In response, lymphocytes attack the foreign protein and unwillingly destroy β cells as collateral damage. It causes an absolute insulin deficiency [11]. It is uncertain what activates the autoimmune response, but some environmental factors as toxins, viral infections, and psychosocial inputs are thought to play a plumbless role [12].

Historically, the usual ratio for T1DM to T2DM has been 1:20. Now it is changing because of expressive increase in the incidence of T2DM in children and young people [11].

1.2 Diabetes mellitus type 2

Type 2 diabetes mellitus (T2DM) or non-insulin-dependent diabetes mellitus (NIDDM) is a syndrome of β cells' dysfunction including relative insulin deficiency associated with insulin resistance [11] and compensatory increases in insulin secretion [13]. It is associated with incorrect sensing of glucose signals by the β cells. T2DM is linked to a stage of insulin resistance. Insulin secreted by the β cells and bound to liver, muscle, and fat cells is subnormally efficacious in carrying out its metabolic action [11]. Generally, T2DM is characterized by the incapability of the pancreatic β cells to secrete appropriate quantities of insulin in order to offset hyperglycemia arising from peripheral insulin resistance and increases hepatic glucose output [14]. It is a multifactorial and complex disorder [13] that is estimated to affect more than 100 million people worldwide [15]. About 80% of all people with diabetes suffer from T2DM [16]. Insulin resistance alone is insufficient to cause diabetes. A progression to overt diabetes required β cells' failure as well [16–18]. Insulin resistance is associated with decrease in insulin receptors in target tissues (muscle, fat, or liver) and insulin receptor kinase activity that causes decrease in glucose transporter 4 (GLUT 4) translocation due to impaired signaling [19]. The onset of T2DM is preceded by an expressive increase in the plasma levels of free fatty acids (FFA) and by sixfold rise in triglyceride (TG) concentration in the pancreatic islets [20]. Chronic exposure to high glucose level and rising FFA concentration is detrimental to β cell function. This situation results in weak glucose-induced insulin secretion and rising level of apoptosis [21].

In spite of the increasing number of T2DM, little is known about the prevention of the disease and its complications at early stages [22]. In this stage insulin-sensitive tissue such as adipose tissue and skeletal muscle become insulin resistant. This causes the development of impaired glucose tolerance, and it can occur over a few years [23].

The pathogenesis of T2DM is complex and is primarily related to gene variation, external and internal environmental factors, abnormal protein modifications, oxidative stress, epigenetic effects, and energy metabolism disorders [24]. It was revealed that also the gut microbiota has been recognized as a key contributor to T2DM, and T2DM is linked to dysbiosis of the intestinal microbiota [25].

1.3 Obesity in type 2 diabetes

Onset and development of T2DM is commonly incurred by several factors, which are combined with lifestyle, obesity, genetic defects, virus infection, and drugs [16]. Obesity is defined as a pathological excess of body fat that results from a permanent positive energy balance [26]. Persistent positive energy balance is pertinent to increased storage of triglycerides. This expands the adipose depots and increases the proportion of hypertrophied adipocytes [27]. Under condition of obesity, the lipid storage capacity of adipocytes is overcome, resulting in adipocyte-derived fatty acids and cytokines leaking into the circulation [28]. Damaging lipid species accumulates in ectopic tissue causing local inflammation and provides lipotoxicity [29]. Lipotoxicity determines an important link between obesity, insulin resistance, and T2DM. It interprets the harmful cellular effects of chronically increased concentrations of fatty acids and excess lipid accumulation in tissues other than adipose tissue. Excess adiposity is considered to promote the onset and severity of insulin resistance, contributing to emergence and progression of impaired glucose tolerance and T2DM [27].

Obesity-induced insulin resistance accelerates pancreatic islet exhaustion and thus the onset of T2DM [13]. Generally, obesity is a major risk factor for developing T2DM [30]. High-fat diet applied in animal's model that has inclination to DM results in obesity, hyperinsulinemia, and altered glucose homeostasis due to insufficient compensation by the islets [31]. Whereupon it is required in human population suffering T2DM to follow diet regimes and restriction of energy in the food so to maintain glucose concentration in acceptable level. In this case the diet has more considerable impact on diabetic primary complications than genetic predisposition [32]. Genetic disposition to obesity is probably commonly due to the small impingements of a wide selection of genes such as those encoding the beta3adrenoceptor, PPARγ and its co-activator-1, fat mass and obesity-associated gene, and adiponectin and a selection of genes that could potentially influence behavior and hypothalamic hunger-satiety mechanisms [33].

Currently, therapeutic strategies for T2DM are limited. They involve insulin and four main classes of oral antidiabetic agents in order to stimulate pancreatic insulin secretion. However, these agents suffer from generally inadequate efficacy and various adverse effects. So there is the possibility to try new therapeutic agents or treatments, most of them are under preclinical and early clinical stages [34].

2. Zucker diabetic fatty rats

An animal model for biomedical investigation is one in which normative biology, behavior, and pathological process can be studied and in which the phenomenon in one or more respects resembles the same phenomenon in humans [35]. Research in diabetes on humans is not possible or only partially possible. Hence, animal model of DM is very useful and advantageous [36]. Animal disease models are essential tools for studying the pathophysiology of DM enabling therapeutic interventions to be developed [37]. It is true that the present therapeutic approaches to treat DM and obesity, which are saving many lives every day, were invented, validated, and optimized on animal models [38]. When studying T2DM the use of an animal model with a homogenous genetic background is advised [39]. Most of the available models are based on rodents [36]. Rodents are most commonly utilized due to their small size, short generation interval, and easy availability [39] and because of economic consideration [36]. Being mammals, the physiology of rats is similar to humans than nonmammalian species [40]. Nevertheless, nonrodent models of diabetes are needed as a valuable supplement to rodents for both practical and physiological reasons with respect to humans [36]. Many animal models for DM research are obese, reflecting the human condition where obesity is closely related to T2DM development [41]. Animals exhibiting a syndrome of insulin resistance and T2DM reflecting the human disease involve many species with genetic, nutritional, or experimental causation [36]. There are a lot of rodent models available for the research in T2DM, but some of them may not always be satisfactory to mirror human T2DM due to the large heterogeneity in the latter. There are no fully unified classification criteria for this type of animal model. But, the spontaneous type 2 diabetic rodent models are considered the most outstanding and most useful [42].

According to Srinivasan and Ramarao [36], spontaneous diabetic animal models have special advantages and also disadvantages. The advantages are:

a. The development of T2DM is of spontaneous origin involving genetic factors.

b. Animals develop characteristic features resembling human T2DM.

c. Most of inbred animal model in which the genetic background is homogeneous and environmental factors can be controlled allow genetic dissection of this multifactorial disease easy.

d. Variability of results is minimal and it required smaller sample size.

Among the disadvantages are mainly:

a. Highly inbred, homogenous, and mostly monogenic inheritance and development of diabetes are highly genetically determined unlike heterogeneity in humans.

b. Limited availability and expensive for the diabetes study.

c. Mortality due to ketosis problem is high in the case of animals with brittle pancreas and requires insulin treatment in later stage for survival.

d. Require sophisticated maintenance.

One of the rodent models that reflect human form of T2DM is Zucker diabetic fatty (ZDF) rats. ZDF rats as spontaneous diabetic animal model exhibit both the prediabetic and the end stage observed in human T2DM patients [43]. Spontaneously diabetic animals of T2DM may be acquired from the individuals with one or several genetic mutations transmitted from generation to generation or selected from nondiabetic outbred animals by repeated breeding through several generations. The result is that these animals inherited DM either as single or multigene defects. The metabolic particularities result from single gene defect (monogenic) which is due to dominant gene or recessive gene, or it can be of polygenic origin [36].

ZDF rats come from a colony of outbred Zucker rats in the laboratory of Dr. Walter Shaw at Eli Lilly Research Laboratories in Indianapolis (USA) during the years 1974–1975. In early 1981, some animals with diabetic lineage were designated and redefined. An inbred line of ZDF rats was established in 1985. Development to a genetic model was established in 1991 [44]. The Zucker fatty (ZF) rats carry a spontaneous mutation in the leptin receptor gene (fa) [45]. ZF rats resulted from the simple autosomal recessive (fa) gene on chromosome 5 [36]. This mutation causes hyperphagia, early onset of obesity, and insulin resistance [14] along with increased growth of subcutaneous fat depot [46]. At the age of 4 weeks, ZF rats gain weight more rapidly due to increased growth of subcutaneous fat depot, and they have a noticeably higher body weight at about the age of 9 weeks [47]. The hyperphagia and obesity in ZF rats are attributed to hypothalamic defect in leptin receptor signaling that is related to mild hyperglycemia, mild glucose intolerance, insulin resistance, hyperlipidemia, and moderate hypertension [46]. ZF rats have impaired glucose tolerance rather than apparent diabetes [42].

Thereafter, a mutation in ZF strain led to a substrain with an evident diabetic phenotype—the Zucker diabetic fatty (ZDF) rats [42]. ZDF rats are less obese than ZF rats having a decrease beta cell mass which resulted in inability to compensate for severe insulin resistance [48]. The ZDF rats were derived by selective inbreeding of hyperglycemic ZF rats [49] within the first months of life due to leptin receptor defect and a genetically reduced insulin promoter activity [17]. ZF rats maintain normoglycemia despite their obese phenotype, hyperlipidemia, and hyperinsulinemia [42].

The ZDF male rats became an experimental model for type 2 diabetes mellitus (T2DM). They have a predictable progression from prediabetic to diabetic state [50]. The ZDF rats carry a genetic defect in β-cell transcription. It is inherited independently of the leptin receptor mutation and insulin resistance [17]. In prediabetic stage of ZDF rats, there is no change in insulin mRNA levels. But, significant reduction (30–70%) of other islet mRNA levels, such as glucokinase, mitochondrial glycerol-3-phosphate dehydrogenase, voltage-dependent Ca^{2+} and K^+ channels, Ca^{2+}-ATPase, and transcription factor islet-1 may be detected [51]. It is known that FFA-induced suppression of insulin output in prediabetic stage of ZDF rats is conveyed by nitric oxide (NO) [52]. ZDF rats start to develop T2DM as early as 10 weeks of age, reaching 100% incidence at around 20 weeks of age [53]. It is possible to shorten prediabetic state and reach the symptoms of T2DM after high-energy diet. But, the animals receiving this diet are in the risky group because the diabetic state with its complications arrives quickly and rats can perish. In our experiment with ZDF rats, high-energy diet caused ketoacidosis that meant two cases of animal death in 7th week after initializing feeding with this caloric diet [54].

Blood glucose concentrations in ZDF rats usually increase from 7 to 10 weeks of age and impaired glucose tolerance at 5–7 weeks of age. At the age of 12 weeks, glucose intolerance becomes more severe than at 5–7 weeks of age [55]. Chronic and increasing hyperglycemia in ZDF rats is related to the loss of insulin and pancreatic duodenal homeobox (PDX-1) mRNAs. The lack of glucose stimulated insulin secretion. The possible prevention of hyperglycemia could block the deficit in insulin amount and PDX-1 gene expression and improve insulin secretion [56].

Male ZDF rats that are homozygous recessive have nonfunctional leptin receptors (fa/fa) and develop hyperlipidemia, obesity, and hyperglycemia. Rats that are homozygous dominant (+/+) or heterozygous (fa/+) are lean with normoglycemia. They are healthy, display no symptoms of diabetes, and are usually used as age-match control rats in the experiments. In young fa/fa rats, insulin resistance appears which extends to a deployed insulin secretory defect that initiates hyperglycemia and inadequate β-cell compensation [17, 49]. The insulin resistance is a result of a mutant leptin receptor that causes obesity [17].

In ZDF rats, there are sex differences for phenotypes of diet-induced insulin resistance and glucose intolerance. The most affected are male individuals [57]. On normal diet, male rats from ZDF strain develop severe hypoinsulinemia and hyperglycemia by 4 months of age. Female individuals maintain normal level of blood glucose and insulin despite advanced obesity [38]. Female ZDF rats with fa/fa genotype become also obese and insulin resistant, but do not progress to hyperglycemia, except when fed a high-fat diet [58, 59]. The female

ZDF rats develop T2DM just on a diabetogenic diet [60]. Thus, male ZDF rats are widely used as animal models for human T2DM and diabetic nephropathy and neuropathy [58, 61]. Hyperglycemia in diabetic ZDF rats was recorded at 2.5 month of age, and then blood glucose increased and reached an average value of 29.5 ± 0.9 mM at the age of 5 months. At the age of 5 months, fatty ZDF rats developed significant symptoms of thermal hypoalgesia indicated by prolonged response latencies in a tail-flick test. With progressing diabetes, the markers of thermal hypoalgesia increased at the age of 7 months and persisted till the 10th month [10]. The ZDF rats undergo a rapid transition between 10 and 15 weeks of age. At 10 weeks of age, they are insulin resistant, hyperlipidemic, and hyperinsulinemic. But, the high plasma insulin levels are insufficient to control glucose level, and the animals are hyperglycemic. Between 10 and 15 weeks of age, a loss in insulin secretory function occurs which leads to a marked decline in plasma insulin levels along with hyperglycemia [43]. The ability to secrete insulin to compensate peripheral insulin resistance is limited. β cells of ZDF rats are brittle and easily succumb to over-secretion pressure. The primary defect lies not in the ability of β cells to proliferate but rather in an enhanced rate of apoptosis. It shows impaired insulin secretory β-cell response to glucose, while it remains untouched to non-glucose secretogogues like arginine, a phenomenon similar to human T2DM. Downregulation of β-cell GLUT 2 transporters together with impaired insulin synthesis is probably responsible for hyperglycemia in ZDF animals. Decreased glucose transport activity and lowered GLUT 4 levels are present in the skeletal muscle and adipose tissue [34, 36, 48, 62]. Generally, it was reported that in the progression of ZDF rats, the decline of β-cell glucose transporter 2 (GLUT 2) membrane receptors and the incidental loss of muscle glucose transporter 4 (GLUT 4) are responsible for the impaired insulin secretion and subsequent hyperglycemia. The activity of GLUT 4 receptors decreased in adipose tissue and skeletal muscle. This results in reduced β-cell transport ability together with the peripheral insulin resistance [50].

Siwy et al. [63] characterized the strain of ZDF rats as appropriate model for human disease based on urinary peptidomic profiles. In the study the diabetic rats were heavier than lean individuals. Consistent with a diabetic phenotype, ZDF rats were hyperglycemic and dyslipidemic already at early (2 months of age) and more severely at late (8 month of age). Renal function was markedly impaired at 8 months. Proteinuria was present at 2 months and progressively increased at 8 months. At 8 months, ZDF rats' renal histology showed pathological changes, including glomerular sclerosis with thickening of the Bowman capsule and retraction of the tuft, tubular atrophy and dilatation, and hyaline casts. Lean rats did not develop any histopathological changes. Chen and Wang [44] introduced the following pharmacologically related characteristics of ZDF rats: 25–55% reduction of GLUT 4 in the adipose tissue, heart, and skeletal muscle; loss of pancreatic duodenal homeobox gene expression; and free fatty acids and nitric oxide induced suppression of insulin output (**Table 1**).

Obesity of ZDF rats (fa/fa) is caused by hyperphagia [59], and food restriction can counteract or delay development of T2DM [55]. Hyperphagia leads to hyperinsulinemia, which upregulates transcription factors that stimulate lipogenesis. This results in ectopic deposition of triacylglycerol in non-adipocytes, thereby providing fatty acid substrate for pathological non-oxidative metabolism, such as ceramide synthesis [64].

The severity of DM in the adult hyperglycemic ZDF rats is reflected in body weight and food consumption [42]. High-energy diet in animal's models inclined to DM leads to obesity, hyperinsulinemia, and altered glucose metabolism

Main feature	Characteristic	Description
Type of diabetes	T2DM • Characteristic features resembling human type • Associated with obesity • Hyperphagia • Polyuria • Polydipsia • Hyperglycemia • Hyperlipidemia • Hyperinsulinemia • Insulin resistance • Reduction of GLUT 4 in adipose tissue	Development spontaneously involving genetic factors Acceleration of symptoms by high-fat diet Mortality due to ketosis after high-fat diet Brittle pancreas
Genetic feature	Leptin receptors • Defect in leptin signaling • Genetic defect in β cell transcription • Insulin receptor deletion	Homozygous recessive (fa/fa) rats—diabetic homozygous dominant (+/+) and heterozygous (fa/+) remain lean and normoglycemic
Using in research	Mechanism of T2DM Obesity	Minimal variability of results, the possibility to use small size groups
Progression of diabetes	Insulin secretory defects Inadequate β cell compensation	Predictable progression from prediabetic to diabetic state

Table 1.
Major genetic, physiological, and pathophysiological characteristics of ZDF rats.

because of insufficient compensation by the pancreatic islets [31]. In our experiment [54] high-energy diet immediately induced hyperglycemia in ZDF rats, animals developed obesity, and we observed disturbance in some hematology parameters as neutrophils, mean platelet volume (MPV), and platelet count (PLT), which is a possible marker of angiopathy. In the groups of diabetic rats, we observed the significant weight decrease when compared to the control animals. It is probably linked to accelerated switch from prediabetic to diabetic state. It suggests the inability to utilize the calories consumed [42] and the degradation of structural proteins and muscle wasting in diabetic individuals [65]. Hempe et al. [59] found that body weight of ZDF obese rats was higher than the lean rats at the beginning of their study. Later, at around 16 weeks of age, body weight of obese animals started to decline. Lean rats reached the same body weight as obese rats at 25 weeks of age. The decrease of weight in obese diabetic rats on high energy is linked to accelerated switch from prediabetic to diabetic accomplished with associated complications of T2DM. Oyedemi et al. [65] explained this reduction in body weight as degradation of structural proteins and muscle wasting.

Hempe et al. [59] evaluated if nephropathy and neuropathy in ZDF rats are linked to the hyperglycemic state of the rats and are real diabetic late complications or are related to other characteristics of the fa/fa genotype. Good glycemic control may be effective in delaying the neuropathic symptoms in diabetic patients [66].

2.1 Hematological parameters of ZDF rats

In general, diabetic patients have increased values of some hematological parameters as platelet count, mean platelet volume (MPV), and platelet distribution width (PDWc). Platelet activation can result in the generation of vascular disease [67]. Hematological parameters can be altered as a result of infection that occurs during DM [65]. In the human study, the total granulocyte count was increased in diabetic patients. It was confirmed that granulocyte count is associated with T2DM [68]. In our experiment significant increase in granulocyte count was also observed in diabetic ZDF rats in comparison with the control animals [54]. It was also published that increased count of one part of granulocytes (neutrophils) correlated with the rising risk of vascular disease in T1DM [69] with consequences as diabetic angiopathy [70]. Mean corpuscular volume (MCV) and mean corpuscular hemoglobin (MCH) were decreased in the diabetic ZDF rats when compared to the lean control [54]. Similar results are published by Mahmoud [71] in white albino rats with experimentally induced DM and Oyedemi et al. [65] in streptozotocin-induced diabetic Wistar rats. Generally, the decrease of these hematological parameters during the diabetes could be an indicator of abnormal hemoglobin synthesis, failure of blood osmoregulation, and plasma osmolality [72].

Platelets play a critical role in atherogenesis and thrombosis-mediated myocardial ischemia accelerated in diabetic state [73]. They are source of inflammatory mediators [74]. We observed increased values of platelets in diabetic ZDF rats against the lean control [54]. Through inflammatory process during the DM, the platelets are highly activated. Activated platelets presumably support neutrophil activation and recruitment through expressing selectins, inflammatory cytokines, and chemokines [75]. It was proven that platelets and neutrophils regulate and affect each other's functions by platelet-leukocyte contact and releasing soluble effector mediators [76].

The marker of platelet function and activation is hematological parameter—MPV [77]. Increased MPV can be an independent risk factor for arterial thrombotic events such as myocardial infarction and cerebral thromboembolism [78]. Usually diabetic patients have increased MPV values [79] correlated with a large thrombocyte size that are more reactive and aggregable [77] which can upset hemostatic system during the diabetic state [80]. In diabetes the risk of retinopathy onset increases with higher MPV [79, 81]. In our study diabetic ZDF rats had increased MPV values in comparison with the healthy lean control [54]. In diabetic patients there is usually higher value of PDWc—the hematological parameter that presents an indicator of variation in platelets' size and activity [79, 82]. In our previous research [83], we observed that the rise in the secondary symptoms of T2DM complications caused by high-energy diet was accompanied with disturbed hematological parameters. It could be also a potential marker of angiopathy.

3. Other rodent models used in research in diabetes

Rodent animal models for investigation of T1DM are streptozotocin- or alloxan-induced animals, nonobese diabetic (NOD) mouse, and bio-breeding (BB) rat [84]. NOD mice and BB rats are rodent animal model with spontaneous development of T1DM [85]. Rodent models for T2DM include except ZDF rats also Goto-Kakizaki (GK) rats, Otsuka Long-Evans Tokushima Fatty (OLETF) rats, spontaneously diabetic Tori (SDT) rats, Kuo Kondo (KK) mice, ob/ob+/+ mice, and db/db+/+ mice [84]. GK rats are nonobese Wistar substrain which develops T2DM early [86]. Male OLETS rats suffer from diabetes at 18–25 weeks of age. The symptoms include polyphagia, mild obesity, hypertriglyceridemia, hyperinsulinemia, and impaired glucose tolerance in 16 weeks of age [87]. Tori SDT rat is inbred strain of Sprague-Dawley rat. Male individuals have high glucose levels by 20 weeks, pancreatic islet histopathology, hemorrhage in pancreatic islets, and inflammatory cell infiltration with fibroblasts, prior to diabetes glucose intolerance with hypoinsulinemia [88].

KK mice are a polygenic model of obesity and T2DM. They are characterized by insulin resistance, hyperinsulinemia, and hyperphagia [89]. The ob/ob+/+ mice carry a mutation in the leptin gene, manifested as obesity, hyperglycemia, impaired glucose intolerance, and hyperinsulinemia [90]. The db/db+/+ mice have a leptin receptor mutation, are spontaneously hyperphagic, and suffer from obesity, hyperglycemia, hyperinsulinemia, and insulin resistance within the first month of life [91]. The advantages of ZDF rats in diabetes research are mainly due to the fact that it is a spontaneous model for T2DM research. It shows characteristics such as hyperglycemia, obesity, hyperphagia, polyuria, insulin disorders, and dyslipidemia due to the mutation in the leptin receptor gene and provides an appropriate model for common human T2DM. Moreover, these rats are calm and dispassionate; the handling and manipulation with them is comfortable.

4. Conclusion

The ZDF strain is of increasing preclinical interest due to its pathophysiological similarities to human T2DM [92, 93]. They are generally used in studies of diabetes with obesity and cardiovascular complications because of dyslipidemia background [44]. Defective insulin release in ZDF rats could be partially restored by glucagon-like peptide (GLP-1). The action of GLP-1 therapy is mediated through Ca^{2+}-independent signaling pathway in pancreatic islets [44]. The use of rosiglitazone protected ZDF rats against the loss of β-cell mass through sustaining cell proliferation, and blocking increased β cells' death [94]. Metformin prevented hyperglycemia in ZDF rats aged between 6 and 12 weeks. This compound significantly reduced free fatty acid level and triglycerides. It delayed the onset of DM which is linked to the improvement in β cell functions, on a par with the lipotoxicity hypothesis for adipogenic diabetes [95]. Some experimental interventions provided on ZDF rats are shown in **Table 2**. In general, animal model for DM research is required and needed to uncover and understand the pathophysiology of disease. This is the key to the development of new therapies and treatment [96].

Today, the number of patients suffering from DM is increasing. The most common form of DM is T2DM. It is a genetic disease demonstrating insulin insufficiency. Therefore the research on this disease is deepening and required. Due to its complex, complicated, multifactorial heterogeneous disease resulting from both environmental factors and genetic responsiveness, accurate animal model that can mirror human T2DM symptoms and complication is required. Presently, the spontaneous T2DM rodent model for research in DM and obesity is ZDF rats. This strain shows characteristics such as obesity, hyperglycemia, insulin disorders, and dyslipidemia due to the mutation in the leptin receptor gene and provides an appropriate model for common human T2DM.

Source	Aim of the study	Treatment	Results and conclusion of the study
Tanaka et al. [97]	To assess if the use of antioxidants prevents glucose toxicity and ameliorates the progression of DM	6 weeks of age till 12 weeks of age, antioxidants—N-acetyl-L-cysteine, aminoguanidine, ZDF rats	• The treatment with antioxidants can partially prevent the progressive β cells' dysfunction
Wasan et al. [43]	To examine the effect of organic vanadium compounds	3-week treatment with the insulin-enhancing agent—vanadium compounds, ZDF	• Increase in plasma homocysteine and cysteine level
Hempe et al. [59]	If the complications in the kidney and nerves correspond to human diabetic complications	Food restriction or pioglitazone (peroxisome proliferator-activated receptor gamma - PPARγ agonist) treatment, ZDF rats	• Food restriction delayed (not prevented) the onset of DM for 8–10 weeks and pioglitazone prevented the development of DM • ZDF rat is a good model for diabetic nephropathy, but alterations in nerve functions were not diabetes-related
Siwy et al. [63]	Evaluation of the similarity between ZDF rats and T2DM in humans	24 hours study, ZDF rats 2 month and 8 month of age	• ZDF rats may be more suitable to study the macrovascular branch within the pathophysiologic cascade of diabetic angiopathies, but it is not a good model for microvascular disease
Wang et al. [42]	Determination whether salsalate, a salicylate with anti-inflammatory properties, is effective in mitigating DM progression	Chronic administration of salsalate from 5 weeks of age to 24 weeks of age, ZDF rats	• The therapy is effective in particular animal model; it may only be effective in a subpopulation of humans with the disease
Kim et al. [98]	Investigation of therapeutic effect of resistance training on T2DM	8 weeks of resistance training, ZF and ZDF rats	• Regular resistance training initiated at the onset of DM improved glucose tolerance and GLUT 4 expression
Ďuračka et al. [99]* Tvrdá et al. [100]*	The effect of bee bread on the oxidative profile of testicular tissue and fertility in diabetic rats	3 months therapy with bee bread in the dose of 250 mg/kg/day, ZDF rats	• Significant increase of total antioxidant capacity of in testicular tissue lysate • Bee bread effectively protected proteins against oxidative damage • Bee bread provided substantial protection against testicular oxidative stress
Soltesova Prnova et al. [10]	Observation of the effect of Cemtirestat on symptoms of peripheral diabetic neuropathy	2 months treatment with Cemtirestat in doses 2.5 and 7.5 mg/kg/day, ZDF rats	• Partial inhibition of sorbitol accumulation in red blood cells and the sciatic nerve • Decrease in plasma level of TBARS • Normalization of peripheral neuropathy symptoms
Álvarez-Cilleros et al. [101]	To examine potential antidiabetic properties of cocoa	10 weeks treatment with cocoa-rich diet, ZDF rats	• Improvement in glucose homeostasis and insulin resistance, protection of renal structure and functionality

Source	Aim of the study	Treatment	Results and conclusion of the study
Capcarova et al. [102]*	To effect of bee bread on DM complications	4 months therapy with bee bread in the dose of 700 mg/kg/ day, ZDF rats	• Treatment of hyperglycemia, used as the prevention of DM in young age

**Experiments realized at the Department of Animal Physiology, Faculty of Biotechnology and Food Sciences, Slovak University of Agriculture in Nitra, Slovak Republic.*

Table 2.
The investigation of some therapeutic strategies in DM research using ZDF rats.

In general, in the future research, many novel strategies in treatment of DM will be surveyed, and the use of ZDF rats in these experiments will be worthy to study.

Acknowledgements

This study was supported from the APVV grant no 15-0229, KEGA 024SPU-4/2018 and VEGA grant no 1/0144/19.

Conflict of interest

The authors declare no conflict of interest.

Author details

Marcela Capcarova* and Anna Kalafova
Faculty of Biotechnology and Food Sciences, Department of Animal Physiology, Slovak University of Agriculture in Nitra, Nitra, Slovak Republic

*Address all correspondence to: marcela.capcarova@uniag.sk

References

[1] Gavard JA, Lustman PJ, Clouse RE. Prevalence of depression in adults with diabetes: An epidemiological evaluation. Diabetes Care. 1993;**16**(8):1167-1178. DOI: 10.2337/diacare.16.8.1167

[2] Ogurtsova K, Da Rocha Fernandez JD, Huang Y, Linnenkamp U, Guariguata L, Cho NH, et al. IDF diabetes atlas: Global estimates for prevalence of diabetes for 2015 and 2040. Diabetes Research and Clinical Practise. 2017;**128**:40-50. DOI: 10.1016/j.diabres.2017.03.024

[3] Cho NH, Shaw JE, Karuranga S, Huang Y, da Rocha Fernandes JD, Ohlrogge AW, et al. IDF diabetes atlas: Global estimates of diabetes prevalence for 2017 and projections for 2045. Diabetes Research and Clinical Practise. 2018;**138**:271-281. DOI: 10.1016/j.diabres.2018.02.023

[4] King AJF. The use of animal models in diabetes research. British Journal of Pharmacology. 2012;**166**:877-894. DOI: 10.1111/j.1476-5381.2012.01911.x

[5] Pang YL, Hu JW, Liu GL, Lu SY. Comparative medical characteristics of ZDF-T2DM rats during the course of development to late stage disease. Animal Models and Experimental Medicine. 2018;**1**:203-211. DOI: 10.1002/ame2.12030

[6] Nakagawa T, Tanabe K, Croker BP, Johnson RJ, Grant MB, Kosugi T, et al. Endothelial dysfunctions of diabetes. Journal of the American Medical Association. 2004;**292**:2495-2499. DOI: 10.1038/nrneph.2010.152

[7] Brownlee M. Biochemistry and molecular cell biology of diabetic complications. Nature. 2001;**414**:813-820. DOI: 10.1038/414813a

[8] Feldman EL, Nave KA, Jensen TS, Bennet DLH. New horizons in diabetic neuropathy: Mechanisms, bioenergetics, and pain. Neuron. 2017;**93**(6):1296-1313. DOI: 10.1016/j.neuron.2017.02.005

[9] Tomlinson DR, Gardiner NJ. Glucose neurotoxicity. Nature Reviews. Neuroscience. 2008;**9**(1):36-45. DOI: 10.1038/nrn2294

[10] Soltesova Prnova M, Svik K, Bezek S, Kovacikova L, Karasu C, Stefek M. 3-mercapto-5H-1,2,4-triazino[5,6-b]indole-5-acetic acid (Cemtirestat) alleviates symptoms of peripheral diabetic neuropathy in Zucker diabetic fatty (ZDF) rats: A role of aldose reductase. Neurochemical Research. 2019;**44**:1056-1064. DOI: 10.1007/s11064-019-02736-1

[11] Robertson RP, Harmon JS. Diabetes, glucose toxicity, and oxidative stress: A case of double jeopardy for the pancreatic islet β cell. Free Radical Biology & Medicine. 2006;**41**:177-184. DOI: 10.1016/j.freeradbiomed.2005.04.030

[12] Akerblom HK, Knip M. Putative environmental factors in type 1 diabetes. Diabetes/Metabolism Reviews. 1998;**14**:31-36. DOI: 10.1002/(sici)1099-0895(199803)14:1<31::aid-dmr201>3.3.co;2-1

[13] Kahn SE, Hull RL, Utzschneider KM. Mechanisms linking obesity to insulin resistance and type 2 diabetes. Nature. 2006;**444**:840-846. DOI: 10.1038/nature05482

[14] Garnett KE, Chapman P, Chambers JA, Waddell ID, Boam DSW. Differential gene expression between Zucker fatty rats and Zucker diabetic fatty rats: A potential role for the immediate-early gene Egr-1 in regulation of beta cell proliferation. Journal of Molecular Endocrinology. 2005;**35**:13-25. DOI: 10.1677/jme.1.01792

[15] Gutterman DD. Vascular dysfunction in hyperglycemia: Is

protein kinase C the culprit? Circulation Research. 2002;**90**(1):5-7. DOI: 10.1161/res.90.1.5

[16] Miyake K, Yang W, Hara K, Yasuda K, Horikawa Y, Osawa H, et al. Construction of a prediction model for type 2 diabetes mellitus in the Japanese population based on 11 genes with strong evidence of the association. Journal of Human Genetics. 2009;**54**(4):236-241. DOI: 10.1038/jhg.2009.17

[17] Griffen SC, Wang J, German MS. A genetic defect in beta-cell gene expression segregates independently from the fa locus in the ZDF rat. Diabetes. 2001;**50**:63-68. DOI: 10.2337/diabetes.50.1.63

[18] Polonsky KS, Sturis J, Bell GI. Seminars in medicine of the Beth Israel hospital, Boston. Non-insulin-dependent diabetes mellitus- a genetically programmed failure of the beta cell to compensate for insulin resistance. The New England Journal of Medicine. 1996;**334**:777-783. DOI: 10.1056/NEJM199603213341207

[19] Lencioni C, Lupi R, Del Prato S. Beta-cell failure in type 2 diabetes mellitus. Current Diabetes Record. 2008;**8**:179-184. DOI: 10.1007/s11892-008-0031-0

[20] Lee Y, Hirose H, Zhou YT, Esser V, McGarry JD, Unger RH. Increased lipogenic capacity of the islets of obese rats: A role in the pathogenesis of NIDDM. Diabetes. 1997;**46**:408-413. DOI: 10.2337/diab.46.3.408

[21] Prentki M, Joly E, El-Assad W, Roduit R. Malonyl-CoA signalling, lipid partitioning, and glucolipotoxicity: Role in beta-cell adaptation and failure in the etiology of diabetes. Diabetes. 2002;**51**:S405-S413. DOI: 10.2337/diabetes.51.2007.S405

[22] Gheni G, Yokoi N, Beppu M, Yamaguchi T, Hidaka A, Hoshino Y, et al. Characterization of the prediabetic state in a novel rat model of type 2 diabetes, the ZFDM rat. Journal Diabetes Research. 2015;**2015**:1-8. Article ID 261418. DOI: 10.1155/2015/261418

[23] DeFronzo R. The triumvirate: Beta cell, muscle, liver. A collusion responsible for NIDDM. Diabetes. 1988;**37**:667-685. DOI: 10.2337/diab.37.6.667

[24] Uchigata Y. Pathogenesis, diagnosis, and treatment of type 2 diabetes in adolescence. Nihon Rinsho. 2016;**74**:517-522

[25] Yano JM, Yu K, Donaldson GP, Shastri GG, Ann P, Ma L, et al. Indigenous bacteria from the gut microbiota regulate host serotonin biosynthesis. Cell. 2015;**161**:264-276. DOI: 10.1016/j.cell.2015.02.047

[26] Hill J, Wyatt HR, Peters JC. Energy balance and obesity. Circulation. 2012;**126**:126-132. DOI: 10.1161/CIRCULATIONAHA.111.087213

[27] Day C, Bailey CJ. Obesity in the pathogenesis of type 2 diabetes. British Journal of Diabetes and Vascular Disease. 2011;**11**(2):55-61. DOI: 10.1177/1474651411407418

[28] Guilherme A, Virbasius JV, Puri V, Czech MP. Adipocyte dysfunctions linking obesity to insulin resistance and type 2 diabetes. Nature Reviews. Molecular Cell Biology. 2008;**9**:367-377. DOI: 10.1038/nrm2391

[29] Feldstein AE, Nathan W, Werneburg AC, Guicciardi ME, Bronk SF, Rydzewski R, et al. Free fatty acids promote hepatic lipotoxicity by stimulating TNF-α expression via a lysosomal pathways. Hepatology. 2004;**40**:185-194

[30] Lumeng CN, Saltiel AR. Inflammatory links between obesity

and metabolic disease. The Journal of Clinical Investigation. 2011;**121**:2111-2117. DOI: 10.1172/JCI57132

[31] Winzell MS, Ahren B. The high-fat diet-fed mouse: A model for studying mechanisms and treatment of impaired glucose tolerance and type 2 diabetes. Diabetes. 2004;**54**(Suppl 3):S215-S219. DOI: 10.2337/diabetes.53.suppl_3.s215

[32] Izumi Y, Ishibashi G, Nakanishi Y, Kikunaga S. Beneficial effect of 3% milled-rice on blood glucose level and serum lipid concentrations in spontaneously non-insulin-dependent diabetic rats. Journal of Nutritional Science and Vitaminology. 2007;**53**: 400-409. DOI: 10.3177/jnsv.53.400

[33] O'Rahilly S. Human genetics illuminates the paths to metabolic disease. Nature. 2009;**402**:307-314. DOI: 10.1038/nature08532

[34] Bailey CJ. Drugs on the horizon for diabesity. Current Diabetes Reports. 2005;**5**:353-359. DOI: 10.1007/s11892-005-0093-1

[35] Chatzigeorgiou A, Halapas A, Kalafatakis K, Kamper E. The use of animal models in the study of diabetes mellitus. In Vivo. 2009;**23**:245-258

[36] Srinivasan K, Ramarao P. Animal models in type 2 diabetes research: An overview. The Indian Journal of Medical Research. 2007;**125**:451-472

[37] Engel H, Xiong L, Reichenberger MA, Germann G, Roth C, Hirche C. Rodent models of diet-induced type 2 diabetes mellitus: A literature review and selection guide. Diabetes and Metabolic Syndrome: Clinical Research and Reviews. 2019;**13**:195-200. DOI: 10.1016/j.dsx.2018.07.020

[38] Kleinert M, Clemmensen C, Hofmann SM, Moore MC, Renner S, Woods SC, et al. Animals models of obesity and diabetes mellitus. Nature. 2018;**140**:140-184. DOI: 10.1038/nrendo.2017.161

[39] Reed MJ, Meszaros K, Entes LJ, Claypool MD, Pinkett JG, Gadbois TM. A new rat model of type 2 diabetes: The fat-fed, streptozotocin-treated rat. Metabolism. 2000;**49**(11):1390-1394. DOI: 10.1053/meta.2000.17721

[40] Rees DA, Alcolado JC. Animal models of diabetes mellitus. Diabetic Medicine. 2005;**22**:359-370. DOI: 10.1111/j.1464-5491.2005.01499.x

[41] Fang JY, Lin CH, Huang TH, Chuang SY. In vivo rodent models of type 2 diabetes and their usefulness for evaluating flavonoid bioactivity. Nutrients. 2019;**11**(3):23pp. DOI: 10.3390/nu11030530

[42] Wang X, DuBois DC, Sukumaran S, Ayyar V, Jusko WJ, Almon RR. Variability in Zucker diabetic fatty rats: Differences in disease progression in hyperglycemic and normoglycemic animals. Diabetes, Metabolic Syndrome and Obesity: Targets and Therapy. 2014;**7**:531-541. DOI: 10.2147/DMSO.S69891

[43] Wasan KM, Risovic V, Yuen VG, McNeill JH. Differences in plasma homocysteine levels between Zucker fatty and Zucker diabetic fatty rats following 3 weeks oral administration of organic vanadium compounds. Journal of Trace Elements in Medicine and Biology. 2006;**19**:251-258. DOI: 10.1016/j.jtemb.2005.10.001

[44] Chen D, Wang MW. Development and application of rodent models for type 2 diabetes. Diabetes, Obesity & Metabolism. 2005;**7**:307-317. DOI: 10.1111/j.1463-1326.2004.00392.x

[45] Philips MS, Liu Q, Hammond HA, Dugan V, Hey PJ, Caskey CJ, et al. Leptin receptor missense mutation in the fatty Zucker rat. Nature Genetics. 1996;**13**:18-19. DOI: 10.1038/ng0596-18

[46] Durham HA, Truett GE. Development of insulin resistance and hyperphagia in Zucker fatty rats. American Journal of Physiology. Regulatory, Integrative and Comparative Physiology. 2006;**210**:652-658. DOI: 10.1152/ajpregu.00428.2004

[47] McCaleb ML, Sredy J. Metabolic abnormalities of the hyperglycemic obese Zucker rat. Metabolism. 1992;**41**(5):522-525. DOI: 10.1016/0026-0495(92)90212-s

[48] Pick A, Clark J, Kubstrup C, Levisetti M, Pugh W, Bonner-Weir S, et al. Role of apoptosis in failure of beta-cell mass compensation for insulin resistance and beta-cell defects in the male Zucker diabetic fatty rat. Diabetes. 1998;**47**:358-364. DOI: 10.2337/diabetes.47.3.358

[49] Etgen GJ, Oldham BA. Profiling of Zucker diabetic fatty rats in their progression to the overt diabetic state. Metabolism. 2000;**49**:684-688. DOI: 10.1016/S0026-0495(00)80049-9

[50] Slieker LJ, Sundell KL, Heath WF, Osborne HE, Blue J, Manetta J, et al. Glucose transporter levels in tissues of spontaneously diabetic Zucker fa/fa (ZDF/drt) and viable yellow mouse (Avy/a). Diabetes. 1992;**41**:187-193. DOI: 10.2337/diab.41.2.187

[51] Tokuyama Y, Sturis J, DePaoli AM, Takeda J, Stoffel M, Tang J, et al. Evolution of beta-cell dysfunction in the male Zucker diabetic fatty rat. Diabetes. 1995;**44**:1447-1457. DOI: 10.2337/diab.44.12.1447

[52] Shimabukuro M, Ohneda M, Lee Y, Unger RH. Role of nitric oxide in obesity-induced beta cell disease. The Journal of Clinical Investigation. 1997;**100**:290-295. DOI: 10.1172/JCI119534

[53] Yokoi N, Hoshino M, Hidaka S, Yoshida E, Beppu M, Hoshikawa R, et al. A novel rat model of type 2 diabetes: The Zucker fatty diabetes mellitus ZFDM rat. Journal Diabetes Research. 2013;**2013**:9. Article ID 103731. DOI: 10.1155/2013/103731

[54] Capcarova M, Kalafova A, Schwarzova M, Soltesova Prnova M, Svik K, Schneidgenova M, et al. The high-energy diet affecting development of diabetes symptoms in Zucker diabetic fatty rats. Biologia. 2018;**73**:659-671. DOI: 10.2478/s11756-018-0076-8

[55] Ohneda M, Inman LR, Unger RH. Caloric restriction in obese pre-diabetic rats prevents beta-cell depletion, loss of beta-cell GLUT 2 and glucose incompetence. Diabetologia. 1995;**38**:173-179. DOI: 10.1007/s001250050267

[56] Harmon JS, Gleason CE, Tanaka Y, Oseid EA, Hunter-Berger KK, Robertson RP. In vivo prevention of hyperglycemia also prevents glucotoxic effects on PDX-1 and insulin gene expression. Diabetes. 1999;**48**:1995-2000. DOI: 10.2337/diabetes.48.10.1995

[57] Nadal-Casellas A, Proenza AM, Llado I, Gianotti M. Sex-dependent differences in rat hepatic lipid accumulation and insulin sensitivity in response to diet-induced obesity. Biochemistry and Cell Biology. 2012;**90**:164-172. DOI: 10.1139/o11-069

[58] Figarolla JL, Loera S, Weng Y, Shanmugam N, Natarajan R, Rahbar S. LR-90 prevents dyslipidaemia and diabetic nephropathy in the Zucker diabetic fatty rat. Diabetologia. 2008;**51**:882-891. DOI: 10.1007/s00125-008-0935-x

[59] Hempe J, Elvert R, Schmidts HL, Kramer W, Herling AW. Appropriateness of the Zucker diabetic fatty rat as a model for diabetic microvascular late complications. Laboratory Animals. 2012;**46**:32-39. DOI: 10.1258/la.2011.010165

[60] Wang Y, Sun G, Sun J, Liu S, Wang J, Xu X, et al. Spontaneous type 2 diabetic rodent models. Journal Diabetes Research. 2013; ID 401723:8. DOI: 10.1155/2013/401723

[61] Brussee V, Guo G, Dong Y, Cheng C, Martinez JA, Smith D, et al. Distal degenerative sensory neuropathy in a long-term type 2 diabetes rat model. Diabetes. 2008;**57**:1664-1673. DOI: 10.2337/db07-1737

[62] Zhang BB, Moller DE. New approaches in the treatment of type 2 diabetes. Current Opinion in Chemical Biology. 2000;**10**:703-710. DOI: 10.1016/s1367-5931(00)00103-4

[63] Siwy J, Zoja C, Klein J, Benigni A, Mullen W, Mayer B, et al. Evaluation of the Zucker diabetic fatty (ZDF) rat as a model for human disease based on urinary peptidomic profiles. PLoS One. 2012;**12**:e51334. DOI: 10.1371/journal.pone.0051334

[64] Unger RH, Orci L. Diseases of liporegulation: New perspective on obesity and related disorders. FASEB Journal. 2001;**15**:312-321. DOI: 10.1096/fj.00-0590

[65] Oyedemi SO, Adewusi EA, Aiyegoro OA, Akinpelu DA. Antidiabetic and haematological effect of aqueous extract of stem bark of Afzelia Africana (Smith) on streptozotocin-induced diabetic Wistar rats. Asian Pacific Journal of Tropical Biomedicine. 2011;**1**(5):353-358. DOI: 10.1016/S2221-1691(11)60079-8

[66] Waldfogel JM, Nesbit SA, Dy SM, Sharma R, Zhang A, Wilson LM, et al. Pharmacotherapy for diabetic peripheral neuropathy pain and quality of life: A systematic review. Neurology. 2017;**88**(20):1958-1967. DOI: 10.1212/wnl.0000000000003882

[67] Curtis TM, Gardiner TA, Stitt AW. Microvascular lesions of diabetic retinopathy: Clues towards understanding pathogenesis? Eye. 2009;**23**:1496-1508. DOI: 10.1038/eye.2009.108

[68] Gkrania-Klotsas E, Ye Z, Cooper AJ, Sharp SJ, Luben R, Biggs ML, et al. Differential white blood cell count and type 2 diabetes: Systematic review and meta-analysis of cross-sectional and prospective studies. PLoS One. 2010;**5**(10):e13405:11 pp. DOI: 10.1371/journal.pone.0013405

[69] Huang J, Xiao Y, Xu A, Zhou Z. Neutrophils in type 1 diabetes. Journal of Diabetes Investigation. 2016;**7**: 652-663. DOI: 10.1111/jdi.12469

[70] Piwowar A, Knapik-Kordecka M, Warwas M. Concentration of leukocyte elastase in plasma and polymorphonuclear neutrophil extracts in type 2 diabetes. Clinical Chemistry and Laboratory Medicine. 2000;**38**(12):1257-1261. DOI: 10.1515/cclm.2000.198

[71] Mahmoud A. Hematological alterations in diabetic rats – Role of adipocytokines and effect of citrus flavonoids. EXCLI Journal. 2013;**12**:647-657

[72] Stookey JD, Burg M, Sellmeyer DE, Greenleaf JE, Arieff A, Van Hove L, et al. A proposed method for assessing plasma hypertonicity in vivo. European Journal of Clinical Nutrition. 2006;**61**:143-146. DOI: 10.1038/sj.ejcn.1602481

[73] Kraakman MJ, Lee MK, Al-Sharea A, Dragoljevic D, Barrett TJ, Montenont E, et al. Neutrophil-derived S100 calcium-binding proteins A8/A9 promote reticulated thrombocytosis and atherogenesis in diabetes. The Journal of Clinical Investigation. 2017;**127**(6):2133-2147. DOI: 10.1172/jci92450

[74] Ruggeri ZM. Platelets in atherothrombosis. Nature Medicine. 2002;**8**:1227-1234. DOI: 10.1038/nm1102-1227

[75] Semple JW, Italiano JE Jr, Freedman J. Platelets and the immune continuum. Nature Reviews. Immunology. 2011;**11**:264-274. DOI: 10.1038/nri2956

[76] Li N, Hu H, Lindqvist M. Platelet-leukocyte cross talk in whole blood. Arteriosclerosis, Thrombosis, and Vascular Biology. 2000;**20**:2702-2708. DOI: 10.1161/01.atv.20.12.2702

[77] Park Y, Schoene N, Harris W. Mean platelet volume as an indicator of platelet activation: Methodological issues. Platelets. 2002;**13**(5-6):301-306. DOI: 10.1080/0953710022020148332

[78] Khandekar MM, Khurana AS, Deshmukh SD, Kakrani AL, Katdare AD, Inamdar AK. Platelet volume indices in patients with coronary artery disease and acute myocardial infarction: An Indian scenario. Journal of Clinical Pathology. 2006;**59**:146-149. DOI: 10.1136/jcp.2004.025387

[79] Tetikoglu M, Aktas S, Sagdik HM, Yigitoglu ST, Özcura F. Mean platelet volume is associated with diabetic macular edema in patients with type-2 diabetes mellitus. Seminars in Ophthalmology. 2016;**32**(5):651-654. DOI: 10.3109/08820538.2016.1157612

[80] Vinik AI, Erbas T, Park TS, Nolan R, Pittenger GL. Platelet dysfunction in type 2 diabetes. Diabetes Care. 2001;**24**:1476-1485. DOI: 10.2337/diacare.24.8.1476

[81] Citirik M, Beyazyildiz E, Simsek M, Beyazyildiz O, Haznedaroglu IC. MPV may reflect subclinical platelet activation in diabetic patients with and without diabetic retinopathy. Eye. 2015;**29**(3):376-379. DOI: 10.1038/eye.2014.298

[82] Jindal S, Gupta R, Gupta R, Kakkar A, Singh AV, Gupta K, et al. Platelet indices in diabetes mellitus: Indicators of diabetic microvascular complication. Hematology. 2011;**16**:86-89. DOI: 10.1179/102453311x12902908412110

[83] Capcarova M, Schwarzova M, Kalafova A, Soltesova Prnova M, Svik K, Schneidgenova M, et al. Alterations in haematological parameters in diabetic rats. Slovak Journal of Animal Science. 2017;**50**(4):161

[84] Kim JH, Nishina PM, Naggert JK. Genetic models for non insulin dependent diabetes mellitus in rodents. Journal of Basic and Clinical Physiology and Pharmacology. 1998;**9**:325-345. DOI: 10.1515/jbcpp.1998.9.2-4.325

[85] Makino S, Kunimoto K, Munaoko Y, Mizushima Y, Katagiri K, Tochino Y. Breeding of a non-obese diabetic strain of mice. Experimental Animals. 1980;**29**:1-13. DOI: 10.1538/expanim1978.29.1_1

[86] Bedow AA, Samuel VT. Fasting hyperglycemia in the Goto-Kakizaki rat is dependent on corticosterone: A confounding variable in rodent models of type 2 diabetes. Disease Models & Mechanisms. 2012;**5**(5):681-685. DOI: 10.1242/dmm.009035

[87] Kawano K, Hirashima T, Mori S, Saitoh Y, Kurosumi M, Natori T. Spontaneous long-term hyperglycemic rat with diabetic complications, Otsuka Long-Evans Tokushima Fatty (OLETF) strain. Diabetes. 1992;**41**:1422-1428. DOI: 10.2337/diabetes.41.11.1422

[88] Sasase T, Ohta T, Masuyama T, Yokoi N, Kakehashi A, Shinohara M. The spontaneously diabetic Torii rat: An animal model of nonobese type 2 diabetes with severe diabetic complications. Journal Diabetes Research. 2013;**2013**:12. Article ID 976209. DOI: 10.1155/2013/976209

[89] Reddi AS, Camerini-Davalos RA. Hereditary diabetes in the KK mouse: An overview. Advances in Experimental Medicine and Biology. 1988;**246**:7-15. DOI: 10.1007/978-1-4684-5616-5_2

[90] Dubuc PU. The development of obesity, hyperinsulinemia, and hyperglycemia in Ob/Ob mice. Metabolism. 1976;**25**:1567-1574. DOI: 10.1016/0026-0495(76)90109-8

[91] Shariff E. Animal models of non-insulin dependent diabetes. Diabetes & Metabolism. 1992;**8**:179-208. DOI: 10.1002/dmr.5610080302

[92] Jones HB, Nugent D, Jenkins R. Variation in characteristics of islets of Langerhans in insulin-resistant, diabetic and non-diabetic-rat strains. International Journal of Experimental Pathology. 2010;**91**:288-301. DOI: 10.1111/j.1365-2613.2010.00713.x

[93] Kakimoto T, Kimata H, Iwasaki S, Fukunari A, Utsumi H. Automated recognition and quantification of pancreatic islets in Zucker diabetic fatty rats treated with exendin-4. The Journal of Endocrinology. 2013;**216**:13-20. DOI: 10.1530/joe-12-0456

[94] Finegood DT, McArthur MD, Kojwang D. Beta-cell mass dynamics in Zucker diabetic fatty rats. Rosiglitazone prevents the rise in net cell death. Diabetes. 2001;**50**:1021-1029. DOI: 10.2337/diabetes.50.5.1021

[95] Sreenan S, Sturis J, Pugh W, Burant CF, Polonsky KS. Prevention of hyperglycemia in the Zucker diabetic fatty rat by treatment with metformin or troglitazone. The American Journal of Physiology. 1996;**271**:e742-e747. DOI: 10.1152/ajpendo.1996.271.4.e742

[96] Capcarova M, Soltesova Prnova M, Svik K, Schneidgenova M, Slovak L, Kisska P, et al. Rodent animal model for research in diabetes: A mini review. Slovak Journal of Animal Science. 2018b;**51**(3):138-145

[97] Tanaka Y, Gleason CE, Tran POT, Harmon JS, Robertson RP. Prevention of glucose toxicity in HIT-T15 cells and Zucker diabetic fatty rats by antioxidants. Proceedings of the National Academy of Sciences. 1999;**96**:10857-10862. DOI: 10.1073/pnas.96.19.10857

[98] Kim J, Cho MJ, So B, Kim H, Seong JK, Song W. The preventive effects of 8 weeks of resistance training on glucose tolerance and muscle fiber type composition in Zucker rats. Diabetes and Metabolism Research. 2015;**39**:424-433. DOI: 10.4093/dmj.2015.39.5.424

[99] Ďuračka M, Sukova M, Valentova G, Tokarova K, Kalafova A, Schneidgenova M, et al. Perga (bee bread) improves the oxidative profile of testicular tissue in obese diabetic rats. In: Nutricon 2018. Skopje: Consulting and Training Center KEY; 2018. pp. 169-170

[100] Tvrdá E, Ďuračka M, Sukova M, Valentova G, Tokarova K, Kalafova A, et al. The effect of perga on the testicular oxidative profile of Zucker diabetic fatty rats. In: Nutricon 2018. Skopje: Consulting and Training Center KEY; 2018. pp. 167-168

[101] Álvarez-Cilleros D, López-Oliva E, Goya L, Martín MA, Ramos S. Cocoa intake attenuates renal injury in Zucker diabetic fatty rats by improving glucose homeostasis. Food and Chemical Toxicology. 2019;**127**:101-109. DOI: 10.1016/j.fct.2019.03.002

[102] Capcarova M, Kalafova A, Schwarzova M, Schneidgenova M, Soltesova Prnova M, Svik K, et al. Consumption of bee bread influences glycaemia and development of diabetes in obese spontaneous diabetic rats. Biologia. 2019 in press

Chapter 6

ZDF Rats: A Suitable Model to Study Male Reproductive Dysfunction in Diabetes Mellitus Type 2 Patients

Filip Benko, Mária Chomová, Oľga Uličná and Eva Tvrdá

Abstract

This chapter examines the impact of diabetes mellitus type 2 (DM 2) on the vitality of male reproductive cells collected from Zucker diabetic fatty (ZDF) rats which could be a suitable experimental model for simulating this metabolic disorder. Epididymal spermatozoa were subjected to the assessment of motility, membrane integrity, mitochondrial activity, DNA fragmentation, and oxidative profile. Our results show that DM 2 in combination with obesity negatively affects the sperm vitality and increases the chances of oxidative damage to male gametes. In conclusion we may state that DM 2 has a negative impact on the spermatogenic aspect of male fertility and decreases the sperm quality.

Keywords: diabetes mellitus type 2, obesity, ZDF rats, male reproduction, spermatozoa

1. Introduction

Diabetes mellitus is a chronic metabolic dysfunction which involves alterations in insulin production. Pancreatic β cells are primarily responsible for insulin secretion. There is a variety of complications associated with this disease such as chronic hyperglycemia, neuropathy, retinopathy, and cardiovascular diseases. Diabetes mellitus may be divided into three types: diabetes mellitus type 1 (DM1), type 2 (DM2), and type 3 (DM3)—also known as gestational diabetes mellitus. DM1 is based on an immune-mediated destruction of β cells. It is an autoimmune disease, most common in children and young adults. The treatment includes monitoring of blood glucose levels and insulin therapy [1, 2]. DM2 is associated with insulin resistance. The main cause is a reduced sensitivity of affected tissues to the metabolic effects of insulin. The development of the disease is often associated with obesity, which, on the other hand, causes health complications, such as fertility issues. Contraindications associated with DM2 include disorders of the male reproductive system because metabolism of glucose is essential for a correct process of spermatogenesis. DM2 has a negative impact on the sperm quality markers such as motility, DNA integrity, and seminal plasma composition [3, 4]. Development of DM3 is mostly common among women who are overweight or obese in comparison with thin or normal-weight women. This type of diabetes

IntechOpen

may cause higher levels of adipose tissue in the fetus and an increased child birth weight. Also a few studies reported that DM3 is associated with higher levels of abdominal fat and an increased risk for visceral adiposity which could be linked to other consequences following birth such as future development of DM2 and cardiovascular diseases [5, 6].

Another factor that negatively affects male fertility is obesity. Obesity has been shown to negatively affect the male reproductive potential not only by decreasing sperm quality but mainly by altering the germ cell molecular structure in the testes [7]. Animal models play a pivotal role in monitoring and understanding DM pathogenesis due to a combination of their genetic and functional characterization [8]. Suitable models to demonstrate the effects of DM2 on the organism are Zucker diabetic fatty (ZDF) rats. This type of rats had been discovered in 1961 following cross-breeding of Merck (M-strain) and Sherman rats. ZDF rats have the unique ability to simulate symptoms and contraindications of DM2. These rats have insulin resistance caused by the presence of homozygous mutation of the leptin hormone receptor (*fa gene*) which causes obesity and an increased insulin secretion [9, 10]. The aim of this chapter was to evaluate the effect of diabetes mellitus type 2 on the vitality of ZDF rat reproductive cells.

2. Material and methods

2.1 Biological material

For the experiment three separate groups (**Figure 1**) of adult ZDF and Wistar rats at the age of 120 days were used. The animals were obtained from the Institute of Experimental Pharmacology (Slovak Academy of Sciences, Slovakia) and kept in plastic cages at 24 ± 1°C and 12 h light/12 h dark photoperiod. The animals were provided with water ad libitum. Institutional and national guidelines for the care and use of laboratory animals were followed, and all procedures were approved by the State Veterinary and Food Institute of the Slovak Republic (no. 493/18-221/3) and Ethics Committee.

The first group consisting of 15 Wistar rats (15) was the healthy control group. The second group consisted of 15 ZDF rats on a normal diet (ZDN), while the third group comprised 16 ZDF rats on a special high-caloric diet (Purina 5000; ZDF).

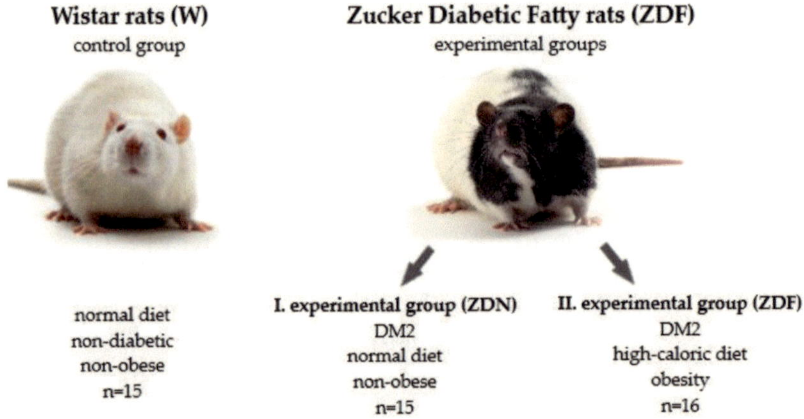

Figure 1.
Distribution of the tested animals (source: Author).

All ZDN and ZDF rats had diabetes confirmed by a blood test. Overall, we analyzed samples from 46 rats.

Following anesthesia and decapitation, epididymes were collected from the rats, cut with a scalpel into smaller fragments, and incubated in phosphate buffer saline (PBS, with Ca^{2+} and Mg^{2+}, Sigma-Aldrich, St. Louis, USA) at 37°C for 15 minutes. After incubation we used the epididymal spermatozoa for further assessment of sperm quality, including motility, membrane and acrosome integrity, DNA fragmentation, mitochondrial activity, lipid peroxidation (LPO), and production of superoxide radicals.

2.2 Sperm motility

Sperm motility is one of the most important parameters examining the fertilizing ability of male gametes. This parameter has an important role in the fertilization process. We examined the motility manually. Rats are known for a high level of motile spermatozoa, generally within a range of 85–96% [11]. After sample collection we applied 10 μl of the epididymal sperm suspension into Makler's counting chamber and counted motile spermatozoa using a light microscope (Olympus, Tokyo, Japan) and a magnification of 40×.

2.3 Membrane integrity

Membrane integrity is a parameter which defines the quality and condition of the reproductive cells. For the determination of membrane integrity, we used a combination of eosin and nigrosin dye. This protocol follows a differential staining method for the analysis of vital and damaged cytoplasmatic membrane in sperm cells of mammals and birds. Eosin is one of the most common dyes to stain the cytoplasm and cytoplasmic proteins of cells. After application of eosin, damaged cytoplasmatic membrane of sperm cells absorbed the dye and changes color into light-red or light-pink. The vital spermatozoa stayed without any change of color. Nigrosin (Sigma-Aldrich, St. Louis, USA) provides background for the smear as a contrast dye for a better differentiation [12–14]. The slides for the analyses were prepared as follows: we applied a drop of semen on the slide and dyed it with 4 μl of eosin solution (Sigma-Aldrich, St. Louis, USA). After application of eosin, we did the same with the nigrosin solution (Sigma-Aldrich, St. Louis, USA), and using another slide glass, we did a smear on the slide. The samples were air-dried at room temperature and observed using a light microscope (Olympus, Tokyo, Japan) and a magnification of 40×. At least 100 cells were evaluated in each slide.

2.4 Acrosome integrity

For the analysis of the acrosome integrity, we used a double fast green–rose bengal stain. After application of 10 μl of semen in a glass slide, we added the same volume of fast green–rose bengal mixture (Sigma-Aldrich, St. Louis, USA), incubated for 60 seconds, and used another slide glass to prepare a smear which was subsequently air-dried at room temperature. With the help of a light microscope (Olympus, Tokyo, Japan), we observed the integrity and compactness of the sperm acrosome. Damage to the acrosome was observed as a disruption of the membrane and cluster of stain localized in the sperm head. We evaluated at least 100 cells in each slide and calculated the percentage of cells with a normal or a damaged acrosome.

2.5 Test of metabolic activity

The mitochondrial toxicity test (MTT test) is a colorimetric test often used for the determination of cell metabolic activity. The main principle of the test is the application of yellow tetrazolium bromide (Sigma-Aldrich, St. Louis, USA) and its subsequent reaction with insoluble mitochondrial succinate dehydrogenase produced by cell mitochondria. This reaction ends with the formation of a blue-violet formazan [15]. The amount of formed formazan is directly proportional to the level of metabolic activity of the cells. We evaluated this test spectrophotometrically at 570 against 620 nm (Multiscan reader, Thermo Fisher, Vantaa, Finland).

2.6 Production of superoxide radicals

For the detection of the superoxide radical concentration, we used the nitroblue tetrazolium or NBT test. The principle of this method is the application of nitroblue tetrazolium (Sigma-Aldrich, St. Louis, USA) to the semen sample. This substance reacts with cellular superoxide to form derivates of formazan. Following washing with PBS and centrifugation (1250 rpm, 5 min), the concentration of formasan derivates was evaluated spectrophotometrically [16] at wavelengths of 620 against 570 nm (Multiscan reader, Thermo Fisher, Vantaa, Finland).

2.7 DNA fragmentation

This type of DNA damage is characterized by both single and double DNA strand breaks. Several types of DNA damage may be observed in mammalian germ cells and are often associated with male infertility. Nowadays various tests are available to detect sperm DNA damage. DNA fragmentation arises from various reasons such as deficiencies in recombination during spermatogenesis, abnormal sperm maturation, abortive apoptosis, or oxidative stress [17, 18]. In our research we used the chromatin-dispersion test by using the Halomax diagnostic kit (Halotech, Madrid, Spain). This kit can analyze the integrity of the DNA molecule and is based on a controlled DNA denaturation process to facilitate the subsequent removal of the proteins contained in each spermatozoon. The main principle of this method is that damaged spermatozoa create halos formed by loops of fragmented DNA at the head of the sperm which are not present in normal spermatozoa. For the evaluation we used a fluorescent microscope (Leica, Holzheim, Germany) and a magnification of 40×. At least 300 cells were evaluated for DNA fragmentation in each slide containing agarose with processed spermatozoa.

2.8 Lipid peroxidation

Lipid peroxidation is a process of membrane lipid degeneration caused by free radicals [19]. The extent of lipid peroxidation in our samples was expressed as the amount of malondialdehyde (MDA) production following the addition of thiobarbituric acid (Sigma-Aldrich, St. Louis, USA) and exposure to heat (100°C, 1 h). The MDA concentration in the samples was evaluated spectrophotomerically at a wavelength of 540 nm (Multiscan reader, Thermo Fisher, Vantaa, Finland).

3. Statistical analysis

The data we obtained were statistically processed using GraphPad Prism (version 6.0 for Windows, GraphPad Software incorporated, San Diego, California,

USA, http://www.graphpad.com/). Differences between the compared groups were statistically evaluated by one-way analysis of variance (ANOVA) and the Tukey comparative test. Statistical significance was assessed at levels ***(P < 0.001), **(P < 0.01), and *(P < 0.05).

4. Results and discussion

When we compared the sperm motility of individual groups (**Figure 2**), we found statistically significant differences (P < 0.001) when comparing the experimental group ZDF with the control W group as well as with the ZDN experimental group. Similarly to our results, Ohta et al. [20] reported a reduction in the sperm motility in rats with obesity in comparison with the control group. They tried to demonstrate the potential link between obesity and a reduction of sperm motility, which ultimately affects the fertility of rats. Simas et al. [21] also reported a decrease of morphologically normal spermatozoa in rats with diabetes. The most common abnormalities found were sperm head deformations and flagella deformities. There is a very close connection between the sperm volume and percentage of motile spermatozoa. Both types of diabetes have a serious impact on the sperm quality, and a lot of studies examine the effect of diabetes on the male fertility. Condorelli et al. [22] showed different pathophysiological effects in type 1 and type 2 of diabetes on the sperm function and quality. The aim of their experiment was to compare DM1 and DM2 patients with healthy and fertile subjects. Male patients suffering from diabetes mellitus type 2 have several contraindications. Among these, the most prominent include low sperm volume, higher concentration of mitochondrial superoxide anions, an increased reactive oxygen species production in the seminal fluid, and lipoperoxidation. The study showed that DM2 caused an inflammatory condition and increased the level of oxidative stress which led to an increased sperm DNA fragmentation and a decreased vitality of spermatozoa. In our experiments we also observed several contraindications in the rats which suffered from diabetes mellitus type 2 including a decreased motility and vitality of sperm cells, a lower mitochondrial activity, followed by higher levels of superoxide production and lipid peroxidation. On the other hand, a combination of DM2 and obesity has more serious consequences to male fertility.

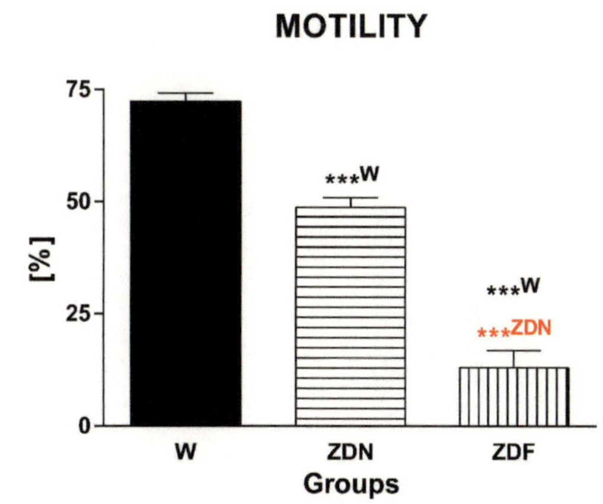

Figure 2.
*Differences in the sperm motility among the observed groups. W, Wistar rats, control group; ZDN, nonobese ZDF rats with diabetes mellitus type 2; ZDF, ZDF rats with obesity and diabetes mellitus type 2. ***(P < 0.001). ^{W}, compared to the Wistar group; ^{ZDN}, compared to the ZDN group.*

Numerous studies have shown the negative impact of obesity on the semen quality. Obesity causes a decrease of sperm concentration and motility and an increased DNA fragmentation index, and it is often associated with erectile dysfunction. Fernandez et al. [23] examined the impact of obesity on the fertility in male Wistar rats which were fed a high-fat diet and compared these with the control group consisting of nonobese rats. Similarly to our study, their results reported that obese animals have a low sperm quality caused by a decreased percentage of spermatozoa with progressive motility which may lead to the development of subfertility. We also recorded higher levels of non-motile spermatozoa in obese rats which were fed with a high-caloric diet.

The evaluation of the membrane integrity of spermatozoa (**Figure 3**) showed statistically significant differences ($P < 0.001$) among the groups. The sperm vitality of the ZDF experimental group was significantly lower than the control W group as well as with the ZDN experimental group. Sabeti et al. [24] also observed statistically significant differences ($P < 0.05$) in the viability between the experimental and control group of rats suffering from diabetes.

The membrane integrity is also affected by the level of cholesterol in the organism. Normal levels of cholesterol are highly important for the membrane fluidity and sperm motility. If there is an increase in the cholesterol levels because of the presence of obesity, then more cholesterol is incorporated into the lipid bilayer of the cells, leading to its adverse effects on the membrane integrity. These facts support the theory that changes in the levels of systemic cholesterol, triglyceride, and free fatty acids are associated with alterations of the cell membrane dynamics, sperm motility, morphology, and susceptibility to DNA damage [25].

As the graph shows, there were no statistically significant differences between any of the studied groups (**Figure 4**). In our opinion this condition could show that diabetes and obesity affect other semen parameters, but the integrity of the acrosome remains unchanged. It is important to know that diabetes mellitus type 2 in combination with obesity does not cause acrosome deformations. This theory was supported by Ding et al. [3] who studied differences between diabetes mellitus type 1 and type 2. In their study, they found that diabetes mellitus type 1 has a markedly impaired glycation process and synthesis of proteins, leading to the development of acrosome integrity disorders, while diabetes mellitus type 2 has less serious consequences to the acrosome. The acrosome integrity was the only parameter which remained stable. It could be caused by the presence of acid hydrolases

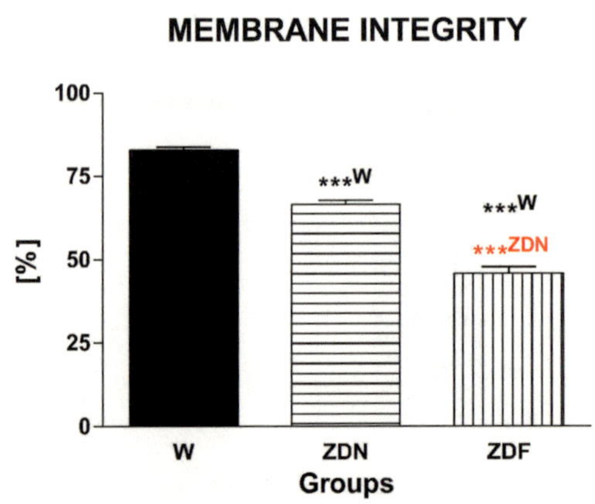

Figure 3.
*Differences in the membrane integrity or viability among the observed groups. W, Wistar rats, control group; ZDN, nonobese ZDF rats with diabetes mellitus type 2; ZDF, ZDF rats with obesity and diabetes mellitus type 2. ***(P < 0.001).* ^{W}*, compared to the Wistar group;* ^{ZDN}*, compared to the ZDN group.*

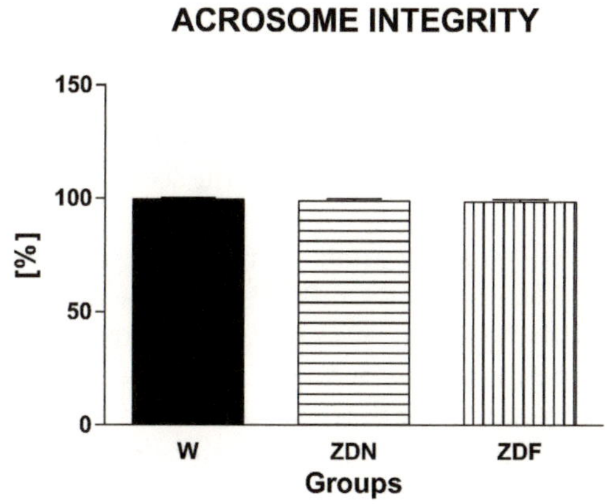

Figure 4.
Acrosome integrity among the observed groups. W, Wistar rats, control group; ZDN, nonobese ZDF rats with diabetes mellitus type 2; ZDF, ZDF rats with obesity and diabetes mellitus type 2. ^{W}, compared to the Wistar group; ^{ZDN}, compared to the ZDN group.

and lysosomal proteins because the acrosome does not need carbohydrates for its function. The presence of polyunsaturated fatty acids in the plasma membrane could exhibit protective effects against the negative activity of ROS and prevent alterations to the acrosome. This theory was also supported by Reddy et al. [26] who tested diabetic-induced reproductive toxicity in male Wistar rats.

The evaluation of the MTT test (**Figure 5**) showed statistically significant differences ($P < 0.5$) between the experimental group ZDF and the experimental group ZDN as well as between the experimental group ZDN and the control group W. Another statistically significant difference ($P < 0.01$) was observed between the ZDN group and the W group. Our results show that the lowest values of mitochondrial activity were observed in the experimental ZDF group. Mitochondrial activity is closely related to the motility as mitochondria are the energy-metabolic center of the cell. We also observed the lowest sperm motility values in ZDF group of rats (**Figure 1**). In the analysis of diabetic rats, Simas et al. [21] reported a significant decrease in the metabolic activity of these rats compared to the control group. Experimental animals had an induced DM1. In all group of rats with diabetes, they observed an increase of inactive mitochondria which caused a

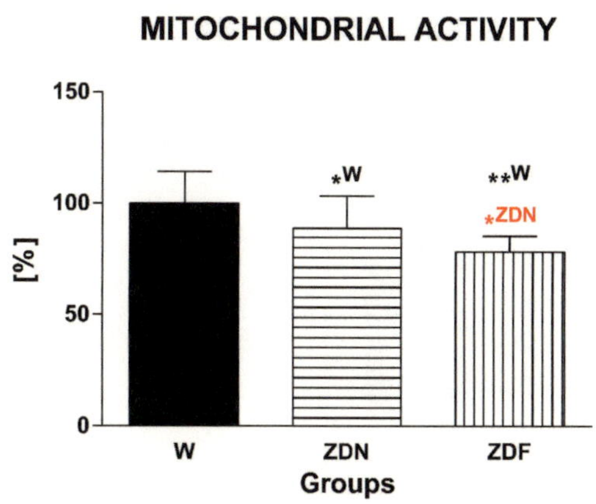

Figure 5.
*Mitochondrial activity among the observed groups. W, Wistar rats, control group; ZDN, nonobese ZDF rats with diabetes mellitus type 2; ZDF, ZDF rats with obesity and diabetes mellitus type 2. **($P < 0.01$); *($P < 0.5$). ^{W}, compared to the Wistar group; ^{ZDN}, compared to the ZDN group.*

decrease of mitochondrial activity. Like Simas et al., we also observed low levels of mitochondrial activity in the ZDF group when compared to other groups. A decrease of mitochondrial activity was also observed in the experimental ZDN group. This phenomenon could be because diabetes mellitus had a negative impact on the metabolism of carbohydrates, which are the source of energy for a proper mitochondrial function.

Results of the superoxide radical production (NBT test) (**Figure 6**) showed statistically significant differences ($P < 0.05$; $P < 0.01$) when we compared both experimental groups with the control group W. In the experimental ZDF group, we recorded the highest superoxide radical production. This finding suggests that the group of ZDF rats is more at risk for the onset of oxidative stress, which negatively affects the viability of spermatozoa produced by these rats. Sabeti et al. [24] demonstrated that the abnormal production and presence of the superoxide radicals in combination with obesity leads to a reduced semen quality and sperm viability. Overproduction of superoxide may also be due to an increase in metabolic functions in order to maintain the homeostasis in obese individuals. Amaral et al. [27] reported no significant differences in the testicular cell concentration between diabetic and nondiabetic rats, but there was a massive decline in the sperm cell concentration due to the influence of hyperglycemia in late stages of spermatogenesis and increased production of reactive oxygen species (ROS) that may lead to the development of oxidative stress. The consequences of oxidative damage include the loss of motility due to lipid peroxidation, induction of DNA damage of the sperm nucleus, and defects in spermatogenesis affecting the final fertilizing potential of sperm cells.

The oxidative environment in testicular tissue of patients with DM2 may result in cellular damage, including lipid peroxidation, formation of carbonyl groups, and DNA damage which causes sperm abnormalities. In fact, this negative condition contributes to a reduction in the sperm motility and viability followed by an increased percentage of abnormalities in the sperm cells of rats [28].

Figure 7 points out the levels of DNA fragmentation between the experimental groups and the control. After evaluation of our results, we found statistically significant differences between the DNA fragmentation following the presence of diabetes mellitus type 2 and obesity. Statistically significant differences ($P < 0.001$) were observed when we compared the experimental ZDF group with the ZDN group and the control W group and also when we compared the experimental ZDN group with the control W group. As seen from our results, a continuous increase in

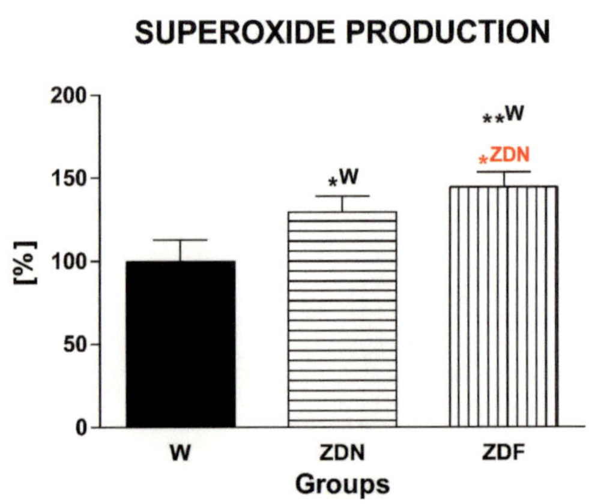

Figure 6.
*Production of superoxide among the observed groups. W, Wistar rats, control group; ZDN, nonobese ZDF rats with diabetes mellitus type 2; ZDF, ZDF rats with obesity and diabetes mellitus type 2. **(P < 0.01); *(P < 0.5). ^{W}, compared to the Wistar group; ^{ZDN}, compared to the ZDN group.*

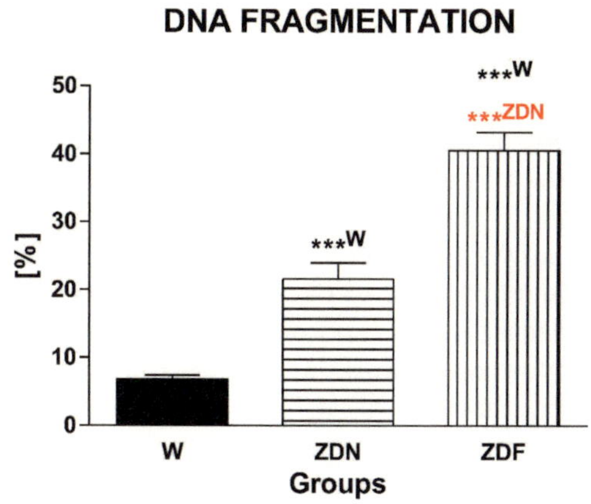

Figure 7.
*Levels of DNA fragmentation among the observed groups. W, Wistar rats, control group; ZDN, nonobese ZDF rats with diabetes mellitus type 2; ZDF, ZDF rats with obesity and diabetes mellitus type 2. *** (P < 0.001). [W], compared to the Wistar group; [ZDN], compared to the ZDN group.*

DNA fragmentation was present, depending on the presence of diabetes mellitus type 2, obesity, or both. Abbasihormozi et al. [29] found the link between leptin receptor mutation and higher levels of DNA fragmentation as a consequence of a higher production of reactive oxygen species. Leptin is an adipose tissue-derived hormone which has an important role in the metabolism and energy homeostasis and ensures a proper function of the neuroendocrine and reproductive system. A direct administration of leptin could affect male infertility by ROS overproduction or hormone profile modulation, but the exact mechanism of the effect of leptin on spermatogenesis is still not understood.

As with the evaluation of the superoxide radical production, we found statistically significant differences ($P < 0.001$) in the case of the concentration of malondialdehyde (**Figure 8**) among both experimental groups (ZDN and ZDF) with the control group (W) and also when comparing the experimental group ZDF with the second experimental group ZDN. Like us, Simas et al. [21] evaluated that there was an increase in the MDA concentration in diabetic rats when compared to all other groups. They also recorded a direct correlation between the superoxide radical production values and MDA concentration. Vignera et al. [30] found a link between

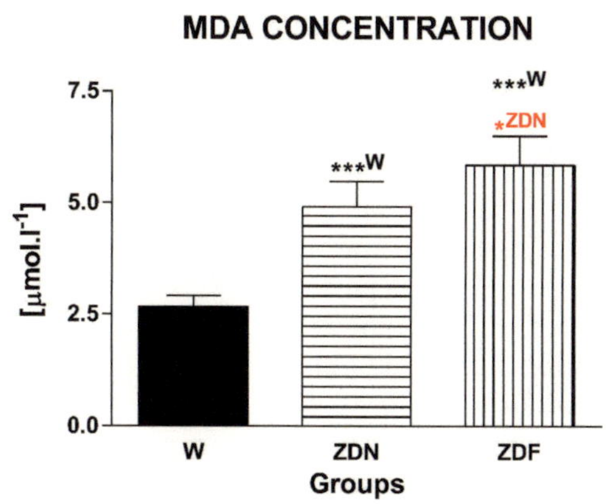

Figure 8.
*Concentration of malondialdehyde (MDA) among the observed groups. W, Wistar rats, control group; ZDN, nonobese ZDF rats with diabetes mellitus type 2; ZDF, ZDF rats with obesity and diabetes mellitus type 2. ***(P < 0.001); *(P < 0.5). [W], compared to the Wistar group; [ZDN], compared to the ZDN group.*

production of MDA and regression of enzymatic activity. The activity of malondialdehyde and caspase-3 were significantly higher, whereas the enzymatic activities of superoxide dismutase and glutathione peroxidase were significantly lower in the diabetic rats than in the control group of healthy rats. Supportive antioxidant treatment could help to increase the antioxidant enzymatic activity and to decrease the activity of malondialdehyde and caspase-3. This treatment also could reduce the apoptosis of germ cells. These facts support the theory about testicular oxidative damage caused by diabetes and subsequent protective effects of antioxidant treatment.

In our opinion the decrease of rat sperm vitality was associated with high levels of body fat because adipose tissue works as an individual endocrine system affecting spermatogenesis and steroidogenesis. On the other hand, diabetes mellitus type 2 affected correct mechanisms of spermatogenesis as carbohydrates are the main source of energy in spermatozoa.

5. Conclusion

The results of our analysis show that diabetes mellitus type 2 has a negative effect on the vitality of rat male reproductive cells. We may conclude that the disease affects sperm motility, viability, and DNA stability negatively. The disease increases the production of superoxide radicals and the concentration of malondialdehyde and also affects the mitochondrial activity of spermatozoa. However, diabetes mellitus type 2 has no effect on the acrosome integrity. At the same time, we observed visible differences in the experimental group of ZDF rats suffering from diabetes mellitus type 2 in combination with obesity in all evaluated parameters. However, differences were also observed in the experimental group of ZDN rats who developed type 2 diabetes mellitus but did not have obesity.

Acknowledgements

This research was created with financial support of the projects APVV-15-0544 and VEGA 1/0314/19.

Conflict of interest

None declared.

Abbreviations

CASA	Computer-assisted sperm analysis
DM1	Diabetes mellitus type 1
DM2	Diabetes mellitus type 2
GDM	Gestational diabetes mellitus
MDA	Malondialdehyde
MTT	Mitochondrial toxicity test
NBT	Nitroblue tetrazolium
ROS	Reactive oxygen species
ZDF	Zucker diabetic fatty rats with high-caloric diet
ZDN	Zucker diabetic fatty rats with normal diet
W	Wistar rats

Author details

Filip Benko[1], Mária Chomová[2], Oľga Uličná[3] and Eva Tvrdá[1*]

1 Department of Animal Physiology, Faculty of Biotechnology and Food Sciences, Slovak University of Agriculture in Nitra, Slovakia

2 Faculty of Medicine, Institute of Medical Chemistry and Clinical Biochemistry, Comenius University in Bratislava, Slovakia

3 Third Intern Clinic, Faculty of Medicine, Comenius University in Bratislava, Slovakia

*Address all correspondence to: evina.tvrda@gmail.com

References

[1] Al-awar A, Kupai M, Veszelka M, Szűcs G, Attieh Z, Murlasits Z, et al. Experimental Diabetes Mellitus in Different Animal Models. Journal Diabetes Research. 2016;**2016**:1-12. DOI: 10.1155/2016/9051426

[2] Siwy J, Zoja C, Klein J, Benigni A, Mullen W, Mayer B, et al. Evaluation of the Zucker diabetic fatty (ZDF) rat as a model for human disease based on urinary peptidomic profiles. PLoS One. 2012;7(12):e51334. DOI: 10.1371/journal.pone.0051334

[3] Ding G, Liu Y, Liu M, Pan J, Guo M, Sheng J, et al. The effects of diabetes on male fertility and epigenetic regulation during spermatogenesis. Asian Journal of Andrology. 2015;**17**:948-953. DOI: 10.4103/1008-682X.150844

[4] Temidayo O, du Plessis S. Diabetes mellitus and male infertility. Asian Pacific Journal of Reproduction. 2018;7:6-14. DOI: 10.4103/2305-0500.220978

[5] Chu S, Callaghan W, Kim S, Schmid C, Lau J, England L, et al. Maternal obesity and risk of gestational diabetes mellitus. Diabetes Care. 2007;**30**:2070-2076. DOI: 10.2337/dc06-2559a

[6] Page K, Romero A, Buchanan T, Xiang A. Gestational diabetes mellitus, maternal obesity, and adiposity in offspring. The Journal of Pediatrics. 2014;**164**(4):807-810. DOI: 10.1016/j.jpeds.2013.11.063

[7] Palmer N, Bakos H, Fullston T, Lane M. Impact of obesity on male fertility, sperm function and molecular composition. Spermatogenesis. 2012;**2**:253-263. DOI: 10.4161/spmg.21362

[8] Teerds KJ, De Rooij DG, Keijer J. Functional relationship between obesity and male reproduction: From humans to animal models. Human Reproduction Update. 2011;**17**:667-683. DOI: 10.1093/humupd/dmr017

[9] Griffen S, Wang J, German M. A genetic defect in beta-cell gene expression segregates independently from the fa locus in the ZDF rat. Diabetes. 2001;**50**:63-68. DOI: 10.2337/DIABETES.50.1.63

[10] Phillips M, Liu Q, Hammond H, Dugan V, Hey P, Caskey C, et al. Leptin receptor missense mutation in the fatty Zucker rat. Nature Genetics. 1996;**13**:18-19. DOI: 10.1038/ng0596-18

[11] Creasy D, Chapin R. Sperm motility. In: Toxicologic Pathology. 3rd ed. Cambridge, Massachusetts, USA: Academic Press; 2013. pp. 2859-2963. DOI: 10.1016/C2010-1-67850-9

[12] Anthony A, Leslie G. Membrane integrity monitoring tests. In: Advanced Membrane Science and Technology for Sustainable Energy and Environmental Applications. Sawston, Cambridge, UK: Woodhead Publishing; 2011. pp. 792-818

[13] Knebel Doeberitz M, Wentzensen N. Cytoplasmatic stain. In: Comprehensice Cytopathology. 3rd ed. Philadelphia, Pennsylvania, USA: Saunders; 2008. pp. 1091-1120. DOI: 10.1016/B978-1-4160-4208-2.X0041-3

[14] Kondracki S, Wysokińska A, Kania M, Górski K. Application of two staining methods for sperm morphometric evaluation in domestic pigs. Journal of Veterinary Research. 2017;**61**:345-349. DOI: 10.1515/jvetres-2017-0045

[15] Bahuguma A, Khan I, Bajpai V, Kang S. MTT assay to evaluate the cytotoxic potential of a drug.

Bangladesh Journal of Pharmacology 2017;**12**:115-118

[16] Esfandiari N, Sharma R, Saleh R, Thomas A, Agarwal A. Utility of the Nitroblue Tetrazolium reduction test for assessment of reactive oxygen species production by seminal leukocytes and spermatozoa. Journal of Andrology. 2013;**24**:862-870. DOI: 10.1002/j.1939-4640.2003.tb03137.x

[17] González-Marín C, Gosálvez J, Roy R. Types, causes, detection and repair of DNA fragmentation in animal and human sperm cells. International Journal of Molecular Sciences. 2012;**13**:14026-14052. DOI: 10.3390/ijms131114026

[18] Enciso M, Sarasa J, Agarwal A, Fernández JL, Gosálvez J. A two-tailed comet assay for assessing DNA damage in spermatozoa. Reproductive BioMedicine Online. 2009;**18**:609-616

[19] Ayala A, Muñoz F, Argüelles S. Lipid peroxidation: Production, metabolism and signaling mechanisms of malondialdehyde and 4-Hydroxy-2-Nonenal. Oxidative Medicine and Cellular Longevity. 2014;**2014**:360438. DOI: 10.1155/2014/360438

[20] Ohta T, Katsuda Y, Miyajima K, Sasase T, Kimura S, Tong B, et al. Gender differences in metabolic disorders and related diseases in spontaneously diabetic Torii-Lepr (fa) rats. Journal Diabetes Research. 2014;**2014**:841957. DOI: 10.1155/2014/841957

[21] Simas J, Mendes T, Paccola C, Vendramini V, Miraglia S. Resveratrol attenuates reproductive alternations in type 1 diabetes-induced rats. Journal of Experimental Pathology. 2017;**98**:312-328. DOI: 10.1111/iep.12251

[22] Condorelli R, La Vignera S, Mongioi L, Alamo A, Calogero A. Diabetes mellitus and infertility: Different pathophysiological effects in type 1 and type 2 on sperm function. Frontiers in Endocrinology. 2018;**9**:268. DOI: 10.3389/fendo.2018.00268

[23] Fernandez C, Bellentani F, Fernandes G, Perobelli J, Favareto A, Nascimento A, et al. Diet-induced obesity in rats leads to a decrease in sperm motility. Reproductive Biology and Endocrinology. 2011;**9**:32. DOI: 10.1186/1477-7827-9-32

[24] Sabeti P, Pourmasumi S, Rahiminia T, Akyash F, Talebi A. Etiologies of sperm oxidative stress. International Journal of Reproductive BioMedicine. 2016;**14**:231-240

[25] Palmer N, Bakos H, Owens J, Setchell B, Lane M. Diet and exercise in an obese mouse fed a high-fat diet improve metabolic health and reverse perturbed sperm function. American Journal of Physiology - Endocrinology and Metabolism. 2012;**302**:768-780. DOI: 10.1152/ajpendo.00401.2011

[26] Reddy K, Narayana Rao M, Murthy J, Reddy P. Lead aggravates the diabetic-induced reproductive toxicity in male Wistar rats. Toxicology Research. 2016;**5**:1465-1476. DOI: 10.1039/c6tx00099a

[27] Amaral S, Moreno A, Santos M, Seica R, Ramalho-Santos J. Effects of hyperglycemia on sperm and testicular cells of Goto-Kakizaki and streptozotocin-treated rat models for diabetes. Theriogenology. 2006;**66**:2056-2067. DOI: 10.1016/j.theriogenology.2006.06.006

[28] Oliviera P, Tomás G, Dias T, Martins A, Rato L, Alves M, et al. White tea consumption restores sperm quality in prediabetic rats preventing testicular oxidative damage. Reproductive Biomedicine Online. 2015;**31**:544-556. DOI: 10.1016/J.RBMO.2015.06.021

[29] Abbasihormozi S, Shahverdi A, Kouhkan A, Cherargi J, Akhlaghi A, Kheimeh A. Relationship of leptin administration with production of reactive oxygen species, sperm DNA fragmentation, sperm parameters and hormone profile in the adult rat. Archives of Gynecology and Obstetrics. 2013;**287**:1241-1249. DOI: 10.1007/s00404-012-2675-x

[30] Vignera S, Condorelli R, Vicari E, D'Agata R, Calogero A. Diabetes mellitus and sperm parameters. Journal of Andrology. 2013;**33**:145-153. DOI: 10.2164/jandrol.111.013193

Chapter 7

Drosophila melanogaster: A Robust Tool to Study Candidate Drug against Epidemic and Pandemic Diseases

Saikat Samadder

Abstract

Drosophila melanogaster is a widely used, dynamic model organism to study various pathogenic diseases observed ubiquitously in the human population. Drosophila, at present, is extensively used to conduct preclinical studies besides its counterpart rodents. The epidemic and pandemic diseases are discussed in this review to demonstrate *Drosophila melanogaster* as a key model. Epidemic and pandemic diseases are still claiming more than 5 million lives every year, and these diseases were well studied in flies. Currently there is no cure for the disease like HIV; the bacterial and fungal infections usually seen in HIV/AIDS patients could be demonstrated elaborately in *Drosophila melanogaster*. Diseases like myocardial infractions and cancer causing viral infection are long term effects of ART (anti-retroviral therapy) that could be experimented in flies. Stable Drosophila S2 cell line, Transgenic flies, transfusion of bacteria and fungi could be implemented to study several infectious diseases and for vaccine development. The latest trends in understanding pathogenic diseases and its potential biochemical markers in flies are discussed in this review to utilize the fruit flies as a functional tool and to explore further it in drug development. The advantages and disadvantages of the fly as a model of infection are discussed along with the epidemiology and the cellular pathophysiology

Keywords: *Drosophila melanogaster*, HIV, Influenza, cholera, tuberculosis, pneumonia, viral diseases, epidemic and pandemic diseases

1. Introduction

1.1 Epidemic disease

The term epidemic is derived from Greek word "epi" meaning "upon" and "demos" meaning "people". It refers to a communicable disease which spreads rapidly in a given population within a very short period of time. Any infectious disease existing in a region does not make it epidemic unless it causes faster mortality. A death rate of around 1.6 folds higher than usual death rate (baseline) caused by a disease in a population within a fixed period could be considered as an epidemic disease. A disease lower than this fold increase, observed in a population could be designated as an outbreak of a disease [1].

Diseases like tuberculosis, hepatitis, yellow fever, chikungunya, ebola virus disease, marburg virus disease, Crimean-Congo haemorrhagic fever, rift valley fever, typhoid fever, Shigellosis, plague, lassa fever, West Nile fever, zika virus disease, meningitis, MERS-CoV, plague, monkeypox, nodding syndrome, nipah virus infection are considered as epidemic diseases as per World Health Organization [2]. Epidemic diseases like plague, small pox and cholera caused unsurpassed deaths in human population till the end of eighteenth century [3].

1.2 Pandemic disease

The term pandemic is derived from Greek word "pan" meaning "all" and "demos" meaning "people". It refers to an epidemic disease which spreads among large population possibly across geographic locations or continents within a short time span [4].

Influenza, along with viral pneumonia, HIV and cholera are considered as pandemic disease and caused millions to die beside high rate of hospitalization and life threatening conditions across the globe [2]. The viral diseases like Influenza, cholera and HIV caused maximum deaths in the twenty-first century [3].

1.3 Vaccination

Vaccines are available for most of the epidemic and pandemic diseases [5]. Vaccination is the most effective prevention technique to suppress the infection in healthy population [6]. However, poor and conflicted regions of Asia and Africa are deprived of these vaccines [7]. World Health Organization plays a major role in epidemic preparedness in these regions and provides extended healthcare facilities during an epidemic outbreak [8].

1.4 Global requirement

Considering, the disease outbreak and its transmission is high in a poor population of developing region [9]. First line and second line antibiotics are the most effective medicines for infected subjects as the vaccines are ineffective after the infection had taken place. First line therapy includes antibiotics are the most commonly prescribed medicines to alleviate the infection process, often not responsive on several types of multi drug resistant infection [10]. Hence it is important to select a cost effective model to screen the first line antibiotics or antivirals.

1.5 Fruit fly as a model organism for drug screening

Today, we need to discover more efficacious antibiotics to fight the infectious diseases. *Drosophila melanogaster* could be a useful model organism to study the infection process and to screen an efficient drug. Due to its shorter lifespan and vast genetic similarity towards vertebrates allows conducting the drug screening experiments. It is reviewed here that *Drosophila melanogaster* was already been used to study infections caused by pandemic and epidemic diseases. But how to utilize the fruit flies to study different infectious disease and techniques to screen a potential drug candidate were not well reviewed.

2. Epidemiology of infectious diseases

2.1 Lower respiratory tract infection epidemiology

As per World Health Organization lower respiratory tract infections are caused mainly by influenza, pneumococcal pneumonia and viral pneumonia are responsible for 3 million deaths [11]. The WHO reported in 2018, 3–5 million cases of Influenza with 290,000–650,000 death cases annually [12]. As per the Global Disease burden (GDB) study report of 2015 there were around 1.5 million deaths in all age groups caused due to pneumococcal pneumonia [13]. SARS (Corona virus) causes viral pneumonia; it is epidemic to more than 30 countries with 8000 reported cases and 774 deaths during the year 2002–2003. MERS is a viral pneumonia causing infections in 688 persons and 282 deaths reported by WHO in 20 countries during 2012 [14].

2.2 Influenza

Influenza originates from *Orthomyxoviridae* family it can be differentiated into three types Influenza pandemic caused by Influenza A/B virus, seasonal Influenza and avian influenza (H5N1) [15]. Influenza virus causes upper respiratory tract infection often found to cause lower respiratory tract infection in association with bacterial co-invasion. The seasonal influenza leads to maximum hospitalization resulting fatality in infants during the seasonal outbreak [16].

2.3 Pneumonia and viral pneumonia

Pneumonia is caused due to several communicable infections usually known as community acquired pneumonia (CAP), often seen in hospitalized patients. Pneumonia can be caused by bacteria like *Streptococcus pneumoniae, Haemophilus influenzae type b (Hib),* and *Chlamydia pneumoniae, Staphylococcus aureus*, *Klebsiella pneumonia* and *Mycoplasma pneumonia.* Viruses like syncytial virus, adenovirus, rhinovirus, metapneumovirus, Influenza A/B viruses, Coronaviruses, parainfluenza virus including MERS and SARS causes viral pneumonia [17]. The viral pneumonia is the influenza often associated with bacterial infection thereby causing fatality better known as superinfection [18].

2.4 Tuberculosis

Tuberculosis is caused by *Mycobacterium tuberculosis*, a gram negative facultative anaerobic bacteria. In 2017 around 10 million people were infected with tuberculosis causing mycobacterium killing 1.6 million peoples across the world [19]. The current estimate of tuberculosis is not significantly different from the 2015 WHO report [20].

2.5 HIV

Currently HIV is the most fatal disease observed in human population across the globe. It caused maximum number of deaths around the world in the last 3 decades. As per the latest WHO report of 2019, HIV/AIDS have claimed more than 35 million deaths till date. Currently 36.9 million (31.1–43.9 million) peoples are living with HIV as of 2017 [21]. Although the rate of infection has decreased in the recent years, still HIV remains a global burden on world economy.

2.6 Diarrhoeal disease

2.6.1 Cholera

The term "cholera" was derived from Sanskrit meaning "stomach disturbance" [22]. Since, early 1800 century cholera outbreak turned out to be pandemic and caused millions to die, altogether six different pandemics took place the seventh started in the year 1961 and is still ongoing [23, 24]. In 2019 WHO report suggests 1.3 million to 4.0 million cases of cholera with an estimated 21,000–143,000 deaths worldwide [25].

2.7 Hepatitis

Viral hepatitis is one of the most life threatening disease, it causes death to 1.4 million peoples across the globe reported in 2018 [26]. Globally around 260 million peoples are infected with HBV and 71 million with HCV infections are reported causing 90% of deaths among viral hepatitis patients [27]. The HBV and HCV has the highest prevalence rate in the global population at present, hepatitis viruses like HAV, HAD and HEV are endemic in many countries [26]. Currently there is no vaccine available for HCV till date.

2.8 Typhoid

The term Typhoid was coined from the Greek word "typhus" which means "Smoky" was used to relate the delirium symptom often associated with typhoid fever [28]. Typhoid fever is caused by gram-negative bacteria known as *Salmonella enterica serovar typhi*. Around 11–21 million cases of typhoid fever outbreak are reported annually, among that it causes death of 128,000–161,000 individuals worldwide [29].

2.9 Malaria

Malaria fever is a severe parasitic disease caused by *Plasmodium falciparum* and *Plasmodium vivax transmits through* female *Anopheles gambiae* mosquitoes. In the year 2017 219 million cases were noted by World Health Organization, this seasonal outbreak of malaria in 87 countries led to 435,000 deaths [30].

2.10 Viral meningitis, viral encephalitis and hemorrhagic fever viruses

Viruses like herpes simplex virus HSV, HIV, mumps virus, measles virus and west Nile virus causes meningitis which causes frequent outbreaks in some regions [31]. Japanese encephalitis virus along with genus *Alphavirus* Togaviridae family viruses are arbovirus (arthropod borne virus) like California encephalitis, Chikungunya, dengue, Eastern equine encephalitis, Powassan, St. Louis encephalitis, Sindbis virus, West Nile, Yellow Fever and Zika virus are capable of causing encephalitis in humans [32, 33]. The viruses capable of causing hemorrhagic fever are dengue virus, rift valley virus, yellow virus, Crimean-Congo Hemorrhagic Fever, Lassa virus, Marburg virus and Ebola virus are epidemic diseases [34].

3. Drosophila model to study highly infectious diseases

There are at present several bacterial, fungal and viral models of infection which were successfully demonstrated to infect flies and used it to understand drug efficacy. Drosophila model of infectious disease could be very low cost model to study drug efficacy in-vivo; it could help to save lives by saving time during an epidemic outbreak. Understanding the disease pathogenesis in humans and drawing out a similar model in *Drosophila melanogaster* would suggest the target genes and proteins responsible for the underlying disease [35].

In the recent past several research works has been conducted to understand the immune system of *Drosophila melanogaster*. At present the immune system of *Drosophila* is a well-studied model to study infectious disease [35]. Adult flies have brain, heart, lung (spiracle), liver (fat body), kidney (renal tubule), GI tract (gut/crop), ovary/testis and versatile circulatory system (hemocyte) [36]. Apart from physiological resemblance Drosophila has 75% genetical similarity with human disease genes, due to this fact genetically tractable model could be generated to what extent is discussed here [35].

4. Host-pathogen interaction

Drosophila melanogaster has a well-built immune system to withstand pathogenic incursion, comprising of cellular, humoral and innate immunity in an effective but in simpler form than humans [37]. However, due to evolutionarily conserved immune pathways found in vertebrates and invertebrates, several components of fly immune system are homologous to humans [38]. The immune activation in flies against pathogens involves processes like recognition, coagulation, melanisation, phagocytosis, apoptosis, regulation of iron metabolism, synthesis of antimicrobial peptides and production of reactive oxygen species [39].

The bacterial and fungal infection leads to the activation of dToll, Imd, Eiger (TNF family homolog) and insulin like receptors (FOXO) in *Drosophila*. The drosophila toll and Imd (immune deficiency) pathways function as innate immunity. Toll receptors in flies play an important part during viral, fungal and bacterial infection. The patterns recognition receptors (PRRs) initiate the signal in fly immune system depending on the type of pathogen upon interaction [40]. Gram positive and gram negative bacterial infection activates peptidoglycan recognition protein SA (PGRP-SA) and Gram-negative binding protein 1 (GNBP1) respectively. PGRP-SA causes proteolytic cleavage of Spatzle upon stimulation of dToll, it mediates downstream signalling of dMyD88, Tube, Pelle, and DIF (dorsalrelated immunity factor) the NF-kB homolog. Imd an intracellular signalling protein located close to the transmembrane PGRP-LE and PGRP-LC proteins, activates Relish protein to trigger autophagy and phagocytosis through ImD regulated genes by rendering cellular immunity against gram negative bacteria [41]. Toll activates the nuclear factor DIF and it promotes humoral immunity in the fat body by producing varieties of anti-microbial peptides AMPs like attacin, cecropin, drosomycin, defensin, metchnikowin, diptericin and drosocin [42].

The fungal pathogen was found to be recognized by GNBP3 along with PGRPSA and GNBP1 it activates the drosophila toll receptors [43]. The Drosophila toll-5 (Tehao) and toll-9 plays major role during fungal infection by inducing Drosomycin gene [44].

During the preliminary stage of viral infection Drosophila toll receptor homolog of human TLR, Imd (TNF-alpha), Domeless (Jak–STAT), and RNAi plays a major role against viral infection these are components of innate immune system [45]. Similar to humans the viral glycoproteins are recognized by toll receptors like toll-4, while toll-7 dependent autophagy observed during viral infection in flies [42, 46]. Jak–STAT and Imd together mediates effective immunity against viral attack in flies [47]. The domeless-hop-stat2 pathway stimulated by upd1/2/3 activates Jak–STAT regulated genes responsible for controlling viral load; it is homologous to mammalian Jak–STAT pathway [48]. The Drosophila P53 and dP38 mediates apoptosis in flies upon stress response generated due to DNA damage, P53 mediated apoptotic genes are regulated by Jak–STAT-MAPK [49]. The dP38 stimulation in flies triggers Unpraired gene (upd protein) a mammalian IL-6 homolog further activates Jak–STAT-*Turandots* pathway which increases tolerance towards the viral invasion [50]. The intrinsic to cell the Dicer2 a viral sensor protein mediates silencing through RNA-induced silencing complex (RISC) dependent RNAi production which inhibits viral components transcription and vago gene activation finally controls viral growth [51, 52]. The anti-viral RNAi are transported from one cell to another through canonical nano-tubes structures [53]. dERK pathway regulates viral infection of flies gut epithelial infection during orally challenge of arbovirus, Sindbis virus and vesicular stomatitis virus [54]. Despite of dynamic immune response against the viral infection viruses like Nora virus, Sigma virus (DmelSV), Drosophila C virus (DCV), and Drosophila X virus (DXV) can cause fatality in flies [55].

5. Markers of infectious diseases

In the recent decades extensive research has been conducted to understand the regulation of immune system in *Drosophila melanogaster*. Using techniques like genome wide screens, *Drosophila* S2 cell line in-vitro models and tissue specific loss of function mutation in transgenic GAL4/UAS fly allows studying selective pathways of immune response [56]. Up-regulation of antimicrobial peptides (AMP) in flies during bacterial and fungal infection was frequently observed, these six AMP genes expression level could be analyzed in flies [42, 57]. ROS level in flies trigger several pathways responsible for tolerance (cell survival) and apoptosis (cell death) could be assayed in virally infected flies [49, 50]. Rescue of diseased transgenic flies upon feeding of desired drug could reveal drug efficacy [56]. Survival of flies would further reveal the effect of drugs during an ongoing pathogenesis [58].

6. Behavioral and physiological characterization of infected flies

6.1 Negative geotaxis assay

Negative geotaxis assay serves the purpose to manifest ongoing pathogenesis inside the live model. It was demonstrated previously that infected flies display significantly lesser motility than healthy flies when exposed under bright light. It could be considered as an important parameter to explain drug efficacy while screening anti-microbial drugs in flies [59].

6.2 Circadian rhythm

Circadian rhythm in flies was studied, the genes timeless or period controls the circadian rhythm of activity-sleep cycle during day-night respectively. It has been observed that infected flies exhibit interrupted circadian control of locomotion thus flies with this deficit shows restlessness at the same time gets lesser sleep than normal flies. The behavioral changes could also be studied in infected flies beside the control/uninfected flies [60].

6.3 Wasting

Wasting is commonly seen physiological changes associated with prolonged diseased condition in humans. Wasting is a common symptom in HIV/AIDS, tuberculosis and cholera patients. Similarly rapid loss in weight could be seen in infected flies prior to its death [61, 62].

7. Factors contributing to suitable infection model

It was previously reported that in order to replicate the outcome of future studies it is important to optimize the lethal dosage selection and the route of inoculum [63]. It is suggested that the selection of microbial strain and gender of flies are two important factors which could potentially impact the findings of future research.

7.1 Route of inoculum

There are two prime techniques for inducing infection in flies, primarily by feeding the flies with the microbes secondly by pricking micro needles dipped in bacterial liquid (inoculum) into fly's abdomen or thorax [62, 63]. Flies could be pricked in the abdomen with micro-needle dipped in the microbial solution, known amount can be useful in pharmacodynamic as well as pharmacokinetic studies [64].

7.2 Flies gender selection

Selecting gender should be considered strictly, few studies do not prefer to report the reason behind choosing the gender male/female type. In a study with *Vibrio cholera* infection narrated that female flies survived approximately 24 h longer than male flies [65].

8. In-vivo models for epidemic and pandemic diseases

The existing models using live bacterial infusion, feeding fungal strains and transgenic flies expressing viral proteins. Under immuno-suppressed condition would serve multiple purposes like studying host pathogen interaction and conducting preclinical trials [62, 66].

8.1 HIV models

Since human viruses do not usually invade insects the use of *Drosophila melanogaster* as a model organism is critical and currently in less usage [67]. In order to establish an HIV model for drug screening in *Drosophila melanogaster* it is important to understand the structural components of human immunodeficiency virus (HIV). The envelope components are comprised of Gp120 and Gp41 encoded by *env* sequence, pol-Gag RNA material encodes for *Matrix, capcid, nuclear capsid, p6, Protease, reverse transcriptase, and Integrase), Vif, Vpr, Nef, vpu, tat,* and *Rev* [68]. Transfection of Tat (transcription activator), Vpu (helps virion budding), Nef (regulator of structural gene expression) and Rev. (Nuclear export protein) in flies (in-vitro/in-vivo) were previously shown, these transfection models could be useful due to the fact that there is no marketed drug to target these viral proteins [69].

The incapacity of Drosophila S2 cells is only associated with the expression of HIV-1 envelope proteins. It is possible to express gycosylated and cleaved Gp120 in S2 cells but fusion with CD4+ receptors of T-helper cells could not be achieved in the model expression system [70]. In another study the expression of Gp120 in drosophila was carried out in S2 cell line, the antigen Gp-120 did not exhibited T-helper cell mediated humoral immune system activation and IgG antibody generation, when introduced in mice [71]. Due to this usual challenge in a different study they expressed HIV-1 virus like proteins in Drosophila S2 cells [72].

The nef transgenic flies exhibited JNK mediated apoptosis further nef inhibits NF-kB necessary for Relish gene activation similarly decreased immune response is common in AIDS patients [73]. In a study transfected viral protein Vpu was shown to cause immune suppression in fat body of flies via toll dependent pathway, in wings the Vpu expression caused apoptosis and hindered wing development, in mammals Vpu is known for causing T-cell lymphocyte death in infected patients [74, 75]. Active microbial invasion in nef flies should be further confirmed before targeting with potent anti-nef drug candidate. The *Rev* transfected S2 cells revealed that expression of Rev. protein directed the translocation of viral mRNA sequence into the cytoplasm, blocked by leptomycin B a secondary metabolite of *Streptomyces* species [76]. Leptomycin B remained unapproved in clinical trials due to high toxicity in cancer patients [77]. The ART drugs like zidovudine, lamivudine, stavudine, didanosine and abacavir were introduced in *D. melanogaster* to study genotoxicity profile [78, 79]. In Drosophila oocytes Tat a nuclear shuttling protein, displayed interaction with tubulin causing dorso-ventral axis mislocalization resulting in delayed microtubule polymerization, similarly tubulin dysfunction causes neurological symptoms observed in HIV+ individuals [80]. The transfected viral proteins in live Drosophila could be used to target drug in a thoughtfully designed model.

Cryptococcosis, Candidiasis and Aspergillosis are common types of fungal infections observed as clinical challenge in HIV-positive patients [81]. Under immunosuppressed condition the invasion of fungi in flies causes fatality. In Drosophila fungal infection could be difficult to achieve as the innate immune system mediates anti-fungal peptide production by haemocyte causing decrease of fungal load and increases fly survival rate [82]. Hence Toll mutant flies were generated and used to induce fungal infection.

Fluconazole and voriconazole showed anti-fungal activity against *Candida albicans* and *Aspergilus fumigatus* respectively in flies [64, 83]. Among *Cryptococcus* species only *Cryptococcus neoformans* is capable of killing flies with mutated toll receptors in drosophila, susceptible to infection acquired from *Cryptococcus* species like *Cryptococcus kuetzingii* or *Cryptococcus laurentii* [83]. Although, the toll mutant flies do not demonstrate an HIV model, it is used to induce fungal infection which can serve as a model for fungal infection in Drosophila for drug screening purpose.

In order to study HPV and EBV there are two model systems to study the effect in flies. In the study with HPV co-expression of viral oncoprotein E6 and human UBE3A did not resulted in tumorigenesis requires Ras or Notch pathway in flies, E6-UBE3A requires insulin receptors for cancer to develop [84]. Upon introducing the BZLF1 gene of EBV led to interaction with shaven gene in flies a homolog of pax gene family of humans responsible for B-cell development [85]. Expression of BRLF1 and BZLF1 genes using *GMR-R* model in Drosophila showed BRLF1 caused overproliferation of cells in flies whereas, BZLF1 resulted in interaction with several tumor suppressor genes and both viral genes showed interactions with core tumor suppressor genes like *reaper, p53, Rab5*, and *Tor* [86]. EVB DNA injection in flies caused Imd mediated pathway to increase diptericin production at the same time hemocyte proliferation and remarkable increase in numbers of hemocyte cells [87]. Human cytomegalovirus derived immediate early gene transfection in flies resulted in embryonic lethality similar to humans [88].

8.2 Influenza infection models

Influenza virus like most other viruses fails to infect the *Drosophila melanogaster*. To construct a suitable model for drug screening was also an important aspect due to high mortality rate caused by flu virus. The influenza virus coat protein consists of hemagglutinin (HA) and neuraminidase (NA). Matrix protein 2 (M2) plays a vital role in maintaining the pH level through proton transport enabling viral uncoating. Expression of M2 protein in flies is achieved through insertion of M2 cDNA sequence in upstream activation sequence (UAS) of pCaSpeR3 p-element insertion vector gave rise to UAS-M2 flies. The crossover between UAS-M2 and C135-Gal4 flies resulted in death at the pupal stage. Therefore the larvae were exposed and not the adult flies to anti-influenza drug amantidine which is a M2 antagonist. Amantadine and several other drugs of its class are not capable of acting against the flu virus due to varying viral strain types. Moving further the flu-fly model of UAS-M2 could be used to study potential anti-influenza drug [56].

8.3 Pneumococcal pneumonia models

There are several pneumococcal pneumonia infection model studied in drosophila using *Streptococcus pneumoniae, Staphylococcus aureus, Pseudomonas aeruginosa* and *Klebsiella pneumonia.* The biofilm formation is widely observed during *Streptococcus pneumoniae* infection in humans [89]. The nasopharyngeal tract is colonized primarily by these gram positive rods bacteria prior to infecting the lower respiratory tract. Similarly the *Pseudomonas aeruginosa* exhibit the biofilm formation during nasocomial infection in humans, a common culprit causing community acquired pneumonia hospitalized in patients [57, 90]. *Pseudomonas aeruginosa* was shown to infect *Drosophila melanogaster* causing its gut epithelium inflammation [57].

Staphylococcus aureus causes osteomyelitis, endocarditis, septicaemia and pneumoniae in humans, it can be selected for mimicking pneumoniae infection in Drosophila [91]. *Staphylococcus aureus* caused rapid death of flies within 48 h due to inoculation of high lethal dosage, extended survival seen upon exposure to antibiotic (tetracycline) [92]. The teichoic acid of peptidoglycan layer in *Staphylococcus aureus* was found to suppress the toll receptors of flies similar to *Streptococcus pneumoniae* toxins autolysin and pneumolysin interacts with toll receptors of macrophages in human [93–95]. *Klebsiella pneumonia* the gram positive bacteria are capable of killing drosophila at higher dose [96]. *Streptococcus pneumoniae* causes the maximum deaths in human causing pneumoniae which could be used as a suitable model for antibiotic screening in *Drosophila melanogaster*.

8.4 Tuberculosis models

There are at present two bacterial models for studying mycobacterium infection in flies, induced by *Mycobacterium marinum* and *Mycobacterium abscessus*. *Mycobacterium marinum* is a non spore forming, non motile, gram positive acid-fast bacillus, which is genetically 99.3% similar *to Mycobacterium tuberculosis* [97, 98]. Vacuole acidification is inhibited by *M. marinum* in drosophila phagocytic cells has been previously identified to be similar with tuberculosis pathogenesis in humans [99, 100]. Tigecycline plus linezolid was shown to have extended fly survival during the *Mycobacterium abscessus* infection. Rifampicin a very potential wide range antibiotics effective to inhibit multi drug resistance tuberculosis (MDRTb), it showed antimycobacterial efficacy in Drosophila infected with *Mycobacterium marinum* [101]. Any potential drug candidate capable of anti-mycobacterial activity can be studied in these models.

8.5 Cholera models

The bacteria *Vibrio cholerae* is a gram-negative and motile bacterium causes diarrheal disease in human. The pathogenesis of *Vibrio cholera* infection in humans was previously reported to be symbolized as comparable disease progression in *Drosophila melanogaster*. Ingestion of cholera bacterium results in lethal infection induced by the toxins in the intestinal cells of the flies. The toxins ingestion could not cause equivalent lethal effect on flies was explained previously. The *V. cholera* infection results in inhibition of adenylyl cyclase, Gsα, or the Gardos K+ channel causing death due to oral ingestion in flies. Clotrimazole a Potassium Calcium-Activated Channel Subfamily N Member 4 (KCNN4) inhibitor exposure increased flies susceptibility to *V. cholera* infection [61]. Quorum sensing is the ability to detect and to respond to a specific density of cell population through gene regulation [102]. *Drosophila melanogaster* initiates quorum sensing during vibrio cholera infection by suppressing succinate (substrate of KEBS cycle) uptake in flies intestine, limiting the wasting process [62]. Quorum sensing enables the bacterium to remain sessile in the flies gut and Vibrio polysaccharide (VPS) gene expression was shown to have increased during *V. cholera* infection of flies [103].

9. Importance of in-vitro model infection in Drosophila

The Drosophila S2 cells were first discovered by I. Schneider in 1972 [104]. S2 cells are derived from primary cell culture of late phase embryo of *Drosophila melanogaster*. S2 cells are macrophage like cells potentially grows in serum free medium as non-adherent suspension. S2 cells can express variety of heterologous proteins, upto 12 proteins could be co-expressed at a time in highly controlled manner, doubling at a rate equivalent to any cell lines derived from human cancerous cell line [105]. These cells do not form coherent clusters with no noisy gene expression profiles by maintaining uniformity during expression and chromosomal aneuploidy gets compensated during expression self adjusted to one gene copy number per cell unlike cancerous lineage [106]. These viable and potent cellular characteristics of S2 cell allows to be chosen for vaccine development, large scale enzyme as well as hormones production similar to Chinese hamster ovary CHO cell lines [104, 107]. The post translational glycosylation process is often not achievable in S2 cells making it disadvantageous [105]. The viral infections models are slow in inducing fatality in immuno-suppressed mutant flies, 50% death occurs after around 18–30 days post infection in live model [44, 108]. Therefore, S2 cell line model could requite certain challenges usually observed during in-vivo infection models.

10. In-vitro model of epidemic and pandemic infectious disease

Using drosophila S2 cell model a study showed that intercellular *Mycobacterium smegmatis* growth inside the host phagosomes is restricted by Rab7, CG8743, and the ESCRT factors [109]. *Cryptococcus neoformans* a fungi responsible for meningoencephalitis infection, S2 cells infected by this fungus up-regulates autophagy initiating proteins like Atg2a, Atg5 and Atg9a beside lysosomal markers like LAMP-1 and cathepsin D [110]. The hepatitis B virus surface antigen (HBsAg) coded by S gene was transfected in S2 cell line gave rise to no variation in expressed protein suggesting S2 cells useful for expression system [111]. The *Plasmodium falciparum* reticulocyte-binding protein homolog 5 (PfRH5) was expressed in S2 cells of Drosophila to produce non-glycosylated variants capable of binding to its receptor in rabbits resulted in IgG production against PfRH5 protein [112]. Highly potential vaccine VAR2CSA against malaria was successfully produced in S2 cells of *Drosophila melanogaster* [113].

11. Viral meningitis, encephalitis and hemorrhagic fever S2 cell line model

Herpes simplex virus was studied in Drosophila S2 cells where transfection of two viral proteins PILRα and gB responsible for binding to mammalian cells were expressed found to be poorly glycosylated [114]. The RNAi pathway was indulged by host cells to inhibit the Dengue virus (Flavivirus family) infection, by knocking down Argonaute (Ago1/2) and Dicer (Dcr1/2) showed sustained viral infection, currently clinical trial is underway NCT00936429 [115, 116]. Japanese encephalitis virus envelope glycoprotein E transfected in Drosophila S2 cells resulted in stable protein expression, this glycoprotein exposure in mice led antibody production against it [117]. Infection of Sindbis virus in live flies led activation of Notch, Jak-STAT and ImD pathway to intervene viral invasion [54]. Notch pathway mediated assimilation of ankyrin, plap, syx13, unc-13, csp, rab1 and rab8 during Sindbis virus infection in S2 cells [115]. The human antibody MR191 specific against Marburg virus was fused with recombinant RAVV GP ectodomain produced in S2 cell line [118]. The Zika virus structural envelope (E) protein were efficiently produced and secreted from transfected Drosophila S2 cell line model [119]. Flies produces RNAi against west Nile virus infection as a result of innate immune response similarly it was seen in S2 cell line, S2 cell lines were used for WNV infection, currently vaccine development NCT01477580 and NCT00707642 is underway [116, 120]. In a study mice were injected with glycoprotein GP of Ebola virus expressed in Drosophila S2 cell line found to produce antibodies against the infused antigen [121] (**Table 1**).

Epidemic/ Pandemic Disease	Microbes	Vaccine/Drugs screened or derived out of fly model (in-vitro/ in-vivo)	References
HIV/AIDS	Human Immuno virus	Leptomycin B (In-Vitro) Unapproved under clinical trials Zidovudine, lamivudine, stavudine, didanosine, Abacavir	[76, 77] [78, 79]
Flu	Influenza A	Amantidine	[56]
Cholera	*Vibrio cholera*	Clotrimazole	[61]
Pneumonia	*Streptococcus pneumoniae*	Tetracycline	[92]
Tuberculosis	*Mycobacterium tuberculosis*	Rifampicin, Tigecycline + Linezolid	[100, 101]
SARS	SARS Corona virus	—	
MERS	MERS corona virus	—	
Measles	Rubeola virus	—	
Typhoid	Salmonella typhi	—	
Hepatitis	Hepatitis A and B	HBsAg expressed in S2 cell line	[111]
Small pox	Variola virus	—	
Malaria	*Plasmodium falciparum*	VAR2CSA/PfRH5 viral protein expressed in S2 cell line	[112, 113]
Zika Fever	Zika virus	Structural envelope (E) protein expressed in S2 cells	[119]
Dengue Fever	Dengue Virus	DEN1-80E expressed in S2 cells	[115, 116]
Encephalitis	Japanese encephalitis virus	JEV E protein expression in S2 cell line	[117]

Epidemic/ Pandemic Disease	**Microbes**	**Vaccine/Drugs screened or derived out of fly model (in-vitro/ in-vivo)**	**References**
Haemorrhagic fever	Ebola virus	glycoprotein GP expressed in S2 cell line	[121]
Haemorrhagic fever	Marburg Virus	MR191 expressed in S2 cell line	[118]
Plague	Yersinia pestis	—	
Yellow fever	Yellow fever virus	—	
West Nile Fever	West Nile Virus	WN-80E expressed in S2 cell line	[116, 120]

Table 1.
List of Drugs/vaccines screened or developed against Infectious diseases in Drosophila melanogaster *as a model organism.*

12. Disadvantages of Drosophila model for drug screening

Drosophila melanogaster being ectothermic organism unlike humans are endothermic homeotherms maintains physiological temperature constantly at 37°C, making it difficult to infect flies with bacteria like *Mycobacterium tuberculosis* grows strictly at 37°C [36, 122]. Several pathogenic viruses capable of infecting humans cannot naturally infect *Drosophila melanogaster* [55, 67]. The fungal dose response in flies is difficult to measure in oral infection model therefore this model is limited to study only the anti-fungal drug efficacy [64]. The presence of symbiotic microbes like Wolbachia a gram negative bacteria associate mostly with drosophila gut, improves the fly immunity against viral infection [123]. Superinfection like viral pneumonia cannot be studied at present to undertake preclinical trial using fly as a model.

13. Future perspective

Irrespective of multiple disadvantages flies could be used for studying drug efficacy. Multi-drug resistance tuberculosis infection could be studied in flies. The ART medication impairs human heart by causing prolonged QT, prolonged arrhythmic condition leads to myocardial infraction, *Drosophila* could be a suitable model to study the effect of anti retroviral therapy on fly heart [124–126]. Shockingly infection induced in flies by vesicular stomatitis virus in toll-7 depleted flies where 50% flies displayed death after 18 days, suggesting HIV infection could also kill toll7 mutant flies, as toll mutant flies displayed fungal invasion, this yet to be confirmed [44]. Alternative to this only viral DNA had been shown in a recent study to evoke immune activation in Drosophila by injecting it in thoracic region [87]. Kaposi sarcoma associated Herpesvirus (KSHV) needs a model which is yet to be studied in flies, however the KSHV viral gene latent nuclear antigen (LANA) interacts with RING3 of human which is homologous to drosophila female sterile homeotic (fsh) has already been identified [127]. Drosophila wound healing an important concern while inducing bacterial infection currently it is well understood and was found regulated by EGFR/ERK pathway essential for tissue repair [128]. The RNAi screens against Dengue or Influenza virus infection in cell culture could not identify Jak–STAT, ImD and toll dependent gene activation suggesting possible alternative pathway associated in infection modulation and no stimulation of inflammatory cytokine activation [40]. The food borne *Salmonella typhimurium* causes stomach flu (*gastroenteritis*) is well studied in Drosophila, but not epidemic pathogen *Salmonella typhi*.

14. Conclusions

Currently the existing models of infection in drosophila are capable of causing infection using viruses, bacteria (gram negative and gram positive) and fungi. These models are of great use since the efficacy of a drug capable of modifying diseased condition could be studied in detail in live *Drosophila* or in-vitro S2 cell. In this detailed review on epidemiology of infectious disease, it could be predicted that infection alone is a threat to overall population imposing death to more than 5 million individuals. Diseases like influenza, HIV, pneumonia, tuberculosis and cholera could be studied in flies. Currently there are 20 diseases which caused epidemic worldwide [2], 13 of the pathogens were studied in *Drosophila melanogaster* and diseases caused by yellow fever virus, Nipah virus, MERS, Hepatitis C virus, Salmonella typhi, Crimean congo hemorrhagic fever virus, chikungunya virus, monkeypox virus, Nipah virus and shigellosis bacteria are yet to be studied in-vitro/in-vivo. These diseases are of pandemic and epidemic criteria it causes huge number of deaths globally. Controlling the epidemic and pandemic diseases should be the main focus of the healthcare sector in the next decade.

Acknowledgements

I would like to thank Professor Sarat Chandra Yenisetti, Nagaland University, India for the effort and advices given for this article. I would like to thank Professor David S. Schneider of Stanford University, USA for clearing doubts regarding tuberculosis and typhoid infection in flies.

Conflict of interest

The author has no conflict of interest.

Notes/Thanks/Other declarations

I would like to thank the IntechOpen Journal for giving 100% waiver to publish this review article. I would like to thank all the researchers for providing their complete articles which are unavailable online.

Author details

Saikat Samadder
Dum Dum Motijheel, Kolkata, India

*Address all correspondence to: saikat.samadder46@gmail.com; saikat.samadder24@gmail.com

References

[1] Green MS, Swartz T, Mayshar E, Lev B, Leventhal A, Slater PE, et al. When is an epidemic an epidemic? Israel Medical Association Journal. 2002;**4**(1):3-6

[2] World Health Organization. Managing epidemics: Key facts about major deadly diseases. 2018. ISBN 978-92-4-156553-0

[3] Hays JN. Epidemics and Pandemics: Their Impacts on Human History. ABC-CLIO. 2005. ISBN 978-1-85109-658-9

[4] World Health Organization. What is a pandemic? Emergencies preparedness, response. 2010. Available at: https://www.who.int/csr/disease/swineflu/frequently_asked_questions/pandemic/en/

[5] Centers for Disease Control and Prevention. Vaccines and Preventable Diseases. List of Vaccines Used in United States. April 13, 2018. [Accessed: 15 August 2019]

[6] Principles of Epidemiology in Public Health Practice, 3rd ed. An Introduction to Applied Epidemiology and Biostatistics. Centers for Disease Control and Prevention. [Retrieved 19 August 2019]

[7] Unicef. Robin Nandy. Immunization under fire. 25 April 2016. [Accessed: 14 August 2019]

[8] Oppenheim B, Gallivan M, Madhav NK, et al. Assessing global preparedness for the next pandemic: development and application of an Epidemic Preparedness Index. BMJ Global Health. 2019;**4**:e001157. DOI: 10.1136/bmjgh-2018-001157

[9] Boutayeb A. The burden of communicable and non-communicable diseases in developing countries. Transactions of the Royal Society of Tropical Medicine and Hygiene. 2010. DOI: 10.1007/978-0-387-78665-0_32

[10] Review of Antibacterial Medicines for the WHO Model List of Essential Medicines 2017 Update. [Accessed: 14 August 2019]

[11] WHO. The top 10 causes of death. Factsheet. 2016. Available at: https://www.who.int/news-room/fact-sheets/detail/the-top-10-causes-of-death [Accessed: 30 June 2019]

[12] WHO. Influenza (seasonal) fact sheet. 2016. Available at: http://www.who.int/mediacentre/factsheets/fs211/en/ [Accessed: 01 July 2019]

[13] Mokdad AH, GBD 2015 LRI Collaborators. Estimates of the global, regional, and national morbidity, mortality, and aetiologies of lower respiratory tract infections in 195 countries: A systematic analysis for the Global Burden of Disease Study 2015. Lancet Infectious Diseases. 2017;**17**(11):1133-1161

[14] Cunha CB, Opal SM. Middle East respiratory syndrome (MERS): A new zoonotic viral pneumonia. Virulence. 2014;**5**(6):650-654

[15] Ziegler T, Mamahit A, Cox NJ. 65 years of influenza surveillance by a World Health Organization-coordinated global network. Influenza Other Respiratory Viruses. 2018;**12**:558-565. DOI: 10.1111/irv.12570

[16] Harish Nair W, Brooks A, Katz M, Roca A. Global burden of respiratory infections due to seasonal infl uenza in young children: A systematic review and meta-analysis. Lancet. 2011;**378**:1917-1930. DOI: 10.1016/S0140-6736(11)61051-9

[17] Al Johani Sameera, Akhter Javed. 2017. Pneumonia of Viral

Etiologies, Contemporary Topics of Pneumonia, Zissis C. Chroneos, IntechOpen. DOI:10.5772/intechopen.71608. Available at: https://www.intechopen.com/books/contemporary-topics-of-pneumonia/pneumonia-of-viral-etiologies

[18] Behrens G, Stoll M. Chapter 4: Pathogenesis and immunology. In: Influenza Report. 2006. ISBN: 3-924774-51-X

[19] WHO. Factsheet. Tuberculosis. 2018. Available at: https://www.who.int/newsroom/facsheets/detail/tuberculosis [Accessed: 01 July 2019]

[20] Raviglione M, Sulis G. Tuberculosis 2015: Burden, challenges and strategy for control and elimination. Infectious Disease Reports. 2016;**8**(2):6570

[21] Progress report on HIV, viral hepatitis and sexually transmitted infections 2019. In: Accountability for the Global Health Sector Strategies, 2016-2021. Geneva: World Health Organization; 2019 (WHO/CDS/HIV/19.7). Licence: CC BY-NC-SA 3.0 IGO

[22] Sen S. Indian cholera: A myth. Indian Journal of History of Science. 2012;**47**(3):345-374

[23] 150 years of cholera epidemiology. Lancet. 2005;**366**(9490):957

[24] Harris JB, La Rocque RC, Qadri F, Ryan ET, Calderwood SB. Cholera. Lancet. 2012;**379**:2466-2476

[25] WHO. Cholera. Fact sheet [Internet]. Available at: https://www.who.int/newsroom/factsheets/detail/cholera [Accessed: 30 June 2019]

[26] Jefferies M, Rauff B, Rashid H, Lam T, Rafiq S. Update on global epidemiology of viral hepatitis and preventive strategies. World Journal of Clinical Cases. 2018;**6**(13):589-599

[27] World Health Organization. Global Hepatitis Report 2017. World Health Organization. 2017. Available at: https://apps.who.int/iris/handle/10665/255016. License: CC BY-NC-SA 3.0 IGO

[28] Ashurst JV, Truong J, Woodbury B. Salmonella Typhi. 2019. Bookshelf ID: NBK519002, PMID: 30085544

[29] WHO. Typhoid. Factsheet. 11 September 2018. Available at: https://www.who.int/immunization/diseases/typhoid/en/ [Accessed: 05 July 2019]

[30] WHO. Malaria. Factsheet. 27 March 2019. Available at: https://www.who.int/news-room/fact-sheets/detail/malaria [Accessed: 05 July 2019]

[31] CDC. Meningitis Home. August 6, 2019. Available at: https://www.cdc.gov/meningitis/viral.html [Accessed: 18 August 2019]

[32] WHO. Health topics encephalitis. Viral Available at: https://www.who.int/topics/encephalitis_viral/en/

[33] New York State Department of Health. Arboviral (Arthropod-borne Viral) Diseases. July 2017. Available at: https://www.health.ny.gov/diseases/communicable/arboviral/fact_sheet.htm

[34] Fernando Cobo Viruses Causing Hemorrhagic Fever. Safety Laboratory Procedures. The Open Virology Journal. 2016;**10**:1-9. DOI: 10.2174/1874357901610010001

[35] Panayidou S, Ioannidou E, Apidianakis Y. Human pathogenic bacteria, fungi, and viruses in Drosophila. Virulence. 2014;**5**(2):253-269. DOI: 10.4161/viru.27524

[36] Tzelepis I, Kapsetaki S-E, Panayidou S, Apidianakis Y. Drosophila melanogaster: A first step and a stepping-stone to anti-infectives. Current Opinion in Pharmacology.

2013;**13**:1-6. DOI: 10.1016/j.coph.2013.08.003

[37] Dionne MS, Schneider DS. Models of infectious diseases in the fruit fly Drosophila melanogaster. Disease Models & Mechanisms. 2008;**1**:43-49. DOI: 10.1242/dmm.000307

[38] Bergman P, Seyedoleslami Esfahani S, Engstrom Y. Drosophila as a model for human diseases—Focus on innate immunity in barrier epithelia. Current Topics in Developmental Biology. 2017;**121**:29-81. DOI: 10.1016/bs.ctdb.2016.07.002

[39] De Gregorio E, Spellman PT, Rubin GM, Lemaitr B. Genome-wide analysis of the *Drosophila* immune response by using oligonucleotide microarrays. Proceedings of the National Academy of Sciences of the United States of America. 2001;**98**(22):12590-12595. DOI: 10.1073/pnas.221458698

[40] Sabin LR, Hanna SL, Cherry S. Innate antiviral immunity in Drosophila. Current Opinion in Immunology. 2010;**22**:4-9

[41] Martin M, Hiroyasu A, Guzman RM, Roberts SA, Goodman AG. Analysis of Drosophila STING reveals an evolutionarily conserved antimicrobial function. Cell Reports. **23**:3537-3550. DOI: 10.1016/j.celrep.2018.05.029

[42] Rutschmann S, Jung AC, Hetru C, Reichhart JM, Hoffmann JA, Ferrandon D. The Rel protein DIF mediates the antifungal but not the antibacterial host defense in Drosophila. Immunity. 2000;**12**(5):569-580. DOI: 10.1016/S1074-7613(00)80208-3

[43] Gottar M, Gobert V, Matskevich AA, Reichhart J-M, Wang C, Butt TM, et al. Dual detection of fungal infections in *Drosophila* through recognition of microbial structures and sensing of virulence factors. Cell. 2006;**127**(7):1425-1437. DOI: 10.1016/j.cell.2006.10.046

[44] Nakamoto M, Moy RH, Xu J, Bambina S, Yasunaga A, Shelly SS, et al. Virus recognition by Toll-7 activates antiviral autophagy in Drosophila. Immunity. 2012;**36**:658-667. DOI: 10.1016/j.immuni.2012.03.003

[45] Lopez WA, Page AM, Ericson BL, Carlson DJ, Carlson KA. Antiviral immunity in the fruit fly, *Drosophila melanogaster*, *Drosophila melanogaster*. In: Perveen FK, editor. Model for Recent Advances in Genetics and Therapeutics. IntechOpen; 2017. DOI: 10.5772/intechopen.69293

[46] Akira S, Uematsu S, Takeuchi O. Pathogen recognition and innate immunity. Cell. 2006;**124**:783-801

[47] Huang Z, Kingsolver MB, Avadhanula V, Hardy RW. An antiviral role for antimicrobial peptides during the arthropod response to alphavirus replication. Pathogenesis and Immunity. DOI: 10.1128/JVI.03360-12

[48] Dostert C, Jouanguy E, Irving P, Troxler L, Galiana-Arnoux D, Hetru C, et al. The Jak-STAT signaling pathway is required but not sufficient for the antiviral response of drosophila. Nature Immunology. 2005;**6**(9):946. DOI: 10.1038/ni1237

[49] Liu B, Behura SK, Clem RJ, Schneemann A, Becnel J, et al. P53-mediated rapid induction of apoptosis conveys resistance to viral infection in *Drosophila melanogaster*. PLoS Pathogens. 2013;**9**(2):e1003137. DOI: 10.1371/journal.ppat.1003137

[50] West C, Silverman N. p38b and JAK-STAT signaling protect against Invertebrate iridescent virus 6 infection in *Drosophila*. PLoS Pathogens. 2018;**14**(5):e1007020. DOI: 10.1371/journal.ppat.1007020

[51] Poirier EZ, Goic B, Tome-Poderti L, Frangeul L, Boussier J, Gausson V, et al. Dicer-2-dependent generation of viral dna from defective genomes of RNA viruses modulates antiviral immunity in insects. Cell Host & Microbe. 2018;**23**:353-365. DOI: 10.1016/j.chom.2018.02.001

[52] Takeuchi O, Akira S. RIG-I-like antiviral protein in flies. Nature Immunology. 2008;**9**(12):1327

[53] Karlikow M, Goic B, Mongelli V, Salles A, Schmitt C, Bonne I, et al. *Drosophila* cells use nanotube-like structures to transfer dsRNA and RNAi machinery between cells. Scientific Reports. **6**:27085. DOI: 10.1038/srep27085

[54] Xu J, Hopkins K, Sabin L, Yasunaga A, Subramanian H, Lamborn I, et al. ERK signaling couples nutrient status to antiviral defense in the insect gut. PNAS. 2013;**110**(37):15025-15030. DOI: 10.1073/pnas.1303193110

[55] Xu J, Cherry S. Viruses and antiviral immunity in *Drosophila*. Developmental and Comparative Immunology. 2014;**42**(1). DOI: 10.1016/j.dci.2013.05.002

[56] Adamson AL, Chohan K, Swenson J, La Jeunesse D. A Drosophila model for genetic analysis of influenza viral/host interactions. Genetics. 2011;**189**:495-506. DOI: 10.1534/genetics.111.132290

[57] Mulcahy H, Sibley CD, Surette MG, Lewenza S. *Drosophila melanogaster* as an animal model for the study of *Pseudomonas aeruginosa* biofilm infections in vivo. PLoS Pathogens. 2011;7:e1002299. DOI: 10.1371/journal.ppat.1002299

[58] Blow NS, Salomon RN, Garrity K, Reveillaud I, Kopin A, Rob Jackson F, et al. *Vibrio cholerae* infection of *Drosophila melanogaster* mimics the human disease cholera. PLoS Pathogens. 2005;**1**(1):e8

[59] Allen JA, Chambers M, Gupta AS, Schneider D. Infection-related declines in chill coma recovery and negative geotaxis in *Drosophila melanogaster*. PLoS One. 2012;**7**(9):e41907. DOI: 10.1371/journal.pone.0041907

[60] Shirasu-Hiza MM, Dionne MS, Pham LN, Ayres JS, Schneider DS. Interactions between circadian rhythm and immunity in *Drosophila melanogaster*. Current Biology. 2007;**17**(10):R354

[61] Dionne MS, Pham LN, Shirasu-Hiza M, Schneider DS. *Akt* and *foxo* dysregulation contribute to infection-induced wasting in *Drosophila*. Current Biology. 2006;**16**(20):1977-1985. DOI: 10.1016/j.cub.2006.08.052

[62] Kamareddine L, ACN W, Vanhove A, Hang S, Purdy A, Kierek-Pearson K, et al. Activation of *Vibrio cholerae* quorum sensing promotes survival of an arthropod host. Nature Microbiology. 2018;**3**. DOI: 10.1038/s41564-017-0065-7

[63] Chambers MC, Jacobson E, Khalil S, Lazzaro BP. Thorax injury lowers resistance to infection in *Drosophila melanogaster*. Infection and Immunity. October 2014;**82**(10):4380-4389. DOI: 10.1128/IAI.02415-14

[64] Chamilos G, Lionakis MS, Lewis RE, Lopez-Ribot JL, Saville SP, Albert ND, et al. *Drosophila melanogaster* as a facile model for large-scale studies of virulence mechanisms and antifungal drug efficacy in *Candida* species. The Journal of Infectious Diseases. 2006;**193**:1014-1022

[65] Berkey CD, Blow N, Watnick PI. Genetic analysis of *Drosophila melanogaster* susceptibility to intestinal *Vibrio cholerae* infection. Cellular Microbiology. 2009;**11**(3):461-474. DOI: 10.1111/j.1462-5822.2008.01267.x

[66] Hughes TT, Allen AL, Bardin JE, Christian MN, Daimon K, Dozier KD, et al. Drosophila as a genetic model for studying pathogenic human viruses. Virology. 2012;**423**:1-5. DOI: 10.1016/j.virol.2011.11.016

[67] Cherry S, Perrimon N. Entry is a rate-limiting step for viral infection in a *Drosophila melanogaster* model of pathogenesis. Nature Immunology. 2004;**5**(1):81-87. DOI: 10.1038/ni1019

[68] Frankel AD, John AT. Young HIV-1: Fifteen proteins and an RNA. Annual Review of Biochemistry. 1998;**67**:1-25

[69] Arts1 EJ, Hazuda DJ. HIV-1 antiretroviral drug therapy. Cold Spring Harbor Perspectives in Medicine. 2012;**2**:a007161

[70] Ivey-Hoyle M, Clark RK, Rosenberg M. The N-terminal 31 amino acids of human immunodeficiency virus type 1 envelope protein gpl20 contain a potential gp4l contact site. Journal of Virology. 1991:2682-2685

[71] Grundner C, Pancera M, Kang J-M, Koch M, Sodroski J, Wyatt R. Factors limiting the immunogenicity of HIV-1 gp120 envelope glycoproteins. Virology. 2004;**330**:233-248

[72] Yang L, Song Y, Li X, Huang X, Liu J, Ding H, et al. HIV-1 virus-like particles produced by stably transfected *Drosophila* S2 cells: A desirable vaccine component. Journal of Virology. 2012;**86**(14):7662-7676

[73] Lee SB, Park J, Jung JU, Chung JK. Nef induces apoptosis by activating JNK signalling pathway and inhibits NF-kB-dependent immune responses in Drosophila. Journal of Cell Science. 2005;**118**:1851-1859. DOI: 10.1242/jcs.02312

[74] Leulier F, Marchal C, Miletich I, Limbourg-Bouchon B, Benarous R, Lemaitre B. Directed expression of the HIV-1 accessory protein Vpu in *Drosophila* fat-body cells inhibits Toll-dependent immune responses. EMBO Reports. 2003;**4**(10)

[75] Marchal C, Vinatier G, Sanial M, Plessis A, Pret A-M, et al. The HIV-1 Vpu protein induces apoptosis in drosophila via activation of JNK signaling. PLoS One. 2012;**7**(3):e34310. DOI: 10.1371/journal.pone.0034310

[76] Fasken MB, Saunders R, Rosenbergi M, David W, Brighty A. Leptomycin B-sensitive homologue of human CRM1 promotes nuclear export of nuclear export sequence-containing proteins in *Drosophila* cells. The Journal of Biological Chemistry. 2000;**275**(3):1878-1886

[77] Klahn P, Fetz V, Ritter A, Collisi W, Hinkelmann B, Arnold T, et al. The nuclear export inhibitor aminoratjadone is a potent effector in extracellular-targeted drug conjugates. Chemical Science. 2019;**10**:5197-5210. DOI: 10.1039/C8SC05542D

[78] Guimaraes NN, Silva CJ, de Andrade HHR, Dihl RR, Lehmann M, Cunha KS. Comparative analysis of genetic toxicity of antiretroviral combinations in somatic cells of Drosophila melanogaster. Food and Chemical Toxicology. 2013;**53**:299-309. DOI: 10.1016/j.fct.2012.12.005

[79] Chen-Chen L, de Jesus Silva Carvalho C, de Moraes Filho AV, Veras JH, Cardoso CG, Bailao EFLC, et al. Toxicity and genotoxicity induced by abacavir antiretroviral medication alone or in combination with zidovudine and/or lamivudine in *Drosophila melanogaster*. Human and Experimental Toxicology. 2019;**38**(4):446-454. DOI: 10.1177/0960327118818248

[80] Piero A, Battaglia SZ, Macchini A, Franca Gigliani A. Drosophila model of

HIV-Tat-related pathogenecity. Journal of Cell Science. 2001;**114**:2787-2794

[81] Marukutira T, Huprikar S, Azie N, Quan S-P, Meier-Kriesche H-U, Horn DL. Clinical characteristics and outcomes in 303 HIV-infected patients with invasive fungal infections: Data from the Prospective Antifungal Therapy Alliance registry, a multicenter, observational study. HIV AIDS (Auckl). 2014;**6**:39-47

[82] Lionakis MS, Lewis RE, May GS, Wiederhold NP, Albert ND, Halder G, et al. *Toll*-deficient *Drosophila* flies as a fast, high-throughput model for the study of antifungal drug efficacy against invasive Aspergillosis and *Aspergillus* virulence. The Journal of Infectious Diseases. 2005;**191**:1188-1195

[83] Apidianakis Y, Rahme LG, Heitman J, Ausubel FM, Calderwood SB, Mylonakis E. Challenge of *Drosophila melanogaster* with *Cryptococcus neoformans* and role of the innate immune response. Eukaryotic Cell. 2004;**3**(2):413-419. DOI: 10.1128/EC.3.2.413-419.2004

[84] Padash Barmchi M, Gilbert M, Thomas M, Banks L, Zhang B, Auld VJ. A *Drosophila* model of HPV E6-induced malignancy reveals essential roles for Magi and the insulin receptor. PLoS Pathogens. 2016;**12**(8):e1005789. DOI: 10.1371/journal.ppat.1005789

[85] Adamson AL, Wright N, LaJeunesse DR. Modeling early Epstein-Barr virus infection in *Drosophila melanogaster*: The BZLF1 protein. Genetics. 2005;**171**:1125-1135. DOI: 10.1534/genetics.105.042572

[86] Adamson A, La Jeunesse D. A study of Epstein-Barr virus BRLF1 activity in a Drosophila model system. The Scientific World Journal. 2012; Article ID 347597, 9 pages. DOI: 10.1100/2012/347597

[87] Sherri N, Salloum N, Mouawad C, Haidar-Ahmad N, Shirinian M, Rahal EA. Epstein-Barr virus DNA enhances diptericin expression and increases hemocyte numbers in *Drosophila melanogaster* via the immune deficiency pathway. Frontiers in Microbiology. 2018;**9**:1268. DOI: 10.3389/fmicb.2018.01268

[88] Steinberg R, Shemer-Avni Y, Adler N, et al. Human cytomegalovirus immediate-early-gene expression disrupts embryogenesis in transgenic Drosophila. Transgenic Research. 2008;**17**:105. DOI: 10.1007/s11248-007-9136-5

[89] Chao Y, Marks LR, Pettigrew EM, Hakansson AP. *Streptococcus pneumoniae* biofilm formation and dispersion during colonization and disease. Frontiers in Cellular and Infection Microbiology. 2015. DOI: 10.3389/fcimb.2014.00194

[90] Bassetti M, Vena A, Croxatto A, Righi E, Guery B. How to manage *Pseudomonas aeruginosa* infections. Drugs in Context. 2018;7:212527. DOI: 10.7573/dic.212527

[91] Apidianakis Y, Pitsouli C, Perrimon N, Rahme L. Synergy between bacterial infection and genetic predisposition in intestinal dysplasia. Proceedings of the National Academy of Sciences of the United States of America. 2009;**106**:20883-20888. DOI: 10.1073/pnas.0911797106

[92] Needham AJ, Kibart M, Crossley H, Ingham PW, Foster SJ. *Drosophila melanogaster* as a model host for *Staphylococcus aureus* infection. Microbiology. 2004;**150**:2347-2355. DOI: 10.1099/mic.0.27116-0

[93] Ragle BE, Karginov VA, Wardenburg JB. Prevention and treatment of *Staphylococcus aureus* pneumonia with a - cyclodextrin derivative. Antimicrobial Agents and Chemotherapy. 2010:298-304. DOI: 10.1128/AAC.00973-09

[94] Kurokawa K, Gong JH, Ryu KH, Zheng L, Chae JH, Kim MS, et al. Biochemical characterization of evasion from peptidoglycan recognition by *Staphylococcus aureus* D-alanylated wall teichoic acid in insect innate immunity. Developmental and Comparative Immunology. 2011;**35**:835-839. DOI: 10.1016/j.dci.2011.03.001

[95] Mook-Kanamori BB, Geldhoff M, van der Poll T, van de Beek D. Pathogenesis and pathophysiology of pneumococcal meningitis. Clinical Microbiology Reviews. 2011;**24**(3):557-591. DOI: 10.1128/CMR.00008-11

[96] Benghezal M, Fauvarque M-O, Tournebize R, Froquet R, Marchetti A, Bergeret E, et al. Specific host genes required for the killing of *Klebsiella* bacteria by phagocytes. Cellular Microbiology. 2006;**8**(1):139-148. DOI: 10.1111/j.1462-5822.2005.00607.x

[97] Pham LN, Dionne MS, Shirasu-Hiza M, Schneider DS. A specific primed immune response in Drosophila is dependent on phagocytes. PLoS Pathogens. 2007;**3**(3):e26. DOI: 10.1371/journal.ppat.0030026

[98] Akram SM, Aboobacker S. Mycobacterium Marinum. Treasure Island (FL): StatPearls Publishing; 2019. Available from: https://www.ncbi.nlm.nih.gov/books/NBK441883/

[99] Sakamoto K. The pathology of *Mycobacterium tuberculosis* infection. Veterinary Pathology;**49**(3):423-439

[100] Dionne MS, Ghori N, Schneider DS. *Drosophila melanogaster* is a genetically tractable model host for *Mycobacterium marinum*. Infection and Immunity. 2003:3540-3550. DOI: 10.1128/IAI.71.6.3540-3550.2003

[101] Chun-Taek O, Moon C, Ok KP, Kwon S-H, Jang J. Novel drug combination for *Mycobacterium abscessus* disease therapy identified in a Drosophila infection model. The Journal of Antimicrobial Chemotherapy. 2014;**69**:1599-1607. DOI: 10.1093/jac/dku024

[102] Whitehead NA, Barnard AML, Slater H, Simpson NJL, Salmond GPC. Quorum-sensing in Gram-negative bacteria. FEMS Microbiology Reviews. 2001;**25**:365-404

[103] Purdy AE, Watnick PI. Spatially selective colonization of the arthropod intestine through activation of *Vibrio cholera* biofilm formation. PNAS. 2011;**108**(49):19737-19742

[104] Adriaan de Jongh W, Salgueiro S, Dyring C. The use of *Drosophila* S2 cells in R&D and bioprocessing. Pharmaceutical Bioprocessing. 2013;**1**(2):197-213

[105] Moraes AM, Jorge SAC, Astray RM, Suazo CAT, Riquelme CEC, Augusto EFP, et al. *Drosophila melanogaster* S2 cells for expression of heterologous genes: From gene cloning to bioprocess development. Biotechnology Advances. 2012;**30**:613-628. DOI: 10.1016/j.biotechadv.2011.10.009

[106] Cherbas L, Willingham A, Zhang D, Yang L, Zou Y, Eads BD, et al. The transcriptional diversity of 25 *Drosophila* cell lines. Genome Research. 2011;**21**(2):301-314. DOI: 10.1101/gr.112961.110

[107] Backovic M, Johansson DX, Klupp BG, Mettenleiter TC, Persson MAA, Rey FA. Efficient method for production of high yields of Fab fragments in Drosophila S2 cells. Protein Engineering, Design & Selection. 2010;**23**(4):169-174. DOI: 10.1093/protein/gzp08

[108] Liu Y, Gordesky-Gold B, Leney-Greene M, Weinbren NL, Tudor M, Inflammation-Induced SC. STING-dependent autophagy restricts Zika

virus infection in the Drosophila brain. Cell Host & Microbe. **24**:57-68. DOI: 10.1016/j.chom.2018.05.022

[109] Philips JA, Porto MC, Wang H, Rubin EJ, Perrimon N. ESCRT factors restrict mycobacterial growth. 3070-3075. PNAS. 2008;**105**(8). DOI: 10.1073.pnas.0707206105

[110] Qin Q-M, Luo J, Lin X, Pei J, Li L, et al. Functional analysis of host factors that mediate the intracellular lifestyle of *Cryptococcus neoformans*. PLoS Pathogens. 2011;**7**(6):e1002078. DOI: 10.1371/journal.ppat.1002078

[111] Jorge SAC, Santos AS, Spina A, Pereira CA. Expression of the hepatitis B virus surface antigen in Drosophila S2 cells. Cytotechnology. 2008;**57**:51-59

[112] Hjerrild KA, Jin J, Wright KE, Brown RE, Marshall JM, Labbe GM, et al. Production of full-length soluble *Plasmodium falciparum* RH5 protein vaccine using a *Drosophila melanogaster* Schneider 2 stable cell line system. Scientific Reports;**6**:30357. DOI: 10.1038/srep30357

[113] de Jongh WA, Resende M d SM, Leisted C, Stroaek A, Berisha B, Nielsen MA, et al. Development of a Drosophila S2 insect-cell based placental malaria vaccine production process. BMC Proceedings. 2013;**7**(Suppl 6):P20. Available at: http://www.biomedcentral.com/1753-6561/7/S6/P20

[114] Fan Q, Bohannon KP, Longnecker R. *Drosophila* Schneider 2 (S2) cells: A novel tool for studying HSV-induced membrane fusion. Virology. 2013;**437**(2):100-109. DOI: 10.1016/j.virol.2013.01.004

[115] Mukherjee S, Hanley KA. RNA interference modulates replication of dengue virus in *Drosophila melanogaster* cells. BMC Microbiology. 2010;**10**:127. Available at: http://www.biomedcentral.com/1471-2180/10/127

[116] Medina LO, Albert TO, Lieberman MM, Wong TAS, Namekar M, Nakano E, et al. A recombinant subunit based Zika virus vaccine is efficacious in non-human primates. Frontiers in Immunology. 2018;**9**:2464. DOI: 10.3389/fimmu.2018.02464

[117] Zhang F, Ma W, Zhang L, Aasa-Chapman M, Zhang H. Expression of particulate-form of Japanese encephalitis virus envelope protein in a stably transfected Drosophila cell line. Virology Journal. 2007;**4**(17). DOI: 10.1186/1743-422X-4-17

[118] King LB, Fusco ML, Flyak AI, Ilinykh PA, Huang K, Gunn B, et al. The Marburgvirus-neutralizing human monoclonal antibody MR191 targets a conserved site to block virus receptor binding. Cell Host Microbe. 2018;**23**(1):101-109.e4. DOI: 10.1016/j.chom.2017.12.003

[119] Qu P, Zhang W, Li D, Zhang C, Liu Q, Zhang X, et al. Insect cell-produced recombinant protein subunit vaccines protect against Zika virus infection. Antiviral Research. 2018 Jun;**154**:97-103. DOI: 10.1016/j.antiviral.2018.04.010 Epub 2018 Apr 14

[120] Chotkowski HL, Ciota AT, Jia Y, Puig-Basagoiti F, Kramer LD, Shi PY, et al. West Nile virus infection of *Drosophila melanogaster* induces a protective RNAi response. Virology. 2008;**377**(1):197-206. DOI: 10.1016/j.virol.2008.04.021

[121] Lai C-Y, Strange DP, Wong TAS, Lehrer AT, Verma S. Ebola virus glycoprotein induces an innate immune response in vivo via TLR4. Frontiers in Microbiology. 2017. DOI: 10.3389/fmicb.2017.01571

[122] Giraldo D, Adden A, Kuhlemann I, Gras H, Bart RH. Geurten correcting locomotion dependent observation biases in thermal preference of *Drosophila*. Scientific Reports. 2019;**9**:3974

[123] Teixeira L, Ferreira A, Ashburner M. The bacterial symbiont *Wolbachia* induces resistance to RNA viral infections in *Drosophila melanogaster*. PLoS Biology. 2008;**6**(12):e1000002. DOI: 10.1371/journal.pbio.1000002

[124] Das M. Cardiac arrhythmias in HIV disease. Cardiovascular Reviews and Reports. 2002;**23**(4):208-212 +226

[125] Dube MP, Lipshultz SE, Fichtenbaum CJ, Greenberg R, Schecter AD, Stacy D. Fisher effects of HIV infection and antiretroviral therapy on the heart and vasculature. Circulation. 2008;**118**:e36e40. DOI: 10.1161/CIRCULATIONAHA.107.189625

[126] Ocorr K, Reeves NL, Wessells RJ, Martin Fink H-S, Chen V, Akasaka T, et al. KCNQ potassium channel mutations cause cardiac arrhythmias in Drosophila that mimic the effects of aging. PNAS. 2007;**104**(10):3943-3948

[127] Platt GM, Simpson GR, Mittnacht S, Schulz TF. Latent nuclear antigen of Kaposi's sarcoma-associated herpesvirus interacts with RING3, a homolog of the Drosophila female sterile homeotic (fsh) gene. Journal of Virology. 1999;**73**(12):9789-9795

[128] Geiger JA, Carvalho L, Campos I, Santos AC, Jacinto A. Hole-in-one mutant phenotypes link EGFR/ERK signaling to epithelial tissue repair in Drosophila. PLoS One. 2011;**6**(11):e28349. DOI: 10.1371/journal.pone.0028349

Chapter 8

Understanding Taste Using *Drosophila melanogaster*

Shivam Kaushik and Pinky Kain

Abstract

Taste is a short-range contact chemosensation required by all animals to detect nutrient rich foods and avoid consuming toxic chemicals. In insects, it is also required to select mates and appropriate oviposition sites. Organization of the fruit fly *Drosophila melanogaster* taste system and availability of experimental tool box, makes *Drosophila* gustatory system an ideal model system for studying the perception of taste and taste elicited behaviors. Like humans, fruit flies also respond to wide range of taste chemical and can differentiate between different taste categories including sweet, bitter, sour, umami and salt. This chapter will present a research progress made in the field of taste using neuroanatomical, genetic, behavioral, molecular and cellular biology techniques in the fruit fly. The compiled survey will provide an outlook of taste research done in fruit fly and its comparison with human taste behavior.

Keywords: *Drosophila*, gustation, neurons, taste receptors, taste behavior

1. Introduction

Animals including *Drosophila melanogaster* use their chemosensory system to monitor the chemical world around them. The chemosensory system includes olfactory system to detect volatile chemicals and gustatory system to detect soluble compounds. The olfactory and visual system helps in food detection and the taste system controls the food acceptance or rejection behavior by helping animals detect nutrient-rich food and avoid toxic substances. The quality and concentration of taste compounds help animals to make such an assessment.

Drosophila and mammals are able to detect basic taste modalities including sugars, bitter compounds, salt, acids, and amino acids [1]. The taste qualities are detected by taste cells present in the periphery. The activation of different taste cells provides a simple mechanism to encode modality. Like mammals' fly taste cells also show dose dependent activation providing the potential to encode different concentrations. The taste system of a fly is distributed over the whole-body, proboscis or labial palps being the main taste organ. It is located on the distal end of the labellum. Like other insects, the taste sensilla are present on labellum, legs, wings and on the female genitalia (**Figure 1**) [2].

The simplicity of the gustatory system of flies provides an ideal situation for comparative studies of taste perception and taste-elicited behaviors. The availability of the experimental tool box including high end imaging of neural circuits in the brain, simple behavioral assays, possibility of electrical recordings and ease

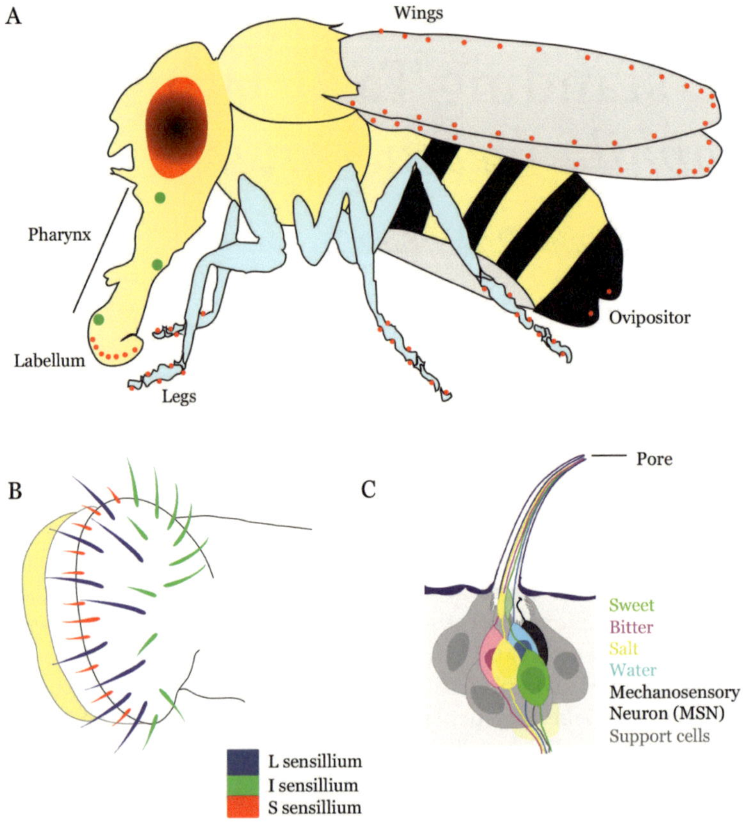

Figure 1.
Taste system of adult Drosophila. (A) Taste neurons are located on different body parts of fly namely labellum, pharynx, legs, wings and ovipositor as shown with red and green dots. (B) Three different types of taste sensillum (Large-L, Intermediate-I and small-S) present on the labellum (mouth part of a fly). (C) Taste sensillum structure showing pore at the tip and different types of taste neurons in a taste sensillum namely sweet (green), bitter (magenta), salt (yellow) and water (sky blue) neurons together with one mechanosensory neuron (black) surrounded by support cells (gray).

of molecular- genetics analyses with the availability of transgenic and mutant flies makes fly a unique system to study taste. In addition, flies share the same molecular logic of taste as mammals.

Different members of gustatory receptor (GRs) genes expressed in gustatory neurons mediate the detection of taste compounds such as sugars and bitter compounds [3–7]. Expression patterns of taste receptors is based largely on transgenic GAL4 expression studies and suggest that different GRs are expressed in overlapping but non-identical subsets of sugar- and bitter-sensing neurons [6–8]. In addition, electrophysiological studies from taste neurons suggest heterogeneity among the responses of individual sugar- or bitter-sensing cells [9–11] suggesting diversity among the peripheral cell types that detect sugars or bitter compounds in *Drosophila.* This organization provides the potential for different taste cell types to be activated by different compounds within a taste modality and the possibility for intra modality discrimination.

This chapter will present a research progress made in the field of taste perception in the fruit fly and will describe the anatomical properties of the *fly* gustatory system. We shall then review taste perception mainly from a molecular genetic perspective that includes the results from behavioral, electrophysiological and imaging analyses. The parallels between the flies and human taste system will provide insight into how the detection of taste compounds regulates feeding decisions.

2. Mammalian taste system

In humans, taste receptors cells (TRC's) helps in the detection of taste stimuli. TRC's are present in taste buds and palate epithelium at the back and sides of the tongue (circumvallate and foliate papillae). The taste buds called fungiform are scattered across the front of the tongue and on the palate. Three morphologically distinct cell types (I, II and III) are present in a taste bud and constitute five functional classes of sensory cells, each specialized to detect one of the five basic taste qualities (bitter, sweet, umami, sour and salty). TRCs are epithelial cells that extend a process to the apical surface of the epithelium, where a taste pore allows direct contact with chemicals in the environment. The life of taste cells is short and they replenish from proliferative basal keratinocytes [12]. TRCs can relay information of taste quality independent of cells relaying other taste qualities [13]. Neurotransmitter receptors are present on taste cells. TRC's release various neurotransmitters to communicate among cells in the taste bud to shape the output of the bud [14]. Vertebrate TRCs do not possess an axon, and instead are innervated by pseudo unipolar neurons whose cell bodies reside in the petrosal and geniculate ganglia. The chorda tympani nerve that (innervates the anterior tongue) contain fungiform papillae and the glossopharyngeal nerve, (innervates the posterior tongue and most of the palate) carry most of the taste information. Neurons from taste ganglia project to the nucleus of the solitary tract, and from there information is relayed to the gustatory cortex [15].

3. Gustatory system of *Drosophila*

Although the same taste preferences are shared between *Drosophila* and mammals, the organization of their gustatory systems are rather different. Unlike humans, flies have wide distribution of taste cells over much of the body including many peripheral organs like labellum, legs, wings and genitalia (**Figure 1**). Such a distribution of taste cells enables the fly to gather contact chemosensory information from many reference points that may make contact with their body enabling detection of potential calorie rich foods or toxic compounds [16]. The presence of taste-sensing cells in other tissues provides the safety benefits allowing evaluation of chemicals without the potential hazard of accidental ingestion. The gustatory sensillum or taste bristle are the main sensory unit of all taste organs housing two to four primary gustatory receptor neurons (GRNs) as well as a single mechanosensory neuron (MSN) [2]. The labellum is the main taste organ in *Drosophila* located at the end of the proboscis (equivalent to human tongue). Labial palps contain 31 bristles (sensilla) each that are arranged in a stereotyped pattern. The sensilla are morphologically classified into three types long, intermediate and short (L, I, and S type) based on their shape and location (**Figure 1**) [2, 17, 18]. L- and S-type sensillum house dendrites of four GRNs, and the I type are associated with two GRNs. Electrophysiological investigations suggest each GRN is thought to respond exclusively to either sugar, water, low salt concentration, or high salt concentration and bitter compounds [11, 19–22]. The terminal pore at the tip of the taste bristle (**Figure 1**) allows taste stimuli access to the dendrite of the GRN, which extends into the bristle shaft [23]. In addition to the peripheral taste sensillum on the palps, legs, and wings, taste neurons are also semi-internally or internally located. The first group consists of row of taste pegs that line the inside of the labial palps and are exposed to foods when the fly 'opens' its palps and readies itself for 'sucking up' foods. The internally located group consists of three sensillum clusters that line the pharynx (**Figure 1**). They allow re-evaluation of the food as it passes and enters the esophagus and the digestive system. Peripheral labial palp GRNs, the internal sensillum, and some leg GRNs project their axons to the sub esophageal zone (SEZ), whereas the wing and a minority of leg GRNs project to the thoracic ganglion (**Figure 2**) [24, 25].

A single MSN and several support cells are also present in the taste sensilla together with the GRNs (**Figure 1**) [26]. These MSNs translate mechanical forces into electrical signals and mediate hearing, positional awareness, and the coordination of movements [27, 28]. The MSNs sense the hardness and viscosity of food [29] similar to the ability of the human tongue to determine the consistency and texture of foods.

3.1 Gustatory receptors of *Drosophila*

Like mammals, GRs in *Drosophila* also detect taste compounds. *Drosophila* utilize ion channels (Ionotropic receptors) to detect salts and sour (acid) compounds. Putative gustatory receptors in *Drosophila* and mammals were discovered almost simultaneously and detect sugars and bitter substances. Mammalian taste receptors belong to the large super family of GTP-binding (G) protein-coupled receptors (GPCRs), but fly GRs share no significant sequence similarity with them [4, 30–38]. A total of 68 *Gr* genes were found in *Drosophila* by analyzing the *Drosophila* genome database using algorithms that identify multi-transmembrane proteins or by performing reiterated Basic Local Alignment Search Tool searches with *Drosophila* olfactory receptor proteins as query sequences [4, 32, 38, 39]. The *Gr* genes are remarkably diverse having similarity between most receptor pairs only 20% or less

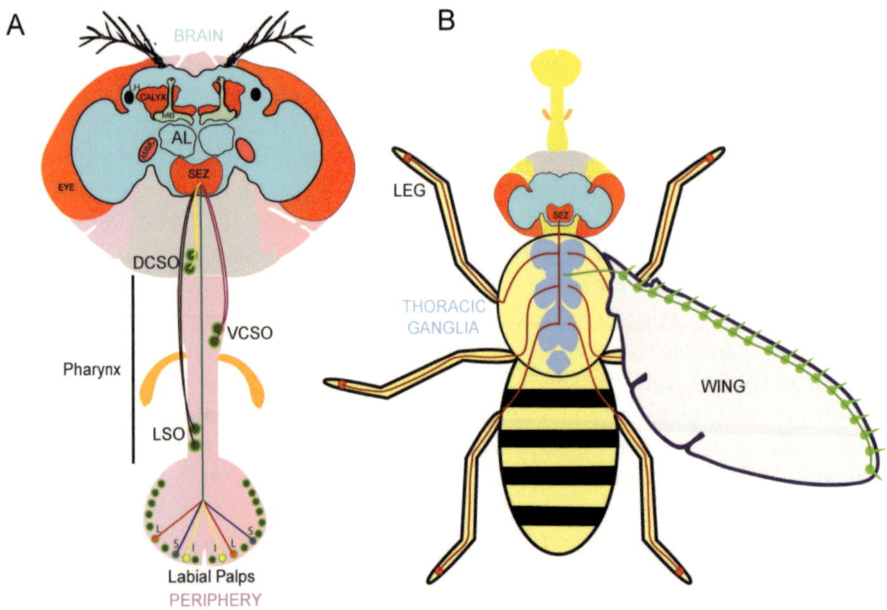

Figure 2.
(A) Taste neurons send projections from the periphery (neurons from different taste sensilla L,I and S) including pharyngeal areas (LSO-labral sense organ, VCSO-ventral cibarial sensory organ, and DCSO-dorsal cibarial sensory organ) to sub esophageal zone (SEZ) in the brain which is the main taste processing center. The taste information flow from SEZ to the antennal and mechanosensory motor center (AMMC). Mushroom body (MB), calyx and lateral horn (LH) are learning and memory centers. Antennal lobes (AL) get information from the olfactory neurons present on antennal surface in Drosophila. (B) GRNs from the legs of an adult fly (Red) send projections to SEZ and some GRNs project to thoracic ganglia only (shown in light blue). GRNs from the wing margins send axonal projections to thoracic ganglia (taste cells present in the wing shown in Green).

(at the amino acid sequence level). There are several gene clusters containing up to six genes, exhibit significantly higher similarity to each other (up to 70%). *Gr* genes with greater than 30% sequence similarity have been grouped into several subfamilies [32]. The domain that is most conserved among all *Gr* genes is located in the region encoding the putative seventh transmembrane domain at the carboxy terminus, a domain that is also shared with the olfactory receptors (ORs) genes this domain was used as a signature motif in one study that lead to the discovery of the fly taste receptors [32].

GRs within a subfamily detect structurally similar taste compounds. For example, the sugar receptor *Gr5a* subfamily consisting of *Gr5a* encoding a trehalose receptor [3, 40, 41] *Gr61a*, and *Gr64a–f* share sequence similarity in the range of 60% and detect diverse sugars [18]. Bitter compounds cover a vast chemically much more diverse structural space than sugars and majority of remaining GRs are devoted to the detection of bitter-tasting and toxic compounds.

Well established GAL4/UAS system, transgenic expression methods helped visualizing the expression of various *Gr* genes [42]. *Gr* gene promoters drive the expression of the yeast transcription factor GAL4 (regulate genes induced by galactose) and GAL4 in turn activates transcription with high specificity via the cis-regulatory element upstream activating sequence (UAS), cloned upstream of green fluorescence protein (GFP), or β-galactosidase reporter genes [42]. Expression analyses of *Gr* genes suggest that they are expressed in distinct subpopulations of GRNs, supporting their role as chemosensory receptors and providing first insights into their complex cellular expression [6, 7, 32, 38].

3.2 Types of taste receptors in mammals and *Drosophila*

The three tastes, bitter, sweet and umami taste are mediated by taste-specific GPCRs, which are expressed in distinct subsets of taste receptor cells in mammals [13]. These three taste employ a canonical G protein phosphoinositide-based pathway, where receptors activate a taste cell-specific G protein that activates PLCβ2, generating second messengers IP_3, DAG and H^+. IP_3 acts on the IP_3 receptor (IP_3R_3) to release Ca^{2+} from intracellular stores, and Ca^{2+} gates the membrane channel TRPM5 [43].

Drosophila taste receptors bear no sequence relationship to mammalian taste receptors. The majority of bitter and sweet taste receptors in insects are members of a large protein superfamily of gustatory receptors [4, 32, 38, 39]. The 68 members of *Drosophila* taste receptors have seven transmembrane domains, but they share no sequence relationship to GPCRs. Rather, they are distantly related to *Drosophila* olfactory receptors (ORs), which have an opposite membrane topology from GPCRs and form ligand-gated ion channels [44–46]. Insect GRs including *Drosophila* have an inverted topology like ORs relative to GPCRs [47] and may form ionotropic receptors [48].

3.2.1 Sweet receptors

Highly concentrated sugars (100–500 mM), artificial sweeteners, and small number of sweet-tasting proteins elicit the sweet taste in mammals. The heterodimer of T1R2 and T1R3 constitutes the sweet receptor [49]. Animals also sense energy-rich foods and various sugars through a mechanism similar to that used by pancreatic β-cells to detect blood glucose [50]. According to this hypothesis, the metabolism of sugars by sweet cells produces ATP, which closes ATP-sensitive K^+ channels leading to membrane depolarization [50].

Flies are attracted to many of the same sugars as humans [9, 51] although they respond most robustly to disaccharides (such as sucrose and maltose) and oligosaccharides [8]. The fly sweet receptors belong to the same superfamily of receptors that includes most of the bitter receptors. In adult flies the three key receptors required for sensing sugars, except for fructose, are Gr5a, Gr64a and Gr64f [8, 40, 52–54]. These three receptors are co-expressed in the sugar-responsive GRNs in the labellum, along with five other related GRs that comprise the *Gr*-Sugar (*Gr*-S) clade [8, 52].

Gr5a and Gr64a sense structurally different sugars. Gr64a participates in the response to sucrose and maltose [8, 52], while Gr5a detect trehalose and melezitose [8, 40, 41]. Gr64f might act as a co-receptor for the responses for all sugars tested except fructose, and functions in concert with Gr5a and Gr64a [53]. Gr43a is the only receptor known to detect fructose [55].

3.2.2 Bitter receptors

Bitter taste allows animals to detect toxins in the environment and avoid consuming them. Compounds such as caffeine, cycloheximide (a protein synthesis inhibitor), denatonium (added to rubbing alcohol to discourage consumption), and quinine (a component of tonic water) taste bitter to humans, mice and flies. In vertebrates, bitter chemicals are detected by a small family of receptors (T2Rs), which are structurally related to rhodopsin, and range in number from 3 to 49, depending on the species [31, 34, 56]. In general, each bitter responsive taste receptor cell expresses multiple types of bitter receptors [57], but not all bitter receptors are expressed by every bitter cell [58], leading formally to the possibility that there are subclasses of bitter cells, as is the case in flies [59]. The chemical receptive field of the bitter receptors fall into two classes—“specialists” that detect one or a few bitter chemicals and “generalists” that detect many [60].

In contrast to vertebrate bitter detection, flies employ a much more complex strategy to sample bitter chemicals. In flies, bitter sensitive GRNs have distinct sensitivities. Based on the response profile to a panel of 16 bitter compounds, the L-, I- and S-type sensilla on the labella are classified into five groups, four of which are sensitive to bitter chemicals (**Figure 1**) [59]. Out of the four, two groups are narrowly tuned to distinct sets of bitter compounds (I-a and I-b). The other two groups respond broadly to bitter tastants but have variable patterns of activity (S-a, S-b). Analysis with a larger panel of bitter compounds may reveal more additional subgroups.

In flies, 33 out of 38 *Gr* genes that express in the labellum are found to be localized to bitter GRNs [59]. The roles of only a few of the bitter GRs have been dissected genetically so far. A minimum of 28 *Grs* can be expressed by some GRNs in the labellum in adult fly. One of the larval GRN classes expresses at least 17 *Grs* [59, 61]. Many GRs act as co-receptors responding to large numbers of aversive chemicals. *Gr32a*, *Gr33a* and *Gr66a* are needed for detection of most bitter chemicals [62, 63]. These three *Grs* with additional *Grs* (*Gr89a* and *Gr39a.a*) are expressed in all bitter responsive GRNs making this group of five GRs to be the “core-bitter GRs” [59]. Other GRs are very narrowly tuned and confer ligand specificity. These receptors are critical in defining the chemical specificity of the GRs, in combination with other GRs. Different combinations of complex sets of GR receptors may explain how a limited number of bitter GRs confer the capacity to respond to a vast collection of structurally diverse bitter compounds. Three TRP channels expressed in the labellum GRNs also contribute to the sensation of aversive compounds through mechanisms that are independent of GRs. TRPA1 show behavioral avoidance to aristolochic acid [64], a related TRPA channel Painless, is required for the behavioral avoidance to isothiocyanates (AITC;

wasabi) [65] and TRP-Like (TRPL) is both necessary and sufficient to confer sensitivity to camphor [66].

3.2.3 Salt taste receptors

Moderate levels of salt is necessary to maintain the important physiological functions such as muscle contraction, action potentials and many other functions while excessive salt intake is deleterious and can lead to hypertension. Salty taste is elicited by Na^+ concentrations ranging from 10 to 500 mM. In humans, salt taste is amiloride-insensitive. The amiloride sensitive component of salt taste is selective for Na^+ and Li^+ over other monovalent cations such as K^+, is sensitive to low concentrations of salts (<100 mM), and is generally appetitive [67]. Amiloride-sensitive salt taste occurs only in the front of the tongue [68]. Based on taste nerve recordings, there is a population of broadly tuned high-salt fibers that are insensitive to amiloride and activated by KCl and NaCl [69]. These fibers innervate both the front and back of the tongue, in contrast to the amiloride-sensitive fibers that innervate only the front of the tongue. Epithelial Na^+ channels (ENaCs) are composed of three subunits—α, β and γ and α subunit is absolutely essential and forms part of the pore [70]. ENaC α has been suggested to be a component of the low salt sensor since a taste-cell specific knockout eliminates sensitivity and behavioral attraction to low concentrations of salt [71].

The cells that mediate the behavioral responses to high salts are not specifically dedicated to sensing high salt, but instead comprise at least two populations of cells with previously identified functions in sensing bitter and sour [72]. Inactivation of TRPM5 or PLCβ2, expressed by bitter cells, eliminates a component of the high salt response, while silencing PKD2L1-expressing sour cells eliminates the remaining components [72]. Remarkably, mice in which PKD2L1-expressing cells are silenced and TRPM5 is inactivated find high salt concentrations attractive, presumably due to activation of the amiloride-sensitive ENaC channels by high salt [72].

Salt taste preferences in *Drosophila* are similar to those in mammals. Both larvae and adult fruit flies prefer low-salt foods, while they reject high-salt concentrations. Two ENaC channels family members, *ppk11* and *ppk19* are reported to be expressed in the terminal organ and required for sensing low salt [73] in *Drosophila* larvae. However, these channels do not appear to function in the salt response in adults [74]. A member of the ionotropic glutamate receptor (IR) family member, Ir76b, is required for low salt sensing in adult flies [74]. IRs were identified originally as a new class of olfactory receptor [75]. However, several IRs are also expressed in GRNs [76]. Ir76b is expressed in GRNs distinct from those that respond to sugars and bitter compounds, and the Ir76b GRNs extend their projections into a unique region of the SEZ [74]. Most recently combined activity of most of all GRN classes encoding salt taste has been proposed [77].

3.2.4 Sour taste

Acidic pH and organic acids such as acetic acid evokes sour taste in the tongue. A subset of taste receptor cells in the tongue and palate epithelium that respond to acidic pH and weak organic acids with electrical activity detects the sour taste [78, 79]. PKD2L1-expressing cells respond are required for sensory response to acids [72, 78, 80] which is mediated by an unusual proton-selective ion channel [80]. Proton selectivity allows the cells to respond to acids without interference from Na^+, which may vary independently in concentration. The taste of carbonation (CO_2) is also detected by PKD2L1-expressing cells. This response is dependent on a membrane anchored carbonic anhydrase isoform 4, Car4 [81], which interconverts $CO_2 + H_2O$ to $H^+ + HCO_3^-$. How Car4 contributes to the activation of sour cells is still not known.

Fruit flies reject foods that are too acidic and prefer the ones which are slightly acidic, such as carbonated water. Carbonated water triggers Ca^{2+} influx in the region of the SEZ innervated by taste peg GRNs, suggesting these neurons are involved in CO_2 detection [82]. Fruit flies avoid many carboxylic acids with a low pH. Behavioral and physiological analysis reveals that the avoidance to carboxylic acid is mainly mediated by a subset of bitter GRNs [83]. The ionotropic receptor Ir7a has been shown for rejecting foods laced with high levels of acetic acid suggesting flies discriminate foods on the basis of acid composition rather than just pH [84].

3.2.5 Amino acid receptors

Umami (amino acid taste) is the sensation elicited by glutamate. In humans, umami is only elicited by glutamate, while mice are sensitive to a wider range of L-amino acids [1]. Addition of the nucleotides IMP or GMP potentiates the umami response, distinguishing it from a more general sensing of glutamate [85]. T1R1/T1R3 is widely recognized as the umami receptor [1].

Fruit flies can taste amino acids too, although their preference is enhanced when raised on a food source devoid of amino acids [86]. Female fruit flies show greatest preference for cysteine, phenylalanine, threonine and tyrosine, while males prefer leucine and histidine. None of the 18 standard amino acids tested stimulates action potentials in GRNs in sugar responsive sensilla [8] raising the possibility of taste pegs in sensing amino acids. Another amino acid, L-canavanine is toxic and elicits an avoidance response in flies [87] and is sensed by GRNs in a subset of S-type sensilla [88]. Gr8a and Gr66a are both required for L-canavanine avoidance [88]. An ionotropic receptor Ir76b has been shown recently for amino acid taste in flies [89].

Activation of fly GRNs by sweet substances, bitter compounds and the amino acid, L-canavanine occur through direct activation of ion channels and G-protein signaling pathways. G proteins subunits Gγ, $G_o\alpha$, $G_s\alpha$ and $G_q\alpha$ are implicated in sugar signaling [90–93]. PLCβ is an effector for $G_q\alpha$. Knockdown in sugar-responsive GRNs of *plcβ21c* or any of the genes encoding TRPC channels (TRP, TRPL and TRPγ) alters the behavioral response to trehalose [92]. Role of G-protein coupled signaling pathways in sensation of bitter tastants has also been suggested for example AC78C is required for the response to caffeine [94], and the PLCβ encoded by *norpA* is required in *trpA1*-expressing GRNs for the behavioral and electrophysiological responses to the bitter compound, aristolochic acid [64] suggesting a role of Gq/PLC/TRPA1 pathway functions in the detection of aristolochic acid. $G_o\alpha$47A is needed for detection of L-canavanine [95]. The predicted role of G-protein coupled signaling pathways in insect taste is to enhance the responses to low concentrations of ligands, as seen for photo transduction cascade in amplifying the response to a photon of light.

3.3 Taste coding

Taste in flies and mammals use labeled line model of coding (in which each cell represents a distinct taste quality and communicates essentially without interruption to the central nervous system). Single taste neurons in flies can detect multiple taste qualities having the same valence (behavioral output) supported by the results that some GRNs are activated by sugars, and low levels of fatty acids, both of which promote feeding [96] while other GRNs are activated by bitter compounds and high concentrations of salt and suppress feeding [20]. In addition, a subset of bitter GRNs is also activated by low pH carboxylic acids, which are feeding deterrents [83]. A complex model for salt coding in flies that combinatorially integrates inputs from across cell types to afford robust and flexible salt behaviors [77].

The taste system of mice also uses a variant of the labeled line model. In mice, taste receptors are segregated into distinct populations such that bitter, sweet, sour and low concentrations of salt are detected by non-overlapping sets of cells [1, 58]. Whether this principle applies to sweet and umami is presently unclear. Recent evidence suggest that aversive high concentrations of salt are not detected by a separate subset of cells, but are instead detected by the populations of cells that detect bitter and sour [72] suggesting that the mammalian taste system is relatively hard-wired to behavior, as is the case in flies.

3.4 Taste modulation in *Drosophila*

Modulation of taste neuron activity prior to the first relay has been suggested. Presence of multiple molecular and cellular mechanisms by which tastant information is integrated in primary taste neurons has been proposed [97]. Various studies suggest that aversive tastants such as bitter compounds and acids, can inhibit the activity of appetitive taste circuits in adult flies and larvae [83, 98, 99]. The reduction of the firing rate of sweet neurons in mixtures of sucrose and the aversive tastants is independent of the activity of the deterrent neuron [83, 98, 99]. Bitter compounds suppress feeding by activating bitter—GRNs and by inhibiting sugar—sensitive GRNs [11]. The suppression of sugar GRNs depends on a odorant binding proteins" (OBP), OBP49a, which is expressed in gustatory organs or indirectly via GABAergic interneurons that connect bitter taste neuron activity to that of sweet taste neurons [100, 101]. Accessory cells synthesized OBP49a and release it into endolymph fluid bathing the GRNs, which then acts non-cell autonomously on sugar activated GRNs. OBP49a binds directly to bitter compounds, and later interacts with the sugar receptor, Gr64a, on the cell surface of the GRNs to suppress its activity [101]. Such non-cell autonomous modulation of the sugar response ensures that bitter compounds in sugar-laden foods are not consumed. Low concentrations of acid tastants have also been observed to modulate detection of bitter compounds in the context of both sweet and deterrent neurons, suppressing their inhibitory effect in the former and their excitatory effect in the latter [102]. Although the mechanisms by which carboxylic acids or low pH inhibit taste neurons remains to be determined.

Internal state can change the gustatory sensitivity as well: starvation potentiates the responses of sweet GRN and suppresses bitter GRN responses; mating increases taste peg GRN sensitivity to polyamines and behavioral responses to low salt in females; and protein deprivation sensitizes taste peg GRNs to yeast and increases behavioral sensitivity to amino acids [86, 103–108]. Taste neuron sensitivity is also modulated by prior dietary experience. Response to camphor (non-toxic bitter compound) decrease after exposing flies to camphor for long [66]. An E3 ubiquitin ligase-regulated decline in the levels of Trpl caused the change in sensitivity. No calories diet also cause increase activity in the *Gr5a+* sweet taste circuit [104, 109] and reduce sensitivity in the bitter taste circuit [105]. In the former case, dopamine signaling acting on the both primary and secondary neurons in the sweet circuit caused the change in sensitivity [104, 109]. The latter is dependent on sNPF acting via GABAergic interneurons [105]. A significant modulation of fly salt taste behavior by salt deprivation has been shown recently [77].

3.5 Non-canonical taste qualities

3.5.1 Fats

Vertebrates can sense a variety of other important taste qualities such as wetness and fattiness. Olfaction and somatosensation helps in the detection of fats, and they elicit post-ingestive effects that promote consumption. It has been shown that mice prefer water spiked with free fatty acids supports a role for the taste system in detecting this rich source of calories [110]. A fatty acid transporter (CD36) and two fat-sensitive GPCRs—GPR40 and GPR120 are putative receptors for fat taste including K^+ channels that are sensitive to polyunsaturated fatty acids [111]. GPR120 is required for preference to fatty acids in mice [112] and is expressed in human TRCs as well [113].

In flies, sweet GRN activation requires the function of the three *Ionotropic receptor* genes *Ir25a*, *Ir76b* and *Ir56d*. *Ir25a and Ir76b* are expressed in several neurons per sensillum, while *IR56d* expression is restricted to sweet GRNs. *Ir25a* and *Ir76b* mutant flies loose appetitive behavioral responses to fatty acids. The phenotype can be rescued by expression of respective transgenes in sweet GRNs [114].

3.5.2 Calcium taste

Ca^{2+}, an ion is required for a vast array of cellular functions. Ca^{2+}-deprived animals show attraction and Ca^{2+}-sated animals show rejection. The aversive response to Ca^{2+}requires a functioning T1R3 receptor, a subunit of the umami and sweet receptor [115]. In human subjects an attenuation of the taste of Ca^{2+} by the T1R3 blocker lactisole has been shown [116].

Fruit flies avoids toxic levels of calcium. This repulsion is mediated by two mechanisms—activation of a specific class of GRNs that suppresses feeding, and inhibition of sugar-activated GRNs, which normally stimulates feeding. The distaste for Ca^{2+}, and electrophysiological responses to Ca^{2+} require three members of the variant ionotropic receptor family Ir25a, Ir62a and Ir76b. The high concentrations of Ca^{2+} show decrease survival in flies [117].

3.5.3 Water

No water receptor has been identified in vertebrate so far. The somatosensory system of animals can detect wetness across their body and also contribute to the sensing of aqueous solutions in the oral cavity. Various tastes have been ascribed to distilled water, from bitter to salty and sweet. Notably, application of water after exposure to some artificial sweeteners, such as saccharin, elicits a sweet taste [118].

A member of the Degenerin/Epithelial Sodium Channel family, ppk28 (an osmosensitive ion channel) mediates the cellular and behavioral response to water in flies. *ppk28* is expressed in water-sensing neurons and loss of *ppk28* abolishes water sensitivity [119].

3.6 Taste signal processing and taste sensory maps in the *Drosophila* brain

In flies, after evaluation of taste input, the information translates into an appropriate behavioral response such as feeding, cessation of feeding, search for alternative food source, courtship, or egg-laying. Detection of sweet compounds by labellum GRs induces a sucking response and sugar detection by the tarsi induces extension of proboscis. It is a requirement to understand the flow of information from peripheral activation of GRNs to behavioral output to gain insight into the neuronal wiring of the taste at each level of information processing.

Unlike mammalian taste cells, fly GRNs from labellum and pharynx send projections of axons directly to the SEZ area of the brain. GRNs in the ovipositor, wings, and some leg sensilla send projections to the thoracic ganglia [24, 25]. Taste neurons send their axons to loosely defined, widely circumscribed zones in the SEZ or thoracic ganglia [32]. Labial palp *Gr5a* positive GRNs project to large areas in the lateral and anterior region of the SEZ, whereas the *Gr66a*-expressing GRNs project to the medial part of the SEZ [6, 7]. Information from GRNs of the legs activates non-overlapping areas of the SEZ than GRNs of the labial palps, suggesting different behavioral outputs of neurons responding to the same ligand but located in different taste tissues. Functional domains of taste have been mapped in the brain using live flies expressing the calcium-sensitive indicator G-CaMP [120] (G-CaMP protein is a fusion of the calmodulin-binding domain from the myosin chain kinase (M13 peptide), permutated EGFP and the calmodulin) in response to sugars and bitter compounds suggesting different taste compounds activate distinct neural ensembles in the SEZ. In terms of their taste quality, organ location, and in some cases sensillar type, at least 10 categories of patterns have been defined in the SEZ and nine in the thoracic abdominal ganglia. Each category is a unique combination of discrete patterns elements that define taste neurons [97].

The SEZ is a primary gustatory center, the higher brain centers where taste information is conveyed from the SEZ are unknown. Recently, sweet second order projection neurons that relay sweet taste information from the SEZ to the antennal and mechanosensory motor center (AMMC) in the deutocerebrum were described [109]. The results support the role of AMMC (generally receives input from mechanosensory and olfactory neurons) in processing multisensory information. Various other studies have identified interneurons that impinge on taste circuits and feeding behavior routines, including a feeding promoting command neuron, feeding promoting dopaminergic neurons, bitter sensitive projection interneurons, feeding restrain GABAergic neurons and neurons in the ventral nerve cord that balance feeding and locomotion [121–126].

The taste representations in the mushroom bodies (MB) (sites for associative learning) examined recently and found that input to the main calyx continues to be segregated according to taste modality and the location that taste information originates from. The bitter and sweet stimuli activate distinct areas, and stimuli from different taste organs activate partially overlapping but distinct patterns [127]. The information about water and sweet qualities, as well as nutritive and non-nutritive sugars is also separated in MB [128, 129]. Unraveling taste circuits, therefore, will be important not only for understanding how sensory input is translated to behavioral output, but also how taste associations are formed in reward and aversive learning [97].

4. Conclusions

Drosophila Grs, IRs, Trp, and ppk receptors underlie detection of various categories of tastants but a lot remains undetermined about the composition and response properties of taste receptors. How combinations of GRs and IRs belonging to different receptor families (e.g. Gr and IR), coordinate within the neurons that house them is a subject of investigation. Feeding behavior is root cause of metabolic disorders. A better understanding of the biology of metabolic disorders in association with GRs, IRs, Trp, and ppk receptors is a need of the hour because of the burden of metabolic disorders, high incidences of cardiovascular diseases, faster aging, dependence on readymade food and consumption of unhealthy junk food in kids. A better understanding of the IRs and other receptors, neural circuits, higher order neurons involved in taste modulation as well as food regulation could provide us with a better understanding about human metabolic disorders and lead to a subsequent development of treatment strategies ultimately benefitting mankind. Recent reports invite exciting new avenues of investigation to determine the higher brain locations that receive taste input from the AMMC, and to trace the circuits by which information is relayed to motor neurons and neurons of the mushroom body to control feeding behavior and associations with appetitive and aversive learning.

Acknowledgements

This work is supported by Wellcome trust/DBT India Alliance Fellowship (grant number IA/I/15/2/502074) awarded to PK.

Conflict of interest

The authors declare no conflict of interest.

Criteria for authorship

SK and PK both substantially contributed to the conception and design of the work. Both participated in drafting and revising the work, made the figures, wrote the chapter and approved the final version for publication.

Abbreviations

GRs	Gustatory receptors
TRCs	Taste receptors cells
GRNs	Gustatory receptor neurons
MSN	Mechanosensory neurons
SEZ	Sub esophageal zone
IRs	Ionotropic receptors
GPCRs	G-protein coupled receptors
OR genes	Olfactory receptor genes
GFP	Green fluorescence protein
UAS	Upstream activating sequence
TRP	Transient receptor potential
TRPL	Transient receptor potential-like
ENaCs	Epithelial Na^+ channels
OBP	Odorant binding proteins
AMMC	Antennal and mechanosensory motor center
MB	Mushroom bodies
IP_3 receptor	Inositol 1,4,5-triphosphate receptor
DAG	Diacylglycerol
norpA	No receptor potential A
plc	Phospholipase C
PKD2L1	Polycystic kidney disease 2-like 1
Car4	Carbonic anhydrase isoform 4

sNPF	Short neuropeptide F
GABA	Gamma-aminobutyric acid
ATP	Adenosine triphosphate
IMP	Inosine monophosphate
GMP	Guanylate monophosphate
ppk	Pickpocket

Author details

Shivam Kaushik and Pinky Kain*
Regional Centre for Biotechnology, NCR Biotech Science Cluster, Faridabad, Haryana, India

*Address all correspondence to: pinkykain@gmail.com

References

[1] Yarmolinsky DA, Zuker CS, Ryba NJ. Common sense about taste: From mammals to insects. Cell. 2009;**139**:234-244

[2] Stocker RF. The organization of the chemosensory system in *Drosophila melanogaster*: A review. Cell and Tissue Research. 1994;**275**:3-26

[3] Chyb S, Dahanukar A, Wickens A, Carlson JR. *Drosophila* Gr5a encodes a taste receptor tuned to trehalose. Proceedings of the National Academy of Sciences of the United States of America. 2003;**100**(Suppl 2):14526-14530

[4] Clyne PJ, Warr CG, Carlson JR. Candidate taste receptors in *Drosophila*. Science. 2000;**287**:1830-1834

[5] Moon SJ, Kottgen M, Jiao Y, Xu H, Montell C. A taste receptor required for the caffeine response in vivo. Current Biology. 2006;**16**:1812-1817

[6] Thorne N, Chromey C, Bray S, Amrein H. Taste perception and coding in *Drosophila*. Current Biology. 2004;**14**:1065-1079

[7] Wang Z, Singhvi A, Kong P, Scott K. Taste representations in the *Drosophila* brain. Cell. 2004;**117**:981-991

[8] Dahanukar A, Lei YT, Kwon JY, Carlson JR. Two gr genes underlie sugar reception in *Drosophila*. Neuron. 2007;**56**:503-516

[9] Hiroi M, Marion-Poll F, Tanimura T. Differentiated response to sugars among labellar chemosensilla in *Drosophila*. Zoological Science. 2002;**19**:1009-1018

[10] Meunier N, Ferveur JF, Marion-Poll F. Sex-specific non-pheromonal taste receptors in *Drosophila*. Current Biology. 2000;**10**:1583-1586

[11] Meunier N, Marion-Poll F, Rospars JP, Tanimura T. Peripheral coding of bitter taste in *Drosophila*. Journal of Neurobiology. 2003;**56**:139-152

[12] Kapsimali M, Barlow LA. Developing a sense of taste. Seminars in Cell & Developmental Biology. 2013;**24**:200-209

[13] Chandrashekar J, Hoon MA, Ryba NJ, Zuker CS. The receptors and cells for mammalian taste. Nature. 2006;**444**:288-294

[14] Chaudhari N, Roper SD. The cell biology of taste. The Journal of Cell Biology. 2010;**190**:285-296

[15] Smith DV, Davis BJ. Neural representation of taste. In: Finger TE, Silver WL, Restrepo D, editors. The Neurobiology of Taste and Smell. New York: Wiley-Liss. 2000. p. 353-394

[16] Ebbs ML, Amrein H. Taste and pheromone perception in the fruit fly *Drosophila melanogaster*. Pflügers Archiv. 2007;**454**:735-747

[17] Amrein H. Pheromone perception and behavior in *Drosophila*. Current Opinion in Neurobiology. 2004;**14**:435-442

[18] Amrein H, Thorne N. Gustatory perception and behavior in *Drosophila melanogaster*. Current Biology. 2005;**15**:R673-R684

[19] Arora K, Rodrigues V, Joshi S, Shanbhag S, Siddiqi O. A gene affecting the specificity of the chemosensory neurons of *Drosophila*. Nature. 1987;**330**:62-63

[20] Hiroi M, Meunier N, Marion-Poll F, Tanimura T. Two antagonistic gustatory receptor neurons responding to sweet-salty and bitter taste in

Drosophila. Journal of Neurobiology. 2004;**61**:333-342

[21] Naoji F, Hiromasa K, Hiromichi M. Impulse frequency and action potential amplitude in labellar chemosensory neurones of *Drosophila melanogaster*. Journal of Insect Physiology. 1984;**30**:317-325

[22] Rodrigues V, Siddiqi O. Genetic analysis of chemosensory pathway. Proceedings of the Indian Academy of Sciences—Section B. 1978;**87**:147-160

[23] Nayak SV, Singh RN. Sensilla on the tarsal segments and mouthparts of adult *Drosophila melanogaster* Meigen (Diptera: Drosophilidae). International Journal of Insect Morphology and Embryology. 1983;**12**:273-291

[24] Stocker RF, Schorderet M. Cobalt filling of sensory projections from internal and external mouthparts in *Drosophila*. Cell and Tissue Research. 1981;**216**:513-523

[25] Rajashekhar KP, Singh RN. Neuroarchitecture of the tritocerebrum of *Drosophila melanogaster*. The Journal of Comparative Neurology. 1994;**349**:633-645

[26] Falk R, Bleiser-Avivi N, Atidia J. Labellar taste organs of *Drosophila melanogaster*. Journal of Morphology. 1976;**150**:327-341

[27] O'Neil RG, Heller S. The mechanosensitive nature of TRPV channels. Pflügers Archiv. 2005;**451**:193-203

[28] Sukharev S, Corey DP. Mechanosensitive channels: Multiplicity of families and gating paradigms. Science's STKE. 2004;**2004**:re4

[29] Zhang YV, Aikin TJ, Li Z, Montell C. The basis of food texture sensation in *Drosophila*. Neuron. 2016;**91**:863-877

[30] Adler E, Hoon MA, Mueller KL, Chandrashekar J, Ryba NJ, Zuker CS. A novel family of mammalian taste receptors. Cell. 2000;**100**:693-702

[31] Chandrashekar J, Mueller KL, Hoon MA, Adler E, Feng L, Guo W, et al. T2Rs function as bitter taste receptors. Cell. 2000;**100**:703-711

[32] Dunipace L, Meister S, McNealy C, Amrein H. Spatially restricted expression of candidate taste receptors in the *Drosophila* gustatory system. Current Biology. 2001;**11**:822-835

[33] Hoon MA, Adler E, Lindemeier J, Battey JF, Ryba NJ, Zuker CS. Putative mammalian taste receptors: A class of taste-specific GPCRs with distinct topographic selectivity. Cell. 1999;**96**:541-551

[34] Matsunami H, Montmayeur JP, Buck LB. A family of candidate taste receptors in human and mouse. Nature. 2000;**404**:601-604

[35] Max M, Shanker YG, Huang L, Rong M, Liu Z, Campagne F, et al. Tas1r3, encoding a new candidate taste receptor, is allelic to the sweet responsiveness locus sac. Nature Genetics. 2001;**28**:58-63

[36] Montmayeur JP, Liberles SD, Matsunami H, Buck LB. A candidate taste receptor gene near a sweet taste locus. Nature Neuroscience. 2001;**4**:492-498

[37] Sainz E, Korley JN, Battey JF, Sullivan SL. Identification of a novel member of the T1R family of putative taste receptors. Journal of Neurochemistry. 2001;**77**:896-903

[38] Scott K, Brady R Jr, Cravchik A, Morozov P, Rzhetsky A, Zuker C, et al. A chemosensory gene family encoding candidate gustatory and olfactory receptors in *Drosophila*. Cell. 2001;**104**:661-673

[39] Robertson HM, Warr CG, Carlson JR. Molecular evolution of the insect chemoreceptor gene superfamily in *Drosophila melanogaster*. Proceedings of the National Academy of Sciences of the United States of America. 2003;**100**(Suppl 2):14537-14542

[40] Dahanukar A, Foster K, van der Goes van Naters WM, Carlson JR. A Gr receptor is required for response to the sugar trehalose in taste neurons of *Drosophila*. Nature Neuroscience. 2001;**4**:1182-1186

[41] Ueno K, Ohta M, Morita H, Mikuni Y, Nakajima S, Yamamoto K, et al. Trehalose sensitivity in *Drosophila* correlates with mutations in and expression of the gustatory receptor gene Gr5a. Current Biology. 2001;**11**:1451-1455

[42] Brand AH, Perrimon N. Targeted gene expression as a means of altering cell fates and generating dominant phenotypes. Development. 1993;**118**:401-415

[43] Zhang Y, Hoon MA, Chandrashekar J, Mueller KL, Cook B, Wu D, et al. Coding of sweet, bitter, and umami tastes: Different receptor cells sharing similar signaling pathways. Cell. 2003;**112**:293-301

[44] Benton R, Sachse S, Michnick SW, Vosshall LB. Atypical membrane topology and heteromeric function of *Drosophila* odorant receptors in vivo. PLoS Biology. 2006;**4**:e20

[45] Sato K, Pellegrino M, Nakagawa T, Nakagawa T, Vosshall LB, Touhara K. Insect olfactory receptors are heteromeric ligand-gated ion channels. Nature. 2008;**452**:1002-1006

[46] Wicher D, Schafer R, Bauernfeind R, Stensmyr MC, Heller R, Heinemann SH, et al. *Drosophila* odorant receptors are both ligand-gated and cyclic-nucleotide-activated cation channels. Nature. 2008;**452**:1007-1011

[47] Zhang HJ, Anderson AR, Trowell SC, Luo AR, Xiang ZH, Xia QY. Topological and functional characterization of an insect gustatory receptor. PLoS One. 2011;**6**:e24111

[48] Sato K, Tanaka K, Touhara K. Sugar-regulated cation channel formed by an insect gustatory receptor. Proceedings of the National Academy of Sciences of the United States of America. 2011;**108**:11680-11685

[49] Nelson G, Hoon MA, Chandrashekar J, Zhang Y, Ryba NJ, Zuker CS. Mammalian sweet taste receptors. Cell. 2001;**106**:381-390

[50] Yee KK, Sukumaran SK, Kotha R, Gilbertson TA, Margolskee RF. Glucose transporters and ATP-gated K^+ (KATP) metabolic sensors are present in type 1 taste receptor 3 (T1r3)-expressing taste cells. Proceedings of the National Academy of Sciences of the United States of America. 2011;**108**: 5431-5436

[51] Gordesky-Gold B, Rivers N, Ahmed OM, Breslin PA. *Drosophila melanogaster* prefers compounds perceived sweet by humans. Chemical Senses. 2008;**33**:301-309

[52] Jiao Y, Moon SJ, Montell C. A *Drosophila* gustatory receptor required for the responses to sucrose, glucose, and maltose identified by mRNA tagging. Proceedings of the National Academy of Sciences of the United States of America. 2007;**104**:14110-14115

[53] Jiao Y, Moon SJ, Wang X, Ren Q, Montell C. Gr64f is required in combination with other gustatory receptors for sugar detection in *Drosophila*. Current Biology. 2008;**18**:1797-1801

[54] Slone J, Daniels J, Amrein H. Sugar receptors in *Drosophila*. Current Biology. 2007;**17**:1809-1816

[55] Miyamoto T, Slone J, Song X, Amrein H. A fructose receptor functions as a nutrient sensor in the *Drosophila* brain. Cell. 2012;**151**:1113-1125

[56] Shi P, Zhang J. Contrasting modes of evolution between vertebrate sweet/umami receptor genes and bitter receptor genes. Molecular Biology and Evolution. 2006;**23**:292-300

[57] Mueller KL, Hoon MA, Erlenbach I, Chandrashekar J, Zuker CS, Ryba NJ. The receptors and coding logic for bitter taste. Nature. 2005;**434**:225-229

[58] Voigt A, Hubner S, Lossow K, Hermans-Borgmeyer I, Boehm U, Meyerhof W. Genetic labeling of Tas1r1 and Tas2r131 taste receptor cells in mice. Chemical Senses. 2012;**37**:897-911

[59] Weiss LA, Dahanukar A, Kwon JY, Banerjee D, Carlson JR. The molecular and cellular basis of bitter taste in *Drosophila*. Neuron. 2011;**69**:258-272

[60] Behrens M, Meyerhof W. Mammalian bitter taste perception. Results and Problems in Cell Differentiation. 2009;**47**:203-220

[61] Kwon JY, Dahanukar A, Weiss LA, Carlson JR. Molecular and cellular organization of the taste system in the *Drosophila* larva. The Journal of Neuroscience. 2011;**31**:15300-15309

[62] Lee Y, Kim SH, Montell C. Avoiding DEET through insect gustatory receptors. Neuron. 2010;**67**:555-561

[63] Moon SJ, Lee Y, Jiao Y, Montell C. A *Drosophila* gustatory receptor essential for aversive taste and inhibiting male-to-male courtship. Current Biology. 2009;**19**:1623-1627

[64] Kim SH, Lee Y, Akitake B, Woodward OM, Guggino WB, Montell C. *Drosophila* TRPA1 channel mediates chemical avoidance in gustatory receptor neurons. Proceedings of the National Academy of Sciences of the United States of America. 2010;**107**:8440-8445

[65] Al-Anzi B, Tracey WD Jr, Benzer S. Response of *Drosophila* to wasabi is mediated by painless, the fly homolog of mammalian TRPA1/ANKTM1. Current Biology. 2006;**16**:1034-1040

[66] Zhang YV, Raghuwanshi RP, Shen WL, Montell C. Food experience-induced taste desensitization modulated by the *Drosophila* TRPL channel. Nature Neuroscience. 2013;**16**:1468-1476

[67] Brand JG, Teeter JH, Silver WL. Inhibition by amiloride of chorda tympani responses evoked by monovalent salts. Brain Research. 1985;**334**:207-214

[68] Ninomiya Y. Reinnervation of cross-regenerated gustatory nerve fibers into amiloride-sensitive and amiloride-insensitive taste receptor cells. Proceedings of the National Academy of Sciences of the United States of America. 1998;**95**:5347-5350

[69] Breza JM, Contreras RJ. Anion size modulates salt taste in rats. Journal of Neurophysiology. 2012;**107**:1632-1648

[70] Canessa CM, Schild L, Buell G, Thorens B, Gautschi I, Horisberger JD, et al. Amiloride-sensitive epithelial Na^+ channel is made of three homologous subunits. Nature. 1994;**367**:463-467

[71] Chandrashekar J, Kuhn C, Oka Y, Yarmolinsky DA, Hummler E, Ryba NJ, et al. The cells and peripheral representation of sodium taste in mice. Nature. 2010;**464**:297-301

[72] Oka Y, Butnaru M, von Buchholtz L, Ryba NJ, Zuker CS. High salt recruits aversive taste pathways. Nature. 2013;**494**:472-475

[73] Liu L, Leonard AS, Motto DG, Feller MA, Price MP, Johnson WA, et al. Contribution of *Drosophila* DEG/ENaC genes to salt taste. Neuron. 2003;**39**:133-146

[74] Zhang YV, Ni J, Montell C. The molecular basis for attractive salt-taste coding in *Drosophila*. Science. 2013;**340**:1334-1338

[75] Benton R, Vannice KS, Gomez-Diaz C, Vosshall LB. Variant ionotropic glutamate receptors as chemosensory receptors in *Drosophila*. Cell. 2009;**136**:149-162

[76] Croset V, Rytz R, Cummins SF, Budd A, Brawand D, Kaessmann H, et al. Ancient protostome origin of chemosensory ionotropic glutamate receptors and the evolution of insect taste and olfaction. PLoS Genetics. 2010;**6**:e1001064

[77] Jaeger AH, Stanley M, Weiss ZF, Musso PY, Chan RC, Zhang H, et al. A complex peripheral code for salt taste in *Drosophila*. eLife. 2018;**7**:e37167

[78] Huang AL, Chen X, Hoon MA, Chandrashekar J, Guo W, Trankner D, et al. The cells and logic for mammalian sour taste detection. Nature. 2006;**442**:934-938

[79] Huang YA, Maruyama Y, Stimac R, Roper SD. Presynaptic (type III) cells in mouse taste buds sense sour (acid) taste. The Journal of Physiology. 2008;**586**:2903-2912

[80] Chang RB, Waters H, Liman ER. A proton current drives action potentials in genetically identified sour taste cells. Proceedings of the National Academy of Sciences of the United States of America. 2010;**107**:22320-22325

[81] Chandrashekar J, Yarmolinsky D, von Buchholtz L, Oka Y, Sly W, Ryba NJ, et al. The taste of carbonation. Science. 2009;**326**:443-445

[82] Fischler W, Kong P, Marella S, Scott K. The detection of carbonation by the *Drosophila* gustatory system. Nature. 2007;**448**:1054-1057

[83] Charlu S, Wisotsky Z, Medina A, Dahanukar A. Acid sensing by sweet and bitter taste neurons in *Drosophila melanogaster*. Nature Communications. 2013;**4**:2042

[84] Rimal S, Sang J, Poudel S, Thakur D, Montell C, Lee Y. Mechanism of acetic acid gustatory repulsion in *Drosophila*. Cell Reports. 2019;**26**(1432-1442):e4

[85] Yamaguchi S. The synergistic taste effect of monosodium glutamate and disodium 5′-inosinate. Journal of Food Science. 1967;**32**

[86] Toshima N, Tanimura T. Taste preference for amino acids is dependent on internal nutritional state in *Drosophila melanogaster*. The Journal of Experimental Biology. 2012;**215**:2827-2832

[87] Mitri C, Soustelle L, Framery B, Bockaert J, Parmentier ML, Grau Y. Plant insecticide L-canavanine repels *Drosophila* via the insect orphan GPCR DmX. PLoS Biology. 2009;**7**:e1000147

[88] Lee Y, Kang MJ, Shim J, Cheong CU, Moon SJ, Montell C. Gustatory receptors required for avoiding the insecticide L-canavanine. The Journal of Neuroscience. 2012;**32**:1429-1435

[89] Ganguly A, Pang L, Duong VK, Lee A, Schoniger H, Varady E, et al. A molecular and cellular context-dependent role for Ir76b in detection of amino acid taste. Cell Reports. 2017;**18**:737-750

[90] Bredendiek N, Hutte J, Steingraber A, Hatt H, Gisselmann G,

Neuhaus EM. Go alpha is involved in sugar perception in *Drosophila*. Chemical Senses. 2011;**36**:69-81

[91] Ishimoto H, Takahashi K, Ueda R, Tanimura T. G-protein gamma subunit 1 is required for sugar reception in *Drosophila*. The EMBO Journal. 2005;**24**:3259-3265

[92] Kain P, Badsha F, Hussain SM, Nair A, Hasan G, Rodrigues V. Mutants in phospholipid signaling attenuate the behavioral response of adult *Drosophila* to trehalose. Chemical Senses. 2010;**35**:663-673

[93] Ueno K, Kohatsu S, Clay C, Forte M, Isono K, Kidokoro Y. Gsalpha is involved in sugar perception in *Drosophila melanogaster*. The Journal of Neuroscience. 2006;**26**:6143-6152

[94] Ueno K, Kidokoro Y. Adenylyl cyclase encoded by AC78C participates in sugar perception in *Drosophila melanogaster*. The European Journal of Neuroscience. 2008;**28**:1956-1966

[95] Devambez I, Ali Agha M, Mitri C, Bockaert J, Parmentier ML, Marion-Poll F, et al. Galphao is required for L-canavanine detection in *Drosophila*. PLoS One. 2013;**8**:e63484

[96] Wisotsky Z, Medina A, Freeman E, Dahanukar A. Evolutionary differences in food preference rely on Gr64e, a receptor for glycerol. Nature Neuroscience. 2011;**14**:1534-1541

[97] Freeman EG, Dahanukar A. Molecular neurobiology of *Drosophila* taste. Current Opinion in Neurobiology. 2015;**34**:140-148

[98] French AS, Sellier MJ, Ali Agha M, Guigue A, Chabaud MA, Reeb PD, et al. Dual mechanism for bitter avoidance in *Drosophila*. The Journal of Neuroscience. 2015;**35**:3990-4004

[99] Konig C, Schleyer M, Leibiger J, El-Keredy A, Gerber B. Bitter-sweet processing in larval *Drosophila*. Chemical Senses. 2014;**39**:489-505

[100] Chu B, Chui V, Mann K, Gordon MD. Presynaptic gain control drives sweet and bitter taste integration in *Drosophila*. Current Biology. 2014;**24**:1978-1984

[101] Jeong YT, Shim J, Oh SR, Yoon HI, Kim CH, Moon SJ, et al. An odorant-binding protein required for suppression of sweet taste by bitter chemicals. Neuron. 2013;**79**:725-737

[102] Chen Y, Amrein H. Enhancing perception of contaminated food through acid-mediated modulation of taste neuron responses. Current Biology. 2014;**24**:1969-1977

[103] Hussain A, Ucpunar HK, Zhang M, Loschek LF, Grunwald Kadow IC. Neuropeptides modulate female chemosensory processing upon mating in *Drosophila*. PLoS Biology. 2016;**14**:e1002455

[104] Inagaki HK, Ben-Tabou de-Leon S, Wong AM, Jagadish S, Ishimoto H, Barnea G, et al. Visualizing neuromodulation in vivo: TANGO-mapping of dopamine signaling reveals appetite control of sugar sensing. Cell. 2012;**148**:583-595

[105] Inagaki HK, Panse KM, Anderson DJ. Independent, reciprocal neuromodulatory control of sweet and bitter taste sensitivity during starvation in *Drosophila*. Neuron. 2014;**84**:806-820

[106] LeDue EE, Mann K, Koch E, Chu B, Dakin R, Gordon MD. Starvation-induced depotentiation of bitter taste in *Drosophila*. Current Biology. 2016;**26**:2854-2861

[107] Steck K, Walker SJ, Itskov PM, Baltazar C, Moreira JM, Ribeiro C. Internal amino acid state modulates

yeast taste neurons to support protein homeostasis in *Drosophila*. eLife. 2018;**7**:e31625

[108] Walker SJ, Corrales-Carvajal VM, Ribeiro C. Postmating circuitry modulates salt taste processing to increase reproductive output in *Drosophila*. Current Biology. 2015;**25**:2621-2630

[109] Kain P, Dahanukar A. Secondary taste neurons that convey sweet taste and starvation in the *Drosophila* brain. Neuron. 2015;**85**:819-832

[110] Gaillard D, Laugerette F, Darcel N, El-Yassimi A, Passilly-Degrace P, Hichami A, et al. The gustatory pathway is involved in CD36-mediated orosensory perception of long-chain fatty acids in the mouse. The FASEB Journal. 2008;**22**:1458-1468

[111] Liu P, Shah BP, Croasdell S, Gilbertson TA. Transient receptor potential channel type M5 is essential for fat taste. The Journal of Neuroscience. 2011;**31**:8634-8642

[112] Cartoni C, Yasumatsu K, Ohkuri T, Shigemura N, Yoshida R, Godinot N, et al. Taste preference for fatty acids is mediated by GPR40 and GPR120. The Journal of Neuroscience. 2010;**30**:8376-8382

[113] Galindo MM, Voigt N, Stein J, van Lengerich J, Raguse JD, Hofmann T, et al. G protein-coupled receptors in human fat taste perception. Chemical Senses. 2012;**37**:123-139

[114] Ahn JE, Chen Y, Amrein H. Molecular basis of fatty acid taste in *Drosophila*. eLife. 2017;**6**:e30115

[115] Tordoff MG, Shao H, Alarcon LK, Margolskee RF, Mosinger B, Bachmanov AA, et al. Involvement of T1R3 in calcium-magnesium taste. Physiological Genomics. 2008;**34**:338-348

[116] Tordoff MG, Alarcon LK, Valmeki S, Jiang P. T1R3: A human calcium taste receptor. Scientific Reports. 2012;**2**:496

[117] Lee Y, Poudel S, Kim Y, Thakur D, Montell C. Calcium taste avoidance in *Drosophila*. Neuron. 2018;**97**(67-74):e4

[118] Galindo-Cuspinera V, Winnig M, Bufe B, Meyerhof W, Breslin PA. A TAS1R receptor-based explanation of sweet 'water-taste'. Nature. 2006;**441**:354-357

[119] Cameron P, Hiroi M, Ngai J, Scott K. The molecular basis for water taste in *Drosophila*. Nature. 2010;**465**:91-95

[120] Marella S, Fischler W, Kong P, Asgarian S, Rueckert E, Scott K. Imaging taste responses in the fly brain reveals a functional map of taste category and behavior. Neuron. 2006;**49**:285-295

[121] Bohra AA, Kallman BR, Reichert H, VijayRaghavan K. Identification of a single pair of interneurons for bitter taste processing in the *Drosophila* brain. Current Biology. 2018;**28**(847-858):e3

[122] Flood TF, Iguchi S, Gorczyca M, White B, Ito K, Yoshihara M. A single pair of interneurons commands the *Drosophila* feeding motor program. Nature. 2013;**499**:83-87

[123] Huckesfeld S, Peters M, Pankratz MJ. Central relay of bitter taste to the protocerebrum by peptidergic interneurons in the *Drosophila* brain. Nature Communications. 2016;**7**:12796

[124] Mann K, Gordon MD, Scott K. A pair of interneurons influences the choice between feeding and locomotion in *Drosophila*. Neuron. 2013;**79**:754-765

[125] Marella S, Mann K, Scott K. Dopaminergic modulation of sucrose acceptance behavior in *Drosophila*. Neuron. 2012;**73**:941-950

[126] Pool AH, Kvello P, Mann K, Cheung SK, Gordon MD, Wang L, et al. Four GABAergic interneurons impose feeding restraint in *Drosophila*. Neuron. 2014;**83**:164-177

[127] Kirkhart C, Scott K. Gustatory learning and processing in the *Drosophila* mushroom bodies. The Journal of Neuroscience. 2015;**35**:5950-5958

[128] Huetteroth W, Perisse E, Lin S, Klappenbach M, Burke C, Waddell S. Sweet taste and nutrient value subdivide rewarding dopaminergic neurons in *Drosophila*. Current Biology. 2015;**25**:751-758

[129] Lin S, Owald D, Chandra V, Talbot C, Huetteroth W, Waddell S. Neural correlates of water reward in thirsty *Drosophila*. Nature Neuroscience. 2014;**17**:1536-1542

Chapter 9

Spermatogenesis in *Drosophila melanogaster*: Key Features and the Role of the NXF1 (Nuclear Export Factor) Protein

Elena Golubkova, Anna Atsapkina, Anna K'ergaard and Ludmila Mamon

Abstract

Now there is interest in finding factors that, once in the ovum, can affect the development of offspring. The *sbr* (*Dm nxf1*) gene in *D. melanogaster* belongs to the evolutionarily conserved family of nxf (nuclear export factor). It is involved in controlling of male fertility and forming the factor that affects the segregation of maternal and paternal chromosomes after fertilization. The Dm NXF1 (SBR) protein seems to play a role in forming this mysterious factor during the meiotic period of spermatogenesis. Male germ cells develop as a syncytium, and success of spermatid individualization depends, to a great degree, on gene expression in primary spermatocytes. Most transcripts formed in primary spermatocytes are long-lived. The presence of the Dm NXF1 protein in the nucleus as well as in the cytoplasm suggests that it plays a role in the biogenesis of long-lived RNA and in cytoskeletal reorganization.

Keywords: spermatogenesis, meiosis, *Drosophila*, NXF, cytokinesis, RNA-binding protein

1. Introduction

Spermatogenesis is an evolutionary conservative process. In testes, the developing germline cells divide synchronously and with incomplete cytokinesis remaining interconnected by cytoplasmic bridges. Therefore, the germline cells develop as a syncytium and only become separated from each other at the end of spermatogenesis during spermatid individualization. In *Drosophila melanogaster*, the stem cell daughter, the gonialblast, undergoes four rounds of synchronous mitotic divisions to produce 16 precursor cells, called spermatogonia [1–3]. The spermatogonia then become spermatocytes and grow 25 times in size [1, 4]. During the growth phase, the majority of the transcripts for sperm differentiation are expressed and potentially stored until their protein activity is required [5–7]. Many of the transcribed genes are important for meiosis and post meiotic stages [6] or as are suggested to be reserved up to sperm maturing and delivered into the egg at fertilization [8, 9]. After the growth phase, the spermatocytes divide twice by meiosis and then differentiate into 64 haploid spermatid interconnected in a syncytium. At sperm individualization, most unneeded products are pushed to the end of the spermatozoon and an

 IntechOpen

actin structure, termed the investment cone, forms a new membrane [10, 11]. Each cell becomes independent and turns into a mobile spermatozoon. Development of a mature spermatozoon with the change of histones on protamines in nucleus, the building of an acrosome and a tail, the forming of cellular membrane, occurs without transcription owing to translation of the long-lived mRNAs. Such mRNAs are keeping as a part of the RNP-complexes localized in the different cellular compartments. The testis-specific isoform of the Dm NXF1 (SBR) protein can be an important component of these RNP-complexes.

2. Long-lived transcripts and interacting proteins: role in the sperm maturation during spermiogenesis

Most transcripts formed in primary spermatocytes are not translated immediately after exiting from the nucleus. These RNAs are reserved in the ribonucleoprotein protein (RNP) complexes for a long time up to the elongation stage during spermatid individualization [6]. In *D. melanogaster*, 529 of the 553 mRNAs detected in spermatid were transcribed in primary spermatocytes [12].

The existence of the NXF (Nuclear eXport Factor) specialized transport receptors, which are the products of paralogous genes of the NXF family and are expressed predominantly in the testes or the brain of humans and mice, has been associated with the presence of long-lived, temporary non-translatable transcripts [13–17]. The most evolutionary conservative member of this family (NXF1) provides transport of the majority of the mRNAs from the nucleus to the cytoplasm [18–21]. The paralogous members of the NXF family differ from the NXF1 protein by divergent sequences enabling interaction with nucleoporins—proteins of nuclear pore complexes and the domain of LRR (leucine-rich repeats). These sequences are responsible for interaction with the partner proteins forming RNP complexes at different steps of mRNA biogenesis [13–15]. The primary localization of testis-specific NXFs in the cytoplasm suggests their participation in the biogenesis of long-lived mRNAs, which are temporarily untranslated [16, 17].

In *D. melanogaster*, the testis-specific *Nxf* paralogous genes are unknown. Unlike its orthologous in humans and mice, the *Dm nxf1* (*sbr*) gene does have testis-specific products [22, 23]. The short testis-specific *sbr* transcripts use the alternative promoters in intron 3 and do not include exons 1–3. As a result, the testis-specific truncated SBR protein excludes nuclear localization signals (NLS) present in the canonical SBR protein and in the part of the RNA binding domain (RBD)

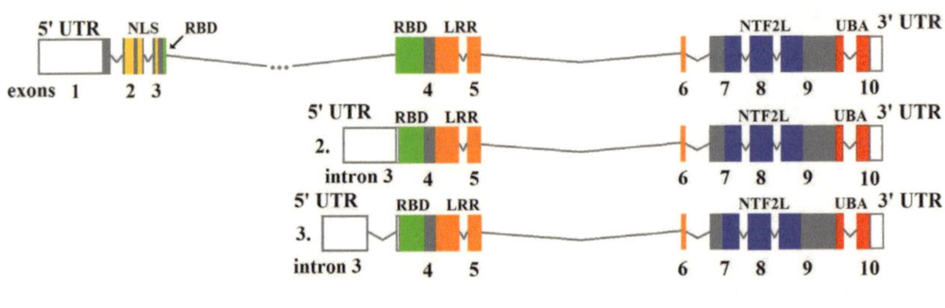

Figure 1.
The testis-specific sbr *mRNAs are detected by 5′ and 3′ RACE-PCR. Boxes indicate exons included in the sbr mRNAs. Exon numbers are signed under each box. RBD – RNA-binding domain; LRR – leucine-rich domain; NTF2L – nuclear transport factor 2 like; UBA – ubiquitin associated domain; NLS – nuclear localization signal. In testes, all* sbr *mRNAs have the shortened 3′ UTR. In all transcripts, length of the 3′ UTR varies from 44 to 124 nucleotides. Two* sbr *mRNAs, using the alternative promoters in intron 3, give rise the shortened protein isoform without NLS and the part of RBD, which are present in the canonical full-length SBR (published in Mamon et al. [24]).*

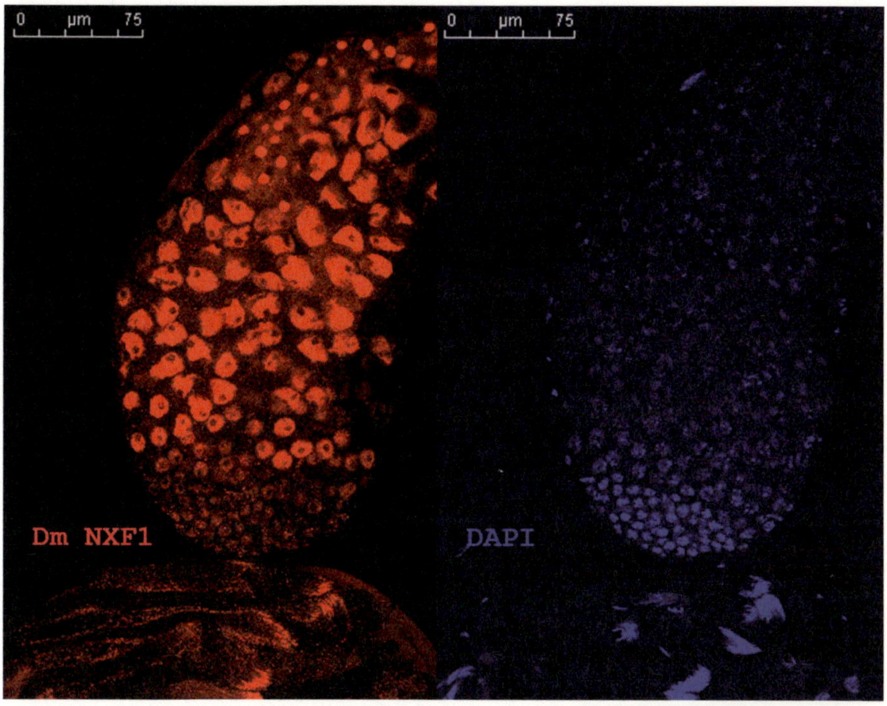

Figure 2.
Adult testis immunofluorescence using anti-SBR (Alexa 546 goat anti-mouse). In the prophase of meiosis (arrowhead) and at the round spermatid stage (arrow)—The SBR proteins are primarily localized in the nucleus. Asterisk denotes the stem cells zone. Scale bar, 75 µm.

(**Figure 1**) [24]. Such truncated SBR form can have other functions in addition to the nuclear RNA export, which are important for spermatogenesis.

During transcriptional activities—the prophase of meiosis and the round spermatid phase—the SBR protein is primarily found in the nucleus (**Figures 2, 3**). This is coordinated with the known universal function of NXF1 proteins in various organisms [18–21]. Unfortunately, antibodies to C-terminal part of the SBR protein cannot contradistinguish full-sized and truncated SBR isoforms.

Nuclei of spermatids in onion stage are brightly painted fluorochromes conjugated with anti-SBR (**Figure 2**). Nebenkern is near each of nucleus and is free from SBR. Only 24 genes are transcribed in *Drosophila* spermatids [12], but spermatid nuclei show the brightest coloring by anti-SBR. Why at onion stage such high concentration of transport receptors for mRNA is necessary. It is possible that spermatid nuclei may be one of the docking place for some long-lived mRNAs because the most dramatic events depending from translation of these mRNAs are related to nucleus and/or structures connected with nucleus during the sperm maturation. In elongated spermatids, the SBR protein is localized on a nuclear surface asymmetrically, only along one of its sides (**Figure 4**). Spermatid elongation is period of intensive nuclear morphogenesis [1]. The nuclear envelope becomes asymmetrical, with nuclear pores being disposed along one side of elongated nucleus. In this region, the dense body consisting from a microtubule (MT) and actin-rich structure is formed and also localized along one side of the nucleus [25]. Nucleoporin binding domains (NTF2L and UBA) in the SBR proteins (**Figure 1**) allow them to participate in the storing of the RNP complexes with the long-lived RNAs connected with the nucleoporins on the nuclear envelope. After the RNP complexes dissociate, the translation of their mRNAs begins close to the nucleus, where there is a need for appropriate proteins. Then the components of the RNP complexes, including the SBR proteins,

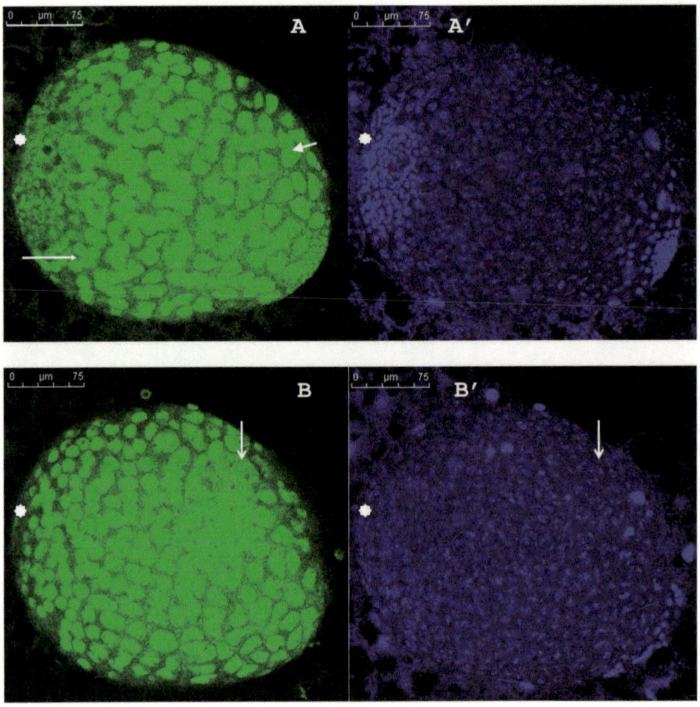

Figure 3.
Prepupal testis immunofluorescence using anti-SBR (Alexa 488 goat anti-mouse – A, B) and DAPI (A', B'). A. Most part of the future gonad is represented of growing spermatocytes, primary spermatocytes, and meiocytes. Initially, the shape of the nucleus is round (A, thin arrow); then the shape becomes irregular (A, short arrow). B. Cyst of 16 meiocytes with the cytoplasmic localization of the SBR protein. Bivalents partially condensed (arrows). Asterisk denotes the stem cells zone. Scale bar, 75 µm.

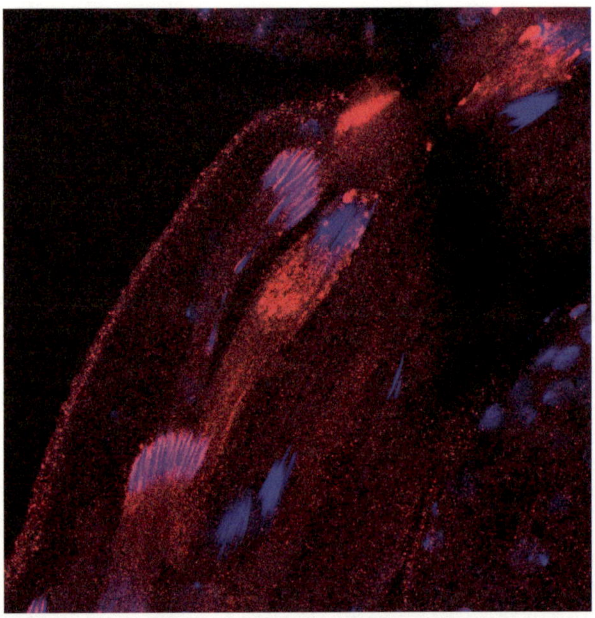

Figure 4.
Spermatids at the elongation or individualization stages in D. melanogaster *males (Alexa Fluor 546). Spermatids at the elongation stage: the SBR protein moves onto one side of the spermatid nuclear. The SBR protein leaves the condensed nuclei and is translocated in large granules into the formed spermatozoon tails at the final stage of spermatid elongation before individualization. 63× magnification.*

relocate in a spermatozoon tail as the granules and are degradate (**Figure 4**) [26, 27]. Proteasome degradation of the proteins is necessary for the formation of mature sperm [28]. SBR (Dm NXF1) protein has UBA-like domain (**Figure 1**), which are involved in a variety of cellular functions and many are connected to the ubiquitin proteasome system, an essential pathway for protein degradation in eukaryotic cells [29, 30]. Proteasome degradation is important for cell cycle progression [31]. Testis-specific proteasome degradation in *D. melanogaster* probably provides effective elimination of SBR during spermatid individualization (**Figure 4**).

It is important that mRNAs encoded by all of the post-meiotically transcribed genes are localized to the extreme distal ends of elongated spermatids. These mRNAs are subdivided into two classes named by comet and cup transcripts [12]. There is not a concentration of SBR to the distal ends of elongated spermatids.

A feature of spermatogenesis is that many genes produce shortened transcripts that encode shortened proteins. These proteins cannot function as the full-sized proteins encoded by the same genes [32]. This is due to the presence of testis-specific promoters or alternative splicing variants [33]. In addition, testis-specific transcripts often use an alternative polyadenylation site located upstream, compared to those in the transcripts of the same gene in other organs [33, 34]. This may be important for the regulation of translation and degradation of transcripts during spermatogenesis. All the transcripts of the *Dm nxf1* (*sbr*) gene have the shortened 3′UTR in testes [22, 23].

3. Allele-specific effects of the *sbr* gene including the male infertility

Mutations of the house-keeping genes are characterized by a broad range of pleiotropic effects. It is no surprise that mutations of the *Dm nxf1* gene lead to both male and female sterility and increase the frequency of the development malformations (FlyBase [35]). That may be the result of disruption of the NXF1 general functions. Allele-specific effects of mutations in the *Dm nxf1* gene suggest that its products may have specialized functions. The organ-specific products of *sbr* can provide such specialization [22–24]. In addition, the source of specialized functions may be the presence of functionally significant sequences in the SBR protein, responsible for interaction with the certain partners. As a result, SBR may be involved in other processes than the nuclear-cytoplasmic export of mRNAs [36, 37]. A mutant product that disrupts the system of macromolecular interactions can have a dominant negative effect, in this case heterozygosity at the null allele will be better than the presence of a mutant protein along with the normal product [38]. Among the mutant alleles of the *sbr (Dm nxf1)* gene with the dominant effects are the formation of three-pole spindles during the first meiotic division in sbr^5/+ females [36] and sterility of sbr^{12}/ *Dp (1;Y)y + v +* males. *Dp (1;Y)y + v +* is the duplication of the segment of the X chromosome with the *sbr+*, y + and v + alleles, translocated on the Y chromosome. The duplication compensates a lethal effect of the different alleles of the *sbr* gene localized on X-chromosome [39]. sbr^{10} is the thermosensitive allele with a block of the heat shock protein synthesis under heat shock (HS) [40]. It is a recessive feature of the sbr^{10} allele [38]. Adult males carrying the sbr^{10} allele are thermosensitive and die from 37°C, 1 h [8]. HS increased the number of abnormalities, of not only paternal but also maternal sex chromosome sets, in the offspring of the sbr^{10} males. Meiocytes were the thermosensitive stage for this unexpected effect [8].

It was interesting to know what feature is characterized the SBR protein distribution in the future gonads of *D. melanogaster* males on the pupal stage. On a pre-pupa stage of development, the future gonad in *D. melanogaster* males is spherical with two poles with dividing cells (**Figure 3**). One pole produces germ-line cells and another—somatic cells of a future gonad. Both poles are quite poor SBR protein. It is as a rule characteristic for SBR in a zone of dividing cells in different tissues. Most part of a future gonad is represented by cysts with 16 spermatocytes. Cysts from 16 meiocytes are outstanding by the highest content of the cytoplasmic SBR protein (**Figure 3B**).

As the growing spermatocytes are characterized by the high level of transcriptional activity, the long-lived RNAs seem to be the most appropriate candidate for the role of factors that, along with the spermatozoon, enter an ovum during fertilization and can participate in chromosome disjunction. Due to the process of translation, a small number of these molecules are sufficient for the regulation of chromosome segregation. This is particularly important during the first hours of embryonal development of *D. melanogaster*, as it takes place in the absence of transcriptional activity of the zygotic genome [41]. It has been demonstrated that along with the spermatozoon, paternal RNAs that may also play a role in fertilization, as well as zygotic and embryonic development, also enter the oocyte [42, 43]. High level of the SBR protein in cytoplasm shows that this protein may play a crucial role in biogenesis of the long-lived RNAs in *D. melanogaster*. The significance of SBR in forming the meiotic spindle in female *D. melanogaster* [36] and for the mitotic divisions in early embryos [44] leave unknown the molecular partners of the SBR protein involved in these processes.

One more target of *sbr*^{12} mutant alleles is the sperm flagellum [26, 45, 46]. The sperm axoneme as a main component of a flagellum defining mobility of a spermatozoon enters the ovum during fertilization [47, 48]. Axoneme is a derivative of a spermatid centroiole, which becomes the basis for the paternal centrosome formation, providing fusion of pronuclei and division of embryonic nuclei [49].

4. The centrosome for male meiosis and building of the axoneme

The centrosome plays a special role in spermatogenesis. It determines the polarity of the stem cells, maintaining contact with the hub and enabling the asymmetric division of stem cells [50]. Male spermatocytes of the wild type contain two centrosomes each. Each centrosome has two orthogonal centrioles that are 10 times larger than those in other cells [51]. Thus, during meiosis I, each of two centrosomes contains a pair of duplicated centrioles.

No centriole duplication occurs before meiosis II. Before chromatid segregation in secondary spermatocytes, each pair of centrioles divides into two single centrioles. Prior to chromatid segregation during meiosis II, one of the centrioles migrates to the opposite pole of a cell. As a result, each spermatozoon inherits only one centriole, which becomes the basal body forming the axoneme [1, 52]. Paternal centrosome participates in the formation of astral microtubules for moving male and female pronuclei toward each other [49]. The role of centrosome RNAs in the asymmetry distribution of cytoplasmic determinants among daughter cells [53] makes plausible the hypothesis that centrosome may be a carrier of the paternal factors, affecting the embryo development.

The axoneme growth during sperm maturation needs translation of the long-lived mRNAs coding the components of axonemal complex. Morphological defects of axoneme are a characteristic dominant manifestation of *sbr*^{12} mutant allele (**Figure 5**) [26, 27].

Translational control is crucial for morphogenical events that take place in the absence of transcription during spermiogenesis [26, 54]. The transcripts encoding proteins required for post-meiotic processes including spermiogenesis are almost all

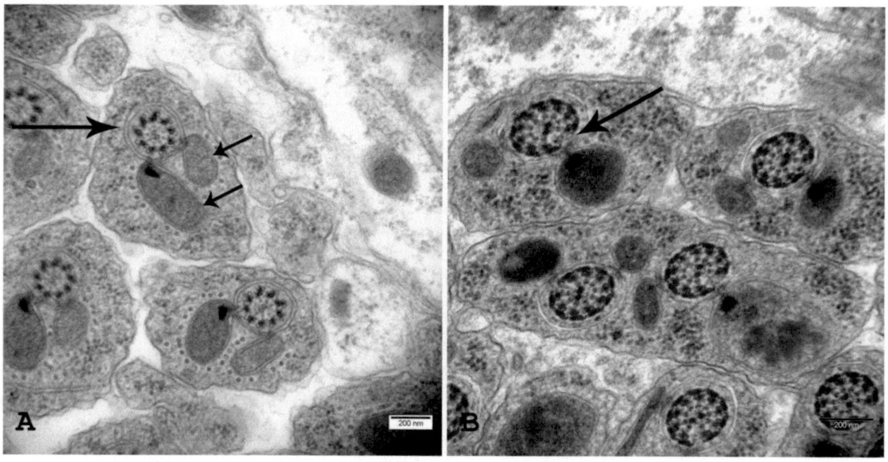

Figure 5.
Electron microscopy images of a cross section of 64 cell cysts. (A) Control (males of Oregon R*). (B) Morphological defects at the spermatid elongation stage in* D. melanogaster sbr12/Dp (1; Y) y + v + *males. Spermatids are often not subdivided into individual cells, and there are anomalies in axoneme structure (long arrow) and in mitochondrial derivatives (short arrow) (published in Mamon et al. [27]). Scale bar, 200 nm.*

abundantly transcribed in spermatocytes [5]. In *D. melanogaster*, elongation of the flagellar axoneme does not require the ubiquitous process of intraflagellar transport [55, 56]. The axoneme growth and spermatid elongation can be carried out at the expense of translation of mRNAs encoding tubulin subunits and other axonemal proteins within the ciliary pocket [25]. The genetic analysis of male fertility has identified numerous genes involved in spermiogenesis control [25, 57]. Spermatid individualization process depends on genes involved in RNAs metabolism [58].

In *sbr12/Dp (1;Y)y + v +* males, the morphology of mitochondrial derivatives and cytokinesis are defective in the elongated spermatids (**Figure 5**). Mitochondria morphogenesis depends on translation efficiency [54], also as a creation of new cellular membrane.

5. Incomplete cytokinesis providing a cellular communication as a feature of male generative cells

Syncytial development during the formation of male generative cells is conservative and is characteristic for different animals [59, 60]. It is assumed that the link between the cells in the syncytium is important primarily for synchronizing differentiation processes [59]. The cytoskeleton and the molecules providing a cellular communication are the main components supporting of equality of each cell in the syncytium.

The cytokinesis of spermatogenic cells is characterized by the formation of ring canals linking the cells that have undergone meiosis or mitosis [61]. During cell division, a contractile ring, made of actin and myosin II filaments, assembles as a result of the interaction between the plus ends of the microtubules and the cell cortex [62]. The contractile ring is double-sided. The microtubules whose minus ends directed toward centrosome on one pole of the dividing cell bind to one side of the contractile ring by their plus ends. The microtubules that have their minus ends turned toward centrosome on the other pole of the dividing cell bind to the opposite side of the contractile ring. Thus, the ring attached equatorially to the membrane of the dividing cell forms an actin-myosin cleavage furrow [60]. The ingression of the cleavage furrow is accompanied by the growth of the daughter cell membranes.

+TIPs—complexes linked to the plus-ends of the microtubules—play a special role in the interaction of the plus-ends of the microtubules with the cell cortex [63]. +TIPs also play an important role in finding and capturing microtubule targets—the cortex and the chromosomes [64, 65]. Localized mRNAs are anchored at the plus ends of microtubules especially in polarized cells [66]. SBR also may be one of the factors that interact with the plus-ends of microtubules.

It is worth noting that the Hs NXF1 (TAP) factor in humans was initially identified *not* as a protein enabling transport of mRNA from the nucleus into the cytoplasm, but as a factor important for cellular adhesion and involved in cell signaling [67]. For this reason, the protein Hs NXF1 was called TAP—Tip-associated protein, where Tip stands for tyrosine kinase-interacting protein.

Tyrosine phosphorylation is important for regulating the assembly and disassembly of the actin cytoskeleton at cell-cell junctions. In many cases, tyrosine phosphorylation in proteins of the junction complex—plakoglobin and β-catenin in adherens junctions—disrupts interactions between cytoskeleton and membrane [68]. It is believed [61] that tyrosine phosphorylation in targets leads to the disassociation of phosphorylated actin from the contractile ring that forms during the cytokinesis of the cells comprising the syncytium during spermatogenesis. This stabilizes the interaction between the walls of the ring and the cell membrane, ends cell division, and forms the ring canal [61]. Building a new cell wall at the cleavage site during cytokinesis requires the involvement of components of the cell membrane and the signaling molecules [69]. A system of transport molecules ensures the delivery of the necessary elements to the region where the cell membrane is being formed [70].

Interactions of NXF1 with cytoskeleton [27] and the cellular membrane [67] may help finding partners of NXF1 in the cytoplasm. Not surprisingly, the closest evolutionary relative of NXF1 is dynein. Dynein plays a significant role in cytokinesis, chromosome segregation, and in enabling the movement of the flagellum of the sperm [70, 71].

6. Homology of proteins of the NXF family and axonemal dyneins as a source of evolutionary relationship

Searching for related proteins and aligning sequences using the multiple sequence alignment tool base has demonstrated that among the proteins that are homologous to the factor NXF1, axonemal dynein of various organisms—especially its light chain—is most frequent. Aligning sequences that exhibit homology demonstrates that the N-terminus of the Dm NXF1 protein, which includes RBD domains and the LRR (leucine-rich repeats), corresponds to the axonemal light chain of dynein in vertebrates (**Figure 6**). At the LRR site, there is a sequence that exhibits a high degree of homology (marked by bold typeface in **Figure 6**), both when comparing different NXF factors (orthologous and paralogous), and comparing the SBR protein with sequences of light chains of dynein (**Figure 6**). The functions of this sequence are yet to be determined.

The dynein complex is a multicomponent system consisting of light, light intermediate, intermediate, and heavy chains of dynein [72]. Interacting with a multitude of protein and ribonucleic partners, the LC8 and TcTex1 light chains of dynein enable a link between different forms of the transported cargo and heavy chains of dynein that function as molecular motors [73, 74]. By interacting with its partner dynacin complex, the dynein complex enables cytoplasmic transport of macromolecular complexes, vesicles, and organelles [75], including particles containing RNAs [74, 76].

DOI: http://dx.doi.org/10.5772/intechopen.90917

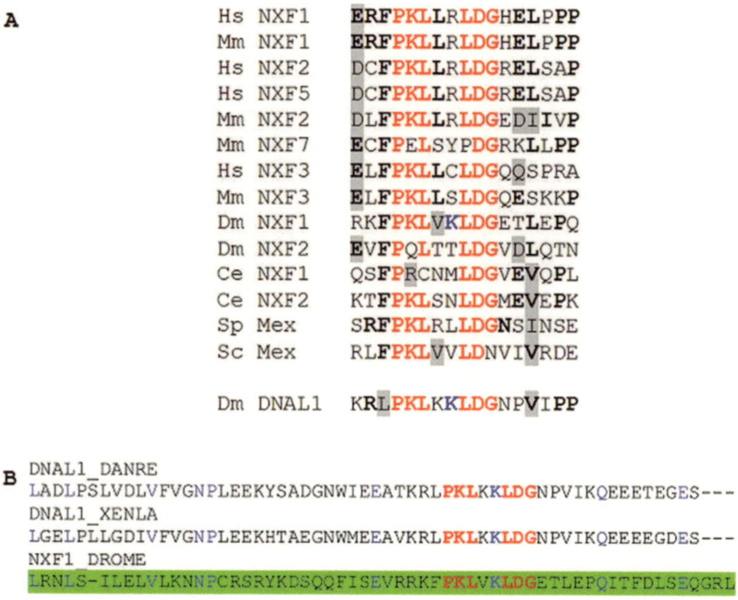

Figure 6.
Results of the alignment of the amino acid sequences of LRR domains of different NXF factors, including SBR, and dynein light chain fragment of D. melanogaster *(DNAL1 Dm) (A). The alignment of the amino acid sequences of dynein light chain fragments (DNAL1) of different organisms (*Danio rerio *– DANRE;* Xenopus laevis *– XENLA) and the LRR domain of SBR (NXF1 DROME) (B). Red letters denote identical amino acids; blue – identical amino acids in the dynein light chain and in the SBR. The bold type notes most often meeting amino acids in corresponding position in protein. Gray color notes interchangeable amino acids. Green color notes sequence of the LRR (leucine rich repeats) domain in SBR protein (http://www.uniprot.org/align/201202072EGVYDYGWG).*

In the fruit fly, the null mutants of the genes that encode the light or heavy chain of dynein have anomalies of wings and bristles, exhibit male and female sterility, and are characterized by disrupted sensory axon trajectories [71], thus demonstrating a similarity to the mutants of the *sbr* gene.

During spermatid individualization, the dynein light chains 1 (DLC1) participate in the assembly of the F-actin [77]. Dynein is necessary for centrosome separation, for forming the division spindle, and for aligning chromosomes equatorially during metaphase [78, 79]. During the interaction with the surface of microtubules, dynein also interacts with MAP proteins [80]. During prometaphase, the DYNLT3 light chain can be observed in kinetochores. During metaphase, dynein moves toward the spindle pole [81, 82]. The dynein complex leaves the kinetochores along with checkpoint proteins. This could be related to the transport of checkpoint proteins (for example, Bub3) from kinetochores to spindle poles [83]. If we recall the homology of the proteins Bub3 and Rae1 [84] and the fact that the protein Rae1, found in RNP complexes that are important for forming the spindle, is capable of direct interaction with NXF1 [85], it is possible to consider these factors as polyfunctional. They are taking part both in nuclear-cytoplasmic transport of macromolecules and in the formation of the mitotic apparatus of the cell.

Because it is the light chain of dynein in the dynein complex that is responsible for interaction with the cargo [72], this resemblance to the N-terminus of the SBR protein suggests that SBR may have the same partners as a dynein complex (that these proteins can have the general partners or the general functions, and they are evolutionary conservative).

7. Conclusion

In *D. melanogaster*, the evolutionarily conservative *sbr (Dm nxf1)* gene is characterized by the testis-specific gene products. The cytoplasmic localization of the SBR protein in meiotic cysts is quite impressive. Some *sbr* mutant alleles cause male sterility and the specific morphological defects of the spermatids at the individualization stage. Transcriptional activity during pre-meiotic and meiotic stages of spermatogenesis is crucial for the sperm motility in *D. melanogaster*. During these stages, the factors are formed that are able to affect the frequency of aneuploidy arising in the offspring after fertilization. The *sbr* gene is an attractive model for investigation of male sterility and the paternal effect on the embryonic development due to the RNA-binding protein SBR involvement in formation and fate of the long-lived RNAs.

Acknowledgements

We are grateful to the staff of KHROMAS (the Center of study of ultrastructure and molecular composition of biological objects at Biological Faculty of St. Petersburg State University) and to its head Elena R. Gaginskaya for their most important assistance in performance of this work.

Funding

The Russian Foundation for Basic Research supported the work (project 19-04-01255).

Declaration of interest

The authors declare that there is no conflict of interest that could be perceived as prejudicing the impartiality of this review.

Author details

Elena Golubkova*, Anna Atsapkina, Anna K'ergaard and Ludmila Mamon
Department of Genetics, St. Petersburg State University, Saint-Petersburg, Russia

*Address all correspondence to: e.golubkova@spbu.ru;
elena_golubkova@mail.ru

References

[1] Fuller MT. Spermatogenesis. In: Bate M, Martinez-Arias A, editors. The Development of *Drosophila melanogaster*. New York: Cold Spring Harbor Laboratory Press; 1993. pp. 71-147. 1564 p

[2] Fuller MT, Spradling AC. Male and female *Drosophila* germline stem cells: Two versions of immortality. Science. 2007;**316**:402-404

[3] Riparbelli MG, Colozza G, Callaini G. Procentriole elongation and recruitment of pericenriolar material and downregulated in cyst cells as they enter quiescence. Journal of Cell Science. 2009;**122**: 3613-3618

[4] Cenci G, Bonaccorsi S, Pisano C, Verni F, Gatti M. Chromatin and microtubule organization during premeiotic, meiotic and early postmeiotic stages of *Drosophila melanogaster* spermatogenesis. Journal of Cell Science. 1994;**107**:3521-3534

[5] Schäfer M, Nayernia K, Engel W, Schäfer U. Translational control in spermatogenesis. Developmental Biology. 1995;**172**:344-352

[6] White-Cooper H. Molecular mechanisms of gene regulation during *Drosophila* stermatogenesis. Reproduction. 2010;**139**:11-21

[7] White-Cooper H, Davidson I. Unique aspects of transcription regulation in male germ cells. Cold Spring Harbor Perspectives in Biology. 2011;**3**(7):a002626

[8] K'ergaard AV, Mamon LA. Hyperthermia of male *Drosophila melanogaster* meiocytes induces abnormalities in both paternal and maternal sex chromosome sets of the offspring. Russian Journal of Genetics. 2007;**43**:1153-1160

[9] Vibranovski MD, Chalopin DS, Lopes HF, Long M, Karr TL. Direct evidence for postmeiotic transcription during *Drosophila melanogaster* spermatogenesis. Genetics. 2010;**186**:431-433

[10] Fabrizio JJ, Hime G, Lemmon SK, Bazinet C. Genetic dissection of sperm individualization in *Drosophila melanogaster*. Development. 1998;**125**:1833-1843

[11] Noguchi T, Miller KG. A role for actin dynamics in individualization during spermatogenesis in *Drosophila melanogaster*. Development. 2003;**130**:1805-1816

[12] Barreau C, Benson E, Gudman nsdottir E, Newton F, White-Cooper H. Post-meiotic transcription in *Drosophila* testes. Development. 2008;**135**: 1897-1902

[13] Herold A, Klymenko T, Izaurralde E. NXF1/p15 heterodimers are essential for mRNA nuclear export in *Drosophila*. RNA. 2001;**7**:1768-1780

[14] Yang J, Bogerd HP, Wang PJ, Page DC, Cullen BR. Two closely related human nuclear export factors utilize entirely distinct export pathways. Molecular Cell. 2001;**8**:397-406

[15] Sasaki M, Takeda E, Takano K, Yomogida K, Katahira J, Yoneda Y. Molecular cloning and functional characterization of mouse *Nxf* family gene products. Genomics. 2005;**85**:641-653

[16] Zhang M, Wang Q, Huang Y. Fragile X mental retardation protein FMRP and the RNA export factor NXF2 associate with and destabilize *Nxf1* mRNA in neuronal cells. Proceedings of the National Academy of Sciences of the United States of America. 2007;**104**:10057-10062

[17] Pan J, Eckardt S, Leu NA, Buffone MG, Zhou J, Gerton GL, et al. Inactivation of *Nxf2* causes defects in male meiosis and age-dependent depletion of spermatogonia. Developmental Biology. 2009;**330**:167-174

[18] Katahira J, Strasser K, Podtelejnikov A, Mann M, Jung JU, Hurt E. The Mex67 pmediated nuclear mRNA export pathway is conserved from yeast to human. The EMBO Journal. 1999;**18**:2593-2609

[19] Bachi A, Braun IC, Rodrigues JP, Pante N, Ribbeck K, von Kobbe C. The C-terminal domain of TAP interacts with the nuclear pore complex and promotes export of specific CTE-bearing RNA substrates. RNA. 2000;**6**:136-158

[20] Herold A, Suyama M, Rodrigues JP, Braun IC, Kutay U, Carmo-Fonseca M, et al. TAP (NXF1) belongs to a multigene family of putative RNA export factors with a conserved modular architecture. Molecular and Cellular Biology. 2000;**20**:8996-9008

[21] Braun IC, Herold A, Rode M, Conti E, Izaurralde E. Overexpression of TAP/p15 heterodimers by passes nuclear retention and stimulates nuclear mRNA export. The Journal of Biological Chemistry. 2001;**276**:20536-20543

[22] Ivankova N, Tretyakova I, Lyozin G, Avanesyan E, Zolotukhin A, Zatsepina OG, et al. Alternative transcripts expressed by *small bristles,* the *Drosophila melanogaster nxf1* gene. Gene. 2010;**458**:11-19

[23] Ginanova V, Golubkova E, Kliver S, Bychkova E, Markoska K, Ivankova N, et al. Testis-specific products of the *Drosophila melanogaster sbr* gene, encoding nuclear export factor 1, are necessary for male fertility. Gene. 2016;**577**:153-160

[24] Mamon L, Ginanova V, Kliver S, Toropko M, Golubkova E. Organ-specific transcripts as a source of gene multifunctionality: Lessons learned from the *Drosophila melanogaster sbr* (*Dm nxf1*) gene. Biomedical Communications. 2019;**62**(2):146-157

[25] Fabian L, Brill JA. *Drosophila* spermatogenesis big things come from little packages. Spermatogenesis. 2012;**2**(3):197-212

[26] Atsapkina AA, Golubkova EV, Kasatkina VV, Avanesyan EO, Ivankova NA, Mamon LA. Peculiarities of spermatogenesis in *Drosophila melanogaster*: Role of main transport receptor of mRNA (Dm NXF1). Cell and Tissue Biology. 2010;**4**(5):429-435

[27] Mamon LA, Ginanova VR, Kliver SF, Yakimova AO, Atsapkina AA, Golubkova EV. RNA-binding proteins of the NXF (nuclear export factor) family and their connection with the cytoskeleton. Cytoskeleton. 2017;**74**:161-169

[28] Zhong L, Belote JM. The testis-specific proteasome subunit Prosα6T of *D. melanogaster* is required for individualization and nuclear maturation during spermatogenesis. Development. 2007;**134**:3517-3525

[29] Suyama M, Doerks T, Braun IC, Sattler M, Izaurralde E, Bork P. Prediction of structural domains of TAP reveals details of its interaction with p15 and nucleoporins. EMBO Reports. 2000;**1**(1):53-58

[30] Madsen L, Schulze A, Seeger M, Hartmann-Petersen R. Ubiquitin domain proteins in disease. BMC Biochemistry. 2007;**8**:S1

[31] Jesenberger V, Jentsch S. Deadly encounter: Ubiquitin meets apoptosis. Nature Reviews Molecular Cell Biology. 2002;**3**:112-121

[32] Kleene KC. Sexual selection, genetic conflict, selfish genes and the atypical patterns of gene expression in spermatogenic cells. Developmental Biology. 2005;**277**:16-26

[33] Kleene KC. A possible meiotic function of the peculiar patterns of gene expression in mammalian spermatogenic cells. Mechanisms of Development. 2001;**106**:3-23

[34] Steger K. Transcriptional and translational regulation of gene expression in haploid spermatids. Anatomy and Embryology. 1999;**199**:471-487

[35] FlyBase. 2019. Available from: https://flybase.org/reports/FBgn0003321

[36] Golubkova EV, Markova EG, Markov AV, Avanesyan EO, Nokkala S, Mamon LA. *Dm nxf1/sbr* gene affects the formation of meiotic spindle in female *Drosophila melanogaster*. Chromosome Research. 2009;**17**:833-845

[37] Kopytova D, Popova V, Kurshakova M, Shidlovskii Y, Nabirochkina E, Brechalov A, et al. ORC interacts with THSC/TREX-2 and its subunits promote Nxf1 association with mRNP and mRNA export in *Drosophila*. Nucleic Acids Research. 2016;**44**:4920-4933

[38] Nikitina EA, Komarova AV, Golubkova EV, Tretyakova IV, Mamon LA. Dominant effects of *l(1)ts403* (*sbr*^{10}) mutation at the disjunction of sex chromosomes in meiosis of *Drosophila melanogaster* females after heat shock. Russian Journal of Genetics. 2003;**39**:269-275

[39] Zhimulev IF, Belyaeva ES, Pokholkova GV, Kotchneva GV, Fomina OV, Bgatov AV, et al. Report of new mutants. Drosophila Information Service. 1982;**58**:210-214

[40] Evgen'ev M, Levin A, Lozovskaya E. The analysis of a temperature-sensitive (ts) mutation influencing the expression of heat shock-inducible genes in *Drosophila melanogaster*. Molecular & General Genetics. 1979;**176**(2):275-280

[41] Edgar BA, Schubiger G. Parameters controlling transcriptional activation during early *Drosophila* development. Cell. 1986;**44**:871-877

[42] Ostermeier GC, Miller D, Huntriss JD, Diamond MP, Krawetz SA. Delivering spermatozoan RNA to the oocyte. Nature. 2004;**429**:154

[43] Miller D. Sperm RNA as a mediator of genomic plasticity. Advances in Biology. 2014;**2014**:179701

[44] Golubkova EV, Nokkala S, Mamon LA. The nuclear export factor gene *small bristles* is involved in the control of early embryonic mitoses in *Drosophila melanogaster*. Drosophila Information Service. 2006;**89**:31-39

[45] Geer BW, Lischwe TD, Murphy KG. Male fertility in *Drosophila melanogaster*: Genetics of the vermilion region. The Journal of Experimental Zoology. 1983;**225**:107-118

[46] Dybas LK, Harden KK, Machnicki JL, Geer BW. Male fertility in *Drosophila melanogaster*: Lesions of spermatogenesis associated with male sterile mutations of the vermilion region. The Journal of Experimental Zoology. 1983;**226**:293-302

[47] Karr T. Intracellular sperm/egg interactions in *Drosophila*: A three-dimensional structural analysis of a paternal product in the developing egg. Mechanisms of Development. 1991;**34**(2-3):101-111

[48] Loppin B, Dubruille R, Horard B. The intimate genetics of *Drosophila* fertilization. Open Biology. 2015;**5**:150076

[49] Callaini G, Riparbelli MG. Fertilization in *Drosophila melanogaster*

centrosome inheritance and organization of the first mitotic spindle. Developmental Biology. 1996;**176**:199-208

[50] Yamashita YM, Mahowald AP, Perlin JR, Fuller MT. Asymmetric inheritance of mother versus daughter centrosome in stem cell division. Science. 2007;**315**:518-521

[51] Gonzalez C, Tavosanis G, Mollinari C. Centrosomes and microtubule organization during *Drosophila* development. Journal of Cell Science. 1998;**111**:2697-2706

[52] Baker JD, Adhikarakunnathu S, Kernan MJ. Mechanosensory-defective, male sterile *unc* mutants identify a novel basal body protein required for cillogenesis in *Drosophila*. Development. 2004;**131**:3411-3422

[53] Lambert JD, Nagy LM. Asymmetric inheritance of centrosomally localized mRNA during embryonic cleavages. Nature. 2002;**420**:682-686

[54] Baker CC, Fuller MT. Translational control of meiotic cell cycle progression and spermatid differentiation in male germ cells by a novel eIF4G homolog. Development. 2007;**134**(15):2863-2869

[55] Han YG, Kwok BH, Kernan MJ. Intraflagellar transport is required in *Drosophila* to differentiate sensory cilia but not sperm. Current Biology. 2003;**13**(19):1679-1686

[56] Sarpal R, Todi SV, Sivan-Loukianova E, Shirolikar S, Subramanian N, Raff EC, et al. *Drosophila* KAP interacts with the kinesin II motor subunit KLP64D to assemble chordotonal sensory cilia, but not sperm tails. Current Biology. 2003;**13**(19):1687-1696

[57] Wakimoto BT, Lindsley DL, Herrera C. Toward a comprehensive genetic analysis of male fertility in *Drosophila melanogaster*. Genetics. 2004;**167**(1):207-216

[58] Sanders C, Smith DP. LUMP is a putative double-stranded RNA binding protein required for male fertility in *Drosophila melanogaster*. PLoS One. 2011;**6**(8):e24151

[59] Fawcett DW. Intercellular bridges. Experimental Cell Research. Supplement. 1961;**8**:174-187

[60] Giansanti MG, Bonnaccorsi S, Bucciarelli E, Gatti M. *Drosophila* male meiosis as a model system for the study of cytokinesis in animal cells. Cell Structure and Function. 2001;**26**:609-617

[61] Hime GR, Brill JA, Fuller MT. Assembly of ring canals in the male germ line from structural components of contractile ring. Journal of Cell Science. 1996;**109**:2779-2788

[62] D'Avino PP, Savoian MS, Glover DM. Cleavage furrow formation and ingression during animal cytokinesis: A microtubule legacy. Journal of Cell Science. 2005;**118**:1549-1558

[63] Howard J, Hyman AA. Dynamics and mechanics of the microtubule plus end. Nature. 2003;**422**:753-758

[64] Schuyler SC, Pellman D. Microtubule "plus-end-tracking proteins": The end is just the beginning. Cell. 2001;**105**:421-424

[65] Mimori-Kiyosue Y, Tsukita S. "Search-and-capture" of microtubules through plus-endbinding proteins (+TIPs). Journal of Biochemistry. 2003;**134**:321-326

[66] Mili S, Macara IG. RNA localization and polarity: From a(PC) to Z(BP). Trends in Cell Biology. 2009;**19**(4):156-164

[67] Yoon DW, Lee H, Seol W, DeMaria M, Rosenzweig M, Jung JU. Tap: A novel cellular protein that

interacts with tip of herpesvirus saimiri and induces lymphocyte aggregation. Immunity. 1997;**6**:571-582

[68] Cowin P, Burke B. Cytoskeleton-membrane interactions. Current Opinion in Cell Biology. 1996;**8**:56-65

[69] Campbell KS, Cooper S, Dessing M, Yates S, Buder A. Interaction of *p59*^{fyn} kinase with dynein light chain, Tctex-1, and colocalization during cytokinesis. Journal of Immunology. 1998;**161**:1728-1737

[70] Ai E, Skop AR. Endosomal recycling regulation during cytokinesis. Communicative & Integrative Biology. 2009;**2**:444-447

[71] Gepner J, Li M-G, Ludmann S, Kortas C, Boylan K, Iyadurai SJP, et al. Cytoplasmic dynein function is essential in *Drosophila melanogaster*. Genetics. 1996;**142**:865-878

[72] Vallee RB, Williams JC, Varma D, Barnhart LE. Dynein: An ancient motor protein involved in multiple modes of transport. Journal of Neurobiology. 2004;**58**:189-200

[73] Tai AW, Chuang JZ, Sung CH. Cytoplasmic dynein regulation by subunit heterogeneity and its role in apical transport. The Journal of Cell Biology. 2001;**153**:1499-1509

[74] Williams JC, Roulhac PL, Roy AG, Vallee RB, Fitzgerald MC, Hendrickson WA. Structural and thermodynamic characterization of a cytoplasmic dynein light chain – Intermediate chain complex. Proceedings of the National Academy of Sciences of the United States of America. 2007;**104**:10028-10033

[75] Puls I, Jonnakuty C, LaMonte BH, Holzbaur EL, Tokito M, Mann E, et al. Mutant dynactin in motor neuron disease. Nature Genetics. 2003;**33**:455-456

[76] Carson JH, Cui H, Barbarese E. The balance of power in RNA trafficking. Current Opinion in Neurobiology. 2001;**11**:558-563

[77] Ghosh-Roy A, Desai BS, Ray K. Dynein light chain 1 regulates Dynamin mediated Factin assembly during sperm individualization in *Drosophila*. Molecular Biology of the Cell. 2005;**16**:3107-3116

[78] Echeverri CJ, Paschal BM, Vaughan KT, Vallee RB. Molecular characterization of the 50-kD subunit of dynactin reveals function for the complex in chromosome alignment and spindle organization during mitosis. The Journal of Cell Biology. 1996;**132**:617-633

[79] Robinson JT, Wojcik EJ, Sanders MA, McGrail M, Hays TS. Cytoplasmic dynein is required for the nuclear attachment and migration of centrosomes during mitosis in *Drosophila*. The Journal of Cell Biology. 1999;**146**:597-608

[80] Dixit R, Ross JL, Goldman YE, Holzbaur ELF. Differential regulation of dynein and kinesin motor proteins by tau. Science. 2008;**319**: 1086-1089

[81] Steuer ER, Wordeman L, Schroer TA, Sheetz MP. Localization of cytoplasmic dynein to mitotic spindles and kinetochores. Nature. 1990;**345**:266-268

[82] Pfarr CM, Coue M, Grissom PM, Hays TS, Porter ME, McIntosh JR. Cytoplasmic dynein is localized to kinetochores during mitosis. Nature. 1990;**345**:263-265

[83] Lo KW-H, Kogoy JM, Pfister KK. The DYNLT3 light chain directly links cytoplasmic dynein to a spindle checkpoint protein, Bub3. The Journal of Biological Chemistry. 2007;**282**:11205-11212

[84] Babu JR, Jeganathan KB, Baker DJ, Wu X, Kang-Decker N, van Deursen JM. Rae1 is an essential mitotic checkpoint regulator that cooperates with Bub3 to prevent chromosome missegregation. The Journal of Cell Biology. 2003;**160**:341-353

[85] Blevins MB, Smith AM, Phillips EM, Powers MA. Complex formation among the RNA export proteins Nup98, Rae1/Gle2, and TAP. The Journal of Biological Chemistry. 2003;**278**:20979-20988

Chapter 10

A Porcine Model of Neonatal Hypoxia-Asphyxia to Study Resuscitation Techniques in Newborn Infants

Megan O'Reilly, Po-Yin Cheung, Tze-Fun Lee and Georg M. Schmölzer

Abstract

Two to three million newborn infants worldwide need extensive cardiopulmonary resuscitation (CPR), and approximately one million of these infants die annually worldwide. Therefore, resuscitation techniques require further refinement to provide better outcomes. To investigate the effectiveness of various interventions and to understand the pathophysiology and pharmacology of neonatal CPR, it is important to have animal models that reliably reproduce features observed in neonates who require resuscitation. Herein, we describe an experimental animal model in newborn piglets that simulates neonatal asphyxia and enables us to examine resuscitation interventions, reoxygenation, and recovery processes. The newborn piglet has several advantages including similar development to a human fetus at 36–38 week's gestation, and comparable body systems and body size, allowing for surgical instrumentation, monitoring, and collection of biological samples. Furthermore, using this model of neonatal asphyxia, we are also able to describe an increasingly important clinical situation in the laboratory setting—pulseless electrical activity (PEA). Since the integration of electrocardiogram into the neonatal resuscitation guidelines, there has been an increased awareness of PEA in newborn infants. The animal model we describe can therefore serve as a valuable tool to bridge the knowledge gap and improve the outcome of asphyxiated newborns in the delivery room.

Keywords: infants, newborn, neonatal resuscitation, asphyxia, heart rate, electrocardiography, auscultation

1. Introduction

Most newborn infants successfully transition from fetal to neonatal life without any help [1]. However, approximately 10–20% of newborns (13–26 million worldwide) need some degree of respiratory support at birth [2–4], which remains the most critical step of neonatal resuscitation. Furthermore, an estimated 0.1% of term infants and up to 15% of preterm infants (2–3 million worldwide) requires extensive cardiopulmonary resuscitation (CPR) at birth,

IntechOpen

which entails chest compressions (CC) and 100% oxygen with or without administration of epinephrine [5–9]. Despite receiving CPR, approximately 1 million newborns die annually worldwide. Even with successful resuscitation, infants receiving extensive CPR in the delivery room have a high incidence of mortality (40–80%) and neurologic morbidity (e.g. 57% hypoxic–ischemic encephalopathy and seizures) [5, 6, 9]. Therefore, resuscitation techniques require further refinement to provide better outcomes. The guidelines for neonatal resuscitation recommended by the American Academy of Pediatrics/American Heart Association Neonatal Resuscitation Program [2–4] are based, in part, on the recognition that the cause of cardiovascular collapse in most newborns is asphyxia. However, in many cases the guidelines rely on data from studies in the adult population and extrapolate it to the neonatal population. Such data may not be entirely applicable to the neonatal population, because the most common cause of cardiovascular collapse in the adult population is primary cardiac compromise/ventricular fibrillation, not asphyxia. Therefore it is imperative that pre-clinical studies with appropriate animal models are carried out to determine the optimal resuscitation techniques before they are translated into the delivery room for newborn infants.

2. Asphyxia at birth

Asphyxia at birth, also known as perinatal asphyxia, is the most common reason that newborn infants fail to make a successful transition to ex-utero life [10]. Asphyxia may occur from several perinatal events, such as failure of placental gas exchange prior to delivery (e.g. placental abruption, uterine rupture, umbilical cord prolapse, chorioamnionitis), or deficient pulmonary gas exchange immediately after birth (e.g. apnea, airway obstruction, respiratory distress syndrome) [10]. Asphyxia is a condition of impaired gas exchange with simultaneous hypoxia and hypercapnia, leading to a mixed metabolic and respiratory acidosis [10]; it depresses myocardial function leading to cardiogenic shock, pulmonary hypertension, mesenteric reperfusion, acute renal failure, and ultimately cardiac arrest. The cascade of hypoxic–ischemic insults results in dysfunction of one or more organ systems in over 80% of asphyxiated newborn infants [11], leading to significant mortality and long-term morbidity. Newborns affected by perinatal asphyxia often present with an inadequate heart rate that does not respond to positive pressure ventilation (PPV). This is due to depressed myocardial function, vasodilation, and very low diastolic blood pressures through which the heart is unable to efficiently contract. Ineffective pumping of enough blood to the lungs inhibits the exchange and consumption of oxygen that is being delivered via PPV [10]. This inevitably leads to the need for CC to mechanically pump the blood through the heart until the myocardium is adequately oxygenated to resume spontaneous contraction and blood circulation [10].

3. Current neonatal resuscitation guidelines

Heart rate (HR) is the most important clinical indicator to evaluate the status of compromised newborns and to guide resuscitation efforts in the delivery room [3]. An increase in the newborn's HR remains the most reliable indicator of adequate ventilation [3]. Until recently, HR assessment in the newborn was achieved via (i) palpation of the umbilical cord, (ii) auscultation of the precordium, and/or

(iii) pulse oximetry [12]. In 2015, the neonatal resuscitation guidelines were updated to integrate the use of electrocardiography (ECG) as a tool for HR assessment immediately after birth [2–4]. This recommendation was based on observational data and small randomized controlled trials showing that ECG displays reliable HR faster than pulse oximetry [2–4]. However, the use of ECG does not replace the need for pulse oximetry, but rather compliment it.

The current neonatal resuscitation guidelines recommend initiation of PPV if the HR is below 100 beats per minute (bpm). If HR does not increase in response to PPV, several ventilation corrective steps are recommended: (i) check the seal of the face mask, (ii) reposition the neonate's head in "sniffing" position, (iii) suction obstructing secretions, (iv) open the mouth to decrease resistance to gas flow, (v) increase the peak inflating pressure, and (vi) establish an advanced airway (intubate or use a laryngeal mask device). If the above ventilation corrective steps fail to improve HR and it decreases to below 60 bpm, CC and 100% oxygen are recommended. If HR persists below 60 bpm despite CC and 100% oxygen, administration of intravenous epinephrine is recommended at a dose of 0.01–0.03 mg/kg. If epinephrine administration is required prior to the establishment of intravenous access, it can be administered endotracheally at a higher dose of 0.05–0.1 mg/kg. The currently recommended technique of delivering CC to a neonate is using a coordinated 3:1 compression-to-ventilation ratio (3:1 C:V). This approach is comprised of 90 CC and 30 ventilation inflations per minute, with a pause after every third CC to deliver one effective ventilation. This technique achieves approximately 120 events per minute. The CC are delivered on the lower third of the sternum and to a depth of approximately one-third of the anterior–posterior chest diameter, and the 2-thumb-encircling hands technique is the preferred method. However the chest compression ratio recommendation is based more on expert opinion and consensus rather than strong scientific evidence, since there is currently very-low-quality evidence to suggest otherwise, according to the guidelines [4].

4. Pulseless electrical activity

Pulseless electrical activity (PEA) is a form of cardiac arrest characterized by cardiac electrical activity, detected by ECG, but with the absence of a detectable pulse [13]. Although PEA is a commonly seen cardiac rhythm in adult and pediatric resuscitations (referred to as a "nonshockable rhythm"), occurring in approximately 32 and 24% of cardiac arrests, respectively [14], its occurrence in neonatal resuscitation is unusual and not widely recognized. In response to the inclusion of ECG in the 2015 neonatal resuscitation guidelines, there has been a rise in awareness of PEA in neonatal resuscitation. ECG displaying PEA could falsely mislead health care providers into overestimating the HR and delay necessary resuscitation techniques. It is possible that PEA may be common in asphyxiated newborns but has been undetected in the clinical setting prior to the recent use of ECG in the delivery room. Recent case reports have raised concerns over the reliability of ECG use during neonatal resuscitation, and the detection of PEA has been cited as a potential limitation of ECG use to guide delivery room resuscitation [15, 16]. Data from studies in the pediatric population indicate decreased survival following resuscitation with PEA events [17, 18], however this is inconsistent throughout the literature. Recent case studies in newborn infants presenting with cardiac arrest with PEA rhythm, as indicated on ECG [15, 16], suggest dire outcomes. Further studies (animal and prospective clinical) are needed to determine the cause and actual incidence of PEA in order to improve the survival of newborns experiencing PEA in the delivery room.

5. Porcine model of hypoxia-asphyxia

Herein we describe an experimental animal model from our research laboratory in newborn piglets that simulates neonatal hypoxia-asphyxia. Using this animal model, we are able to examine the systemic and regional hemodynamic changes during hypoxia-asphyxia, resuscitation interventions, reoxygenation, and the recovery process. The described experimental animal model is a non-survival acute procedure in neonatal pigs, aged between 1 and 3 days old and weighing approximately 1.5–2 kg. The approximate duration of the procedure is between six to eight hours and can be divided into the following sections: (i) anesthesia and surgical instrumentation, (ii) monitoring and stabilization, (iii) hypoxia and asphyxia, (iv) resuscitation intervention, and (v) reoxygenation and recovery.

5.1 Anesthesia and surgical instrumentation

Surgical procedures enable the establishment of mechanical ventilation, arterial and central venous access, and placement of catheters and flow probes for continuously monitoring intravascular pressures and blood flow across the common carotid artery, respectively. Anesthesia is induced using 5% inhaled isoflurane in 100% oxygen (delivered via a nose cone), and is then maintained at 2–3% with fine adjustment by 0.5% as appropriate, depending on the condition of the piglet.

Following induction of anesthesia, an incision is made in the right groin and the right femoral artery and vein are exposed. An area of approximately 1 cm is dissected around each of the vessels, which are then isolated by threading two lengths of suture ties under each vessel. The vessel is ligated distally using a suture tie, a small cut is made in the vessel wall, and then an Argyle catheter (3.5 or 5 French, Covidien, Mansfield, MA) is inserted into the vessel. A double-lumen catheter is used for the femoral vein and is inserted to 15 cm so it is placed close to the right atrium. The venous catheter can be used for medication and maintenance fluid infusions as well as continuous central venous pressure measurement. A single-lumen catheter is used for the femoral artery and is inserted to 5 cm so it is placed at the infra-renal aorta. The single-lumen arterial catheter can be used for continuous mean arterial pressure measurement and blood sampling. The groin incision is then sutured closed. Once vascular access has been established, the inhaled anesthesia can be switched to intravenous anesthesia using morphine and propofol infusions via the venous catheter. This is done after the piglet has been connected to the ventilator machine (see below).

The piglet is then intubated via tracheostomy. A horizontal incision is made at the neck, the trachea is dissected and exposed, and two lengths of suture ties are threaded around the trachea. An endotracheal tube (3.0 or 3.5) is inserted, connected to a ventilator and pressure-controlled ventilation (Acutronic Fabian HFO; Hirzel, Switzerland) is commenced at a respiratory rate of 16–20 breaths/min and pressures of 20/5 cm H_2O. Oxygen saturation is kept within 90–100% by adjusting the fraction of inspired oxygen between 21 and 30%.

The right common carotid artery is dissected and exposed, and one length of suture tie is threaded around to isolate the artery. A real-time ultrasonic flow probe (2 mm; Transonic Systems, Ithaca, New York, USA) is placed around the artery and secured, and ultrasonic gel is placed between the flow probe and artery to allow for optimal signal transduction. The flow probe provides continuous carotid blood flow (CBF) measurement. The neck incision is sutured closed. **Figure 1** shows the surgical instrumentation of the piglet.

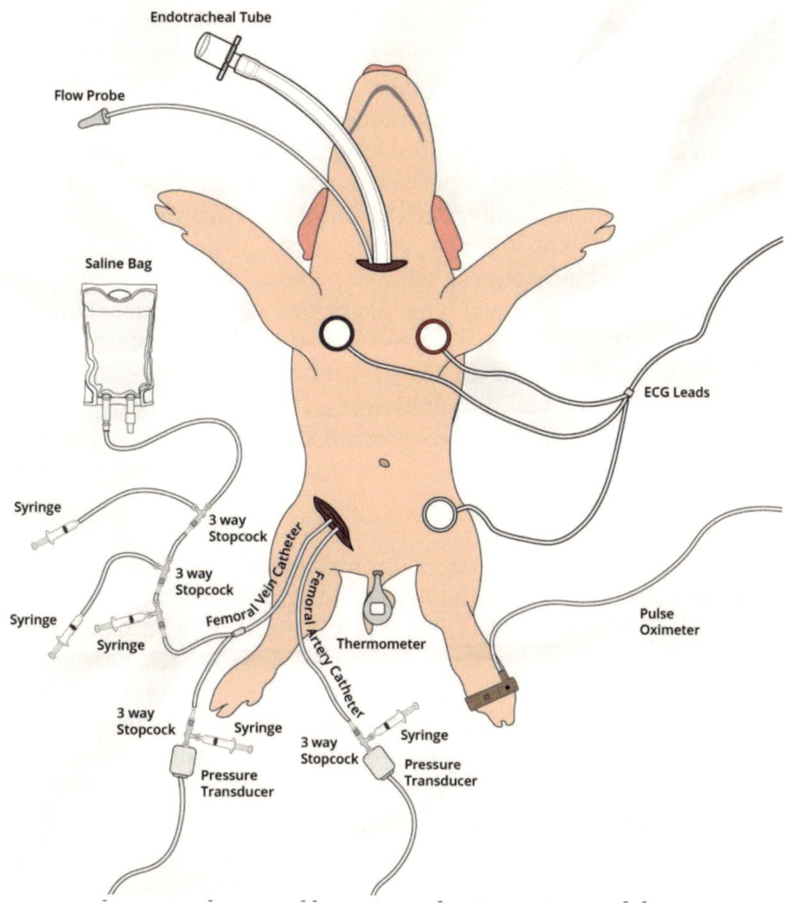

Figure 1.
Schematic of neonatal hypoxia-asphyxia porcine model (copyright https://www.playretain.com).

5.2 Monitoring and stabilization

A pulse oximeter is placed on the piglet's left hind limb for measuring percutaneous oxygen saturation. Continuous monitoring of the HR is achieved by attaching a 3-lead ECG to the piglet's skin (continuously measured and recorded with Hewlett Packard 78833B monitor, Hewlett Packard, Palo Alto, California, USA). Generally, baseline HR is between 150 and 200 bpm. Glucose level and hydration is maintained with an intravenous infusion of 5% dextrose at 10 mL/kg/hour. The piglet's body temperature is maintained at 38–40°C using an overhead warmer and a heating pad. During the experiment, anesthesia is maintained with intravenous propofol (5-10 mg/kg/hour) and morphine (0.1 mg/kg/hour). Additional doses of propofol (1–2 mg/kg) and morphine (0.05–0.1 mg/kg) are given as needed. The anesthetic state of the piglet is regularly monitored throughout the entire experiment using various criteria: neurological (body movements), behavioral (agitation), cardiovascular (tachycardia and hypertension), and respiratory (tachypnoea). The piglet is allowed to stabilize for 1 hour post surgery before the hypoxia-asphyxia protocol is commenced. **Figure 1** shows the placement of the monitoring devices on the piglet's body.

5.3 Hypoxia and asphyxia

The piglet is exposed to severe hypoxemia, which is induced via 30–60 min of normocapnic alveolar hypoxia. The piglet is ventilated with low inspired oxygen concentration delivered by increasing the inhaled concentration of nitrogen gas to induce hypoxemia. The inspired oxygen concentration is adjusted between 10 and 15% to obtain arterial oxygen saturations (SaO_2) of 30–40% and partial pressure of oxygen (PaO_2) of 20–40 mmHg. Arterial blood sampling is conducted to assess the partial pressure of carbon dioxide ($PaCO_2$) and the ventilator rate is then adjusted accordingly.

Hypoxia is followed by asphyxia, which is achieved by disconnecting the ventilator and clamping the endotracheal tube. Asphyxia can be conducted until either bradycardia, asystole or PEA (cardiac arrest). In this experimental animal model, bradycardia is defined as 25% of baseline heart rate, and asystole or PEA is defined as zero CBF and confirmed by auscultation of no HR. Following hypoxia-asphyxia, the resuscitation intervention protocol is commenced.

5.4 Resuscitation intervention

The primary goal of this experimental animal model is to provide a platform to investigate various resuscitation interventions in a pre-clinical scenario. Although the exact details of resuscitation interventions vary, they are predominantly comprised of the following elements: PPV (performed with a Neopuff T-Piece, Fisher and Paykel, Auckland, New Zealand), CC, ventilations, oxygen, and epinephrine administration. The ultimate outcome of the resuscitation intervention is to achieve return of spontaneous circulation (ROSC) in a timely manner, defined as an unassisted HR $\geq$100 bpm for at least 15 s. Section 8 summarizes the various resuscitation interventions published from our research group using this experimental animal model.

5.5 Reoxygenation and recovery

Following the resuscitation intervention and ROSC, the piglet is then reconnected to the ventilator with 100% oxygen briefly, and weaned down to 21% oxygen for the 4-hour post-resuscitation recovery period. At the end of the recovery period, the piglet is euthanized with an intravenous overdose of sodium pentobarbital (120 mg/kg). Tissue samples are collected as required.

6. Pulseless electrical activity in the porcine model of neonatal hypoxia-asphyxia

Using our porcine model of neonatal hypoxia-asphyxia, we are able to describe an increasingly important clinical situation in the laboratory setting. Recent studies from our group have identified the presence of PEA rhythms in nearly half of neonatal pigs that were subjected to hypoxia-asphyxia in animal models of neonatal resuscitation [19–21]. In the study by Patel et al., 43% of piglets (23/54) had no CBF or HR on auscultation but had a HR of 15–80 bpm displayed on ECG, indicating PEA rhythm [20]. Luong et al. reported that 49% of piglets (22/45) presented with PEA rhythms, as indicated by no CBF or HR on auscultation but a HR of 17–75 bpm was displayed on ECG [19]. Solevag et al. also reported that 43% of piglets (9/21) presented with PEA rhythm on ECG, as confirmed by zero CBF and no audible HR/pulse; however, only 56% of piglets with PEA rhythms achieved ROSC compared to 100% of piglets with non-PEA rhythms ($p = 0.02$) [21]. Furthermore, survival to 4-hours post-ROSC occurred in only 33% of PEA piglets versus 58% of non-PEA piglets [21]. These studies indicate that cardiac arrest in the presence of a non-perfusing cardiac rhythm is common in asphyxiated neonatal piglets. Furthermore, this animal data is in agreeance with clinical observations of reduced CPR success in the presence of PEA in the delivery room in newborn infants [15, 16].

Studies from other research groups have also reported on the presence of PEA rhythms in their porcine model [22, 23]. It is important to note, however, that these studies were conducted in older piglets (2-month old "infant/pediatric" pigs) and were subjected to asphyxial cardiac arrest without the preceding hypoxia period. In the study by Lopez-Herce et al., 62% of piglets (44/71) had a PEA rhythm at the time of cardiac arrest [23]. However, there was no significant difference in the rate of ROSC between piglets with PEA rhythm (43%; 19/44) and piglets with non-PEA rhythm (30%; 7/23). Another study by the same research group, Gonzalez et al., also reported the presence of PEA rhythm at the time of cardiac arrest: 45% of piglets (22/49) [22]. Interestingly, the rate of ROSC was greater in piglets with PEA rhythm (45%; 10/22) versus non-PEA rhythm piglets (20%; 4/20) ($p = 0.037$).

The apparent discordance in the rate of ROSC post-PEA event between neonatal piglets [21] and pediatric piglets [22, 23] highlights the need for strong scientific evidence obtained from appropriate neonatal models to further our knowledge of delivery room resuscitation, rather than extrapolating data gained from the pediatric or adult populations. The percentages of PEA in our above-described neonatal model indicate relative consistency and can therefore be generalizable as a methodology. In this neonatal animal model, PEA is confirmed by electrical activity recorded on ECG in combination with no HR/pulse detected by auscultation and pulse oximetry and zero CBF. This animal model is beneficial for research directed at the management of PEA in newborns. Due to the increased awareness of PEA events in newborn infants, it is necessary to further investigate specifically tailored resuscitation techniques or changes in the resuscitation guideline algorithms to improve their survival. This translational model will therefore serve as a valuable tool to bridge the knowledge gap and improve the outcome of newborns that experience PEA in the delivery room.

7. Advantages and limitations of the porcine model of neonatal hypoxia-asphyxia

Owing to its many advantages, the clinically relevant porcine model of neonatal hypoxia-asphyxia has provided a platform to extensively investigate neonatal resuscitation. The newborn piglet is equivalent to a human infant at 36–38 weeks of gestational age, and has a comparable size and weight (1.5–2 kg body weight). This allows for relatively easy instrumentation to invasively monitor hemodynamic and physiological measurements, such as blood pressure and blood gases, as well as the ability to monitor the degree of hypoxia-asphyxia and reoxygenation in the recovery phase. The large size of this animal model (compared to smaller rodent models) allows the repeated collection of biological samples (plasma, whole blood) during the experimental period for biochemical assays. The piglet's cerebral metabolic data and many of the body systems, including cerebrovascular and cardiovascular systems, are also comparable to the human counterparts. This allows for better interpretation of the findings and makes it an exceptional animal model to study resuscitation interventions. The porcine model of neonatal hypoxia-asphyxia closely simulates delivery room events, with the gradual onset of severe hypoxia-asphyxia leading to cardiac arrest. Bradycardia or asystole (cardiac arrest) in newborn infants is usually caused by hypoxia/asphyxia, rather than primary cardiac compromise/ventricular fibrillation observed in adult patients. Furthermore, using our newborn piglet model, we are able to describe an increasingly important clinical situation in the laboratory setting – PEA, which is not well described in newborns in the delivery room. However, the asphyxia model uses piglets that have already undergone the transition from fetal to neonatal circulation and have cleared their lung fluid, which may present as a limitation. Furthermore, our model requires piglets to be intubated with a tightly sealed endotracheal tube to prevent any endotracheal tube leak. This may not occur in the delivery room where infants are either intubated (larynx bypassed, leak present) or receive respiratory support via a facemask, resulting in the possibility of airway obstruction or mask leaks. Nevertheless, many of its advantages make up for the few limitations of the model.

8. Contribution of the porcine model of neonatal hypoxia-asphyxia to current knowledge

The porcine model of neonatal hypoxia-asphyxia has proven to be an invaluable tool through which new resuscitation techniques can be studied pre-clinically. It has also proven to be a crucial element in increasing our understanding of physiological and pharmacological changes surrounding neonatal resuscitation. Below is a summary of studies that have utilized the model to gain further knowledge in various aspects of neonatal resuscitation. Knowledge gained from the below described studies are key in shaping the future neonatal resuscitation guidelines [24, 25].

8.1 Sustained inflations

The current neonatal resuscitation guidelines and the previous guidelines in 2010 [2–4, 26] recommend using a 3:1 C:V ratio when CC are needed, however these recommendations are not based on strong scientific evidence and the most effective C:V ratio in newborns remains controversial. Using our porcine model, Schmölzer et al. investigated an alternative approach to providing ventilation during CPR in the means of sustained inflations (SI) [27]. Rather than the standard coordinated 3:1 C:V technique, Schmölzer et al. proposed that SI during CC would passively deliver an adequate tidal volume into the lungs and improve survival. SI was delivered with a peak inflating pressure of 30 cmH2O for duration of 30 s. During the SI, CC was delivered at a rate of 120/min; SI was interrupted after 30 s for 1 s before a further 30 s of SI was provided [27]. The results showed that piglets resuscitated with SI during CC not only achieved ROSC faster than piglets resuscitated with the standard 3:1 C:V technique, but also had improved hemodynamic recovery and survival [27]. Following that study, Li et al. investigated the optimal rate of CC during SI by comparing CC rates of 90/min and 120/min [28]. Both groups had a similar time to ROSC, survival rates, and hemodynamic and respiratory parameters during CPR, and the hemodynamic recovery in the subsequent 4-hours was also similar in both groups. This leads the authors to suggest that resuscitation with a CC rate of 120/min during SI did not show a significant advantage compared to 90/min and higher CC rates are not necessarily an advantage [28]. To assure this suggestion, another study by Li et al. compared SI with CC at a rate of 90/min to the standard 3:1 C:V technique [29]. Piglets resuscitated with SI during CC at 90/min had significantly improved time to ROSC and also a reduced requirement for 100% oxygen and improved respiratory parameters compared to piglets resuscitated with 3:1 C:V [29]. Mustofa et al. investigated the optimal length of SI during CC by comparing resuscitation with SI duration of either 20 s or 60 s [30]. Using SI duration of 60 s resulted in a similar time to ROSC as SI duration of 20 s, as well as similar survival rate and hemodynamic recovery [30]. Furthermore, Mustofa et al. showed no significant differences in lung and brain pro-inflammatory cytokine concentrations between the SI groups and the 3:1 C:V group, suggesting that the SI technique does not promote more acute brain and lung injuries that the currently practiced technique of 3:1 C:V [30].

8.2 Asynchronous ventilation

Using the porcine model of neonatal hypoxia-asphyxia, Schmölzer et al. investigated a different approach to neonatal resuscitation with asynchronous ventilation during continuous CC; the rationale being that giving continuous CC without pausing for ventilation (as with 3:1 C:V) may avoid interruption in coronary perfusion and may improve minute ventilation during CPR [31]. Piglets were resuscitated with either the standard 3:1 C:V technique or the asynchronous ventilation technique, which delivered continuous CC at a rate of 90/min with asynchronous ventilation at a rate of 30 inflations/min [31]. Both groups had a similar time to ROSC, survival rates, epinephrine and oxygen administration, and hemodynamic and respiratory parameters during CPR; systemic and regional hemodynamic recovery in the subsequent 4-hour recovery period was also similar. This suggests that asynchronous ventilation during continuous CC is not more beneficial to the standard 3:1 C:V technique. In a following study, Patel et al. examined whether the outcome will improve by using different CC rates with asynchronous ventilation, namely 90/min, 100/min, and 120/min [32]. Even though rate and time to ROSC were similar between groups, increasing the CC rate to 120/min with asynchronous ventilation significantly improved hemodynamic recovery, as indicated by CBF, and cerebral and renal perfusion [32].

8.3 Oxygen

Current neonatal resuscitation guidelines recommend the use of 100% oxygen when CC are needed, however this is based on minimal evidence and 100% oxygen is also associated with increased oxidative stress [2–4, 33], and increased morbidity and mortality [34, 35]. Solevåg et al. examined the effect of using 21% oxygen (air) or 100% oxygen during resuscitation using either the 3:1 C:V technique or continuous CC with asynchronous ventilation (rate of 90/min) [36]. Time to achieve ROSC was similar between groups, however resuscitation with air was associated with a higher left ventricular stroke volume after ROSC and less myocardial oxidative stress compared to resuscitation with 100% oxygen [36]. This suggests that air during CC may reduce myocardial oxidative stress and improve cardiac function compared to 100% oxygen. However, the use of continuous CC with asynchronous ventilation in this study was less effective than the standard 3:1 C:V technique [36].

8.4 Chest compressions

Pasquin et al. used the porcine model of neonatal hypoxia-asphyxia to examine different ratios of CC to ventilations; the standard 3:1 C:V technique was compared to a C:V ratio of 2:1 and 4:1 [37]. Time to ROSC, mortality, oxygen requirements, epinephrine administration, and hemodynamic recovery were similar between all groups, indicating no difference in the efficacy of various C:V ratios in asphyxial-induced cardiac arrest of neonatal piglets.

8.5 Respiratory parameters

The purpose of inflations during CC is to deliver an adequate tidal volume to facilitate gas exchange [38], however limited information exists regarding tidal volume delivery during CC. Therefore, Li et al. examined the changes in tidal volume during CC and their effect on lung aeration in the porcine model of hypoxia-asphyxia [39]. Li et al. shows that when resuscitating using the SI with CC technique, passive lung ventilation/aeration can be achieved. In contrast, although use of the 3:1 C:V technique delivered tidal volume, it also resulted in a relative loss of tidal volume per 3:1 C:V cycle of up to 4.5 mL/kg [39]. This suggests that tidal volume delivery is greater when using SI with CC to resuscitate compared to the standard 3:1 C:V technique; this may lead to better alveolar oxygen delivery and lung aeration [39].

Using an objective method to evaluate recovery or predict the outcome of resuscitation may help decision-making during resuscitation. Therefore, Li et al. examined the temporal changes in end-tidal CO_2 ($ETCO_2$), volume of expired CO_2 (VCO_2), and the partial pressure of exhaled CO_2 ($PECO_2$) and their relationship with survivability and hemodynamic changes during CPR in the neonatal porcine model [40]. Li et al. reported that surviving piglets had significantly higher values of $ETCO_2$, VCO_2, and $PECO_2$ during CPR compared to non-surviving piglets, suggesting that continuously monitoring $ETCO_2$, VCO_2, and $PECO_2$ during CC has the potential to be a non-invasive method to indicate ROSC [40]. To further investigate if other parameters could be used as early outcome predictors after CPR, Solevåg et al. examined if cerebral and renal tissue oxygen saturation was different between surviving piglets and non-surviving piglets that were resuscitated after asphyxia-induced cardiac arrest [41]. The relationship of the tissue oxygen saturations with cardiac output, blood pressure, and biochemical variables was also examined [41]. No correlation between cardiac output or blood pressure and cerebral or renal tissue oxygen saturation was observed.

8.6 Hemodynamics

Espinoza et al. examined the changes in HR during adequate PPV following severe bradycardia in the porcine model of hypoxia-asphyxia [42]. The Neonatal Resuscitation Program (NRP) states that if adequate PPV is given for low HR, then the infant's HR should increase within the first 15 s of PPV. However in contrast to the NRP, Espinoza et al. showed that adequate PPV does not increase HR within 15 s of ventilation in piglets with asphyxia-induced bradycardia; after 30 s of PPV only half of piglets had an increase in HR. This study challenges the current NRP statement and suggests that clinicians should not expect an increase in HR after 15 s of PPV if there is severe bradycardia [42].

8.7 Epinephrine

Current neonatal resuscitation guidelines recommend the administration of intravenous epinephrine during if HR persists below 60 bpm despite CC and 100% oxygen [2–4]. However there is currently a lack of data evaluating the hemodynamic effects of epinephrine during neonatal resuscitation. Wagner et al. utilized the porcine model of hypoxia-asphyxia to examine hemodynamic changes after epinephrine administration during resuscitation and compare surviving and non-surviving piglets; epinephrine was administered at a dose of 0.01 mg/kg [43]. Epinephrine had no effect on either HR or cardiac output in survivors versus non-survivors during resuscitation; it did not increase survival rates or ROSC [43].

The abovementioned studies highlight the practicality of this neonatal animal model not only in driving progress in our understanding of neonatal resuscitation, but also in paving the way for new techniques into the delivery room.

9. Conclusions

Animal models that reliably reproduce the events surrounding neonatal resuscitation in the delivery room are imperative to improve the outcome of newborn infants requiring CPR and may also lead to benefits for the pediatric population. Due to its many advantages, the porcine model of neonatal hypoxia-asphyxia is one of the most commonly used large animal models for neonatal resuscitation studies. Not only has this model provided a further understanding of the effects of various resuscitation interventions, but it has also enabled the study of an increasingly important clinical situation in the laboratory setting – pulseless electrical activity. Using this animal model will further accelerate knowledge on neonatal resuscitation that will ultimately benefit patients.

Acknowledgements

We would like to thank the public for donating money to our funding agencies: GMS is a recipient of the Heart and Stroke Foundation/University of Alberta Professorship of Neonatal Resuscitation, a National New Investigator of the Heart and Stroke Foundation Canada and an Alberta New Investigator of the Heart and Stroke Foundation Alberta. The study was supported by a Grant from the SickKids Foundation in partnership with the Canadian Institutes of Health Research (CIHR - Institute of Human Development, Child and Youth Health (IHDCYH)), New Investigator Research Grant Program (Grant number - No. NI17-033) and a Grant-in-Aid from the Heart and Stroke Foundation Canada (Grant Number: G-15-0009284). We would like to acknowledge support from the Women and Children's Health Research Institute, University of Alberta.

Author details

Megan O'Reilly[1,2], Po-Yin Cheung[1,2], Tze-Fun Lee[1,2] and Georg M. Schmölzer[1,2]*

1 Centre for the Studies of Asphyxia and Resuscitation, Royal Alexandra Hospital, Edmonton, Alberta, Canada

2 Department of Pediatrics, Faculty of Medicine and Dentistry, University of Alberta, Edmonton, Alberta, Canada

*Address all correspondence to: georg.schmoelzer@me.com

References

[1] Aziz K, Chadwick M, Baker M, Andrews W. Ante- and intra-partum factors that predict increased need for neonatal resuscitation. Resuscitation. 2008;**79**(3):444-452

[2] Perlman JM, Wyllie J, Kattwinkel J, Wyckoff MH, Aziz K, Guinsburg R, et al. Neonatal resuscitation chapter collaborators. Part 7: Neonatal resuscitation: 2015 international consensus on cardiopulmonary resuscitation and emergency cardiovascular care science with treatment recommendations. Circulation. 2015;**132**(16 Suppl 1): S204-S241

[3] Wyckoff MH, Aziz K, Escobedo MB, Kapadia VS, Kattwinkel J, Perlman JM, et al. Part 13: Neonatal resuscitation: 2015 American Heart Association guidelines update for cardiopulmonary resuscitation and emergency cardiovascular care. Circulation. 2015;**132**(18 Suppl 2):S543-S560

[4] Wyllie J, Perlman JM, Kattwinkel J, Wyckoff MH, Aziz K, Guinsburg R, et al. Neonatal resuscitation chapter collaborators. Part 7: Neonatal resuscitation: 2015 international consensus on cardiopulmonary resuscitation and emergency cardiovascular care science with treatment recommendations. Resuscitation. 2015;**95**:e169-e201

[5] Barber CA, Wyckoff MH. Use and efficacy of endotracheal versus intravenous epinephrine during neonatal cardiopulmonary resuscitation in the delivery room. Pediatrics. 2006;**118**(3):1028-1034

[6] Harrington DJ, Redman CW, Moulden M, Greenwood CE. The long-term outcome in surviving infants with apgar zero at 10 minutes: A systematic review of the literature and hospital-based cohort. American Journal of Obstetrics and Gynecology. 2007;**196**(5):463 e461-463 e465

[7] Shah PS, Shah P, Tai KF. Chest compression and/or epinephrine at birth for preterm infants <32 weeks gestational age: Matched cohort study of neonatal outcomes. Journal of Perinatology. 2009;**29**(10):693-697

[8] Soraisham AS, Lodha AK, Singhal N, Aziz K, Yang J, Lee SK, et al. Neonatal outcomes following extensive cardiopulmonary resuscitation in the delivery room for infants born at less than 33 weeks gestational age. Resuscitation. 2014;**85**(2):238-243

[9] Fischer N, Soraisham A, Shah PS, Synnes A, Rabi Y, Singhal N, et al. The Canadian Neonatal Follow-up Network Canadian Neonatal Network Site Investigators. Extensive cardiopulmonary resuscitation of preterm neonates at birth and mortality and developmental outcomes. Resuscitation. 2019;**135**:57-65

[10] Kapadia V, Wyckoff MH. Chest compressions for bradycardia or asystole in neonates. Clinics in Perinatology. 2012;**39**(4):833-842

[11] Martin-Ancel A, Garcia-Alix A, Gaya F, Cabanas F, Burgueros M, Quero J. Multiple organ involvement in perinatal asphyxia. The Journal of Pediatrics. 1995;**127**(5):786-793

[12] Phillipos E, Solevag AL, Pichler G, Aziz K, van Os S, O'Reilly M, et al. Heart rate assessment immediately after birth. Neonatology. 2016;**109**(2):130-138

[13] Myerburg RJ, Halperin H, Egan DA, Boineau R, Chugh SS, Gillis AM, et al. Pulseless electric activity: Definition, causes, mechanisms, management, and research priorities for the next decade: Report from a national heart, lung, and

blood institute workshop. Circulation. 2013;**128**(23):2532-2541

[14] Nadkarni VM, Larkin GL, Peberdy MA, Carey SM, Kaye W, Mancini ME, et al. National Registry of cardiopulmonary resuscitation investigators. First documented rhythm and clinical outcome from in-hospital cardiac arrest among children and adults. JAMA. 2006;**295**(1):50-57

[15] Luong D, Cheung PY, Barrington KJ, Davis PG, Unrau J, Dakshinamurti S, et al. Cardiac arrest with pulseless electrical activity rhythm in newborn infants: A case series. Archives of Disease in Childhood. Fetal and Neonatal Edition. 2019. pii: fetalneonatal-2018-316087. DOI: 10.1136/archdischild-2018-316087

[16] Sillers L, Handley SC, James JR, Foglia EE. Pulseless electrical activity complicating neonatal resuscitation. Neonatology. 2019;**115**(2):95-98

[17] Donoghue A, Berg RA, Hazinski MF, Praestgaard AH, Roberts K, Nadkarni VM, et al. American Heart Association National Registry of CPR Investigators. Cardiopulmonary resuscitation for bradycardia with poor perfusion versus pulseless cardiac arrest. Pediatrics. 2009;**124**(6):1541-1548

[18] Girotra S, Spertus JA, Li Y, Berg RA, Nadkarni VM, Chan PS. American Heart Association get with the guidelines-resuscitation investigators. Survival trends in pediatric in-hospital cardiac arrests: An analysis from get with the guidelines-resuscitation. Circulation. Cardiovascular Quality and Outcomes. 2013;**6**(1):42-49

[19] Luong DH, Cheung PY, O'Reilly M, Lee TF, Schmölzer GM. Electrocardiography vs. auscultation to assess heart rate during cardiac arrest with pulseless electrical activity in newborn infants. Frontiers in Pediatrics. 2018;**6**:366

[20] Patel S, Cheung PY, Solevag AL, Barrington KJ, Kamlin COF, Davis PG, et al. Pulseless electrical activity: A misdiagnosed entity during asphyxia in newborn infants? Archives of Disease in Childhood. Fetal and Neonatal Edition. 2019;**104**(2):F215-F217

[21] Solevag AL, Luong D, Lee TF, O'Reilly M, Cheung PY, Schmölzer GM. Non-perfusing cardiac rhythms in asphyxiated newborn piglets. PLoS One. 2019;**14**(4):e0214506

[22] Gonzalez R, Urbano J, Botran M, Lopez J, Solana MJ, Garcia A, et al. Adrenaline, terlipressin, and corticoids versus adrenaline in the treatment of experimental pediatric asphyxial cardiac arrest. Pediatric Critical Care Medicine. 2014;**15**(6):e280-e287

[23] Lopez-Herce J, Fernandez B, Urbano J, Mencia S, Solana MJ, del Castillo J, et al. Terlipressin versus adrenaline in an infant animal model of asphyxial cardiac arrest. Intensive Care Medicine. 2010;**36**(7):1248-1255

[24] Schmölzer GM, Pichler G, Solevag AL, Fray C, van Os S, Cheung PY. Collaborators SVt. The surv1ve trial-sustained inflation and chest compression versus 3:1 chest compression-to-ventilation ratio during cardiopulmonary resuscitation of asphyxiated newborns: Study protocol for a cluster randomized controlled trial. Trials. 2019;**20**(1):139

[25] Schmölzer GM. Chest compressions during sustained inflation during cardiopulmonary resuscitation in newborn infants translating evidence from animal studies to the bedside. JACC: Basic to Translational Science. 2019;**4**(1):116-121

[26] Perlman JM, Wyllie J, Kattwinkel J, Atkins DL, Chameides L, Goldsmith JP, et al. Neonatal Resuscitation Chapter Collaborators. Part 11: Neonatal resuscitation: 2010 international

consensus on cardiopulmonary resuscitation and emergency cardiovascular care science with treatment recommendations. Circulation. 2010;**122**(16 Suppl 2): S516-S538

[27] Schmölzer GM, O'Reilly M, Labossiere J, Lee TF, Cowan S, Qin S, et al. Cardiopulmonary resuscitation with chest compressions during sustained inflations: A new technique of neonatal resuscitation that improves recovery and survival in a neonatal porcine model. Circulation. 2013;**128**(23):2495-2503

[28] Li ES, Cheung PY, Lee TF, Lu M, O'Reilly M, Schmölzer GM. Return of spontaneous circulation is not affected by different chest compression rates superimposed with sustained inflations during cardiopulmonary resuscitation in newborn piglets. PLoS One. 2016;**11**(6):e0157249

[29] Li ES, Gorens I, Cheung PY, Lee TF, Lu M, O'Reilly M, et al. Chest compressions during sustained inflations improve recovery when compared to a 3:1 compression:Ventilation ratio during cardiopulmonary resuscitation in a neonatal porcine model of asphyxia. Neonatology. 2017;**112**(4):337-346

[30] Mustofa J, Cheung PY, Patel S, Lee TF, Lu M, Pasquin MP, et al. Effects of different durations of sustained inflation during cardiopulmonary resuscitation on return of spontaneous circulation and hemodynamic recovery in severely asphyxiated piglets. Resuscitation. 2018;**129**:82-89

[31] Schmölzer GM, O'Reilly M, Labossiere J, Lee TF, Cowan S, Nicoll J, et al. 3:1 compression to ventilation ratio versus continuous chest compression with asynchronous ventilation in a porcine model of neonatal resuscitation. Resuscitation. 2014;**85**(2):270-275

[32] Patel S, Cheung PY, Lee TF, Pasquin MP, Lu M, O'Reilly M, et al. Asynchronous ventilation at 120 compared with 90 or 100 compressions per minute improves haemodynamic recovery in asphyxiated newborn piglets. Archives of Disease in Childhood. Fetal and Neonatal Edition. 2019. pii: fetalneonatal-2018-316610. DOI: 10.1136/archdischild-2018-316610

[33] Vento M, Asensi M, Sastre J, Garcia-Sala F, Pallardo FV, Vina J. Resuscitation with room air instead of 100% oxygen prevents oxidative stress in moderately asphyxiated term neonates. Pediatrics. 2001;**107**(4):642-647

[34] Saugstad OD. Resuscitation of newborn infants: From oxygen to room air. Lancet. 2010;**376**(9757):1970-1971

[35] Saugstad OD, Ramji S, Soll RF, Vento M. Resuscitation of newborn infants with 21% or 100% oxygen: An updated systematic review and meta-analysis. Neonatology. 2008;**94**(3):176-182

[36] Solevag AL, Schmölzer GM, O'Reilly M, Lu M, Lee TF, Hornberger LK, et al. Myocardial perfusion and oxidative stress after 21% vs. 100% oxygen ventilation and uninterrupted chest compressions in severely asphyxiated piglets. Resuscitation. 2016;**106**:7-13

[37] Pasquin MP, Cheung PY, Patel S, Lu M, Lee TF, Wagner M, et al. Comparison of different compression to ventilation ratios (2: 1, 3: 1, and 4: 1) during cardiopulmonary resuscitation in a porcine model of neonatal asphyxia. Neonatology. 2018;**114**(1):37-45

[38] Kattwinkel J, Perlman JM, Aziz K, Colby C, Fairchild K, Gallagher J, et al. Part 15: Neonatal resuscitation: 2010 american heart association guidelines for cardiopulmonary resuscitation and emergency cardiovascular care. Circulation. 2010;**122**(18 Suppl 3): S909-S919

[39] Li ES, Cheung PY, O'Reilly M, Schmölzer GM. Change in tidal volume during cardiopulmonary resuscitation in newborn piglets. Archives of Disease in Childhood. Fetal and Neonatal Edition. 2015;**100**(6):F530-F533

[40] Li ES, Cheung PY, O'Reilly M, LaBossiere J, Lee TF, Cowan S, et al. Exhaled co2 parameters as a tool to assess ventilation-perfusion mismatching during neonatal resuscitation in a swine model of neonatal asphyxia. PLoS One. 2016;**11**(1):e0146524

[41] Solevag AL, Lee TF, Lu M, Schmölzer GM, Cheung PY. Tidal volume delivery during continuous chest compressions and sustained inflation. Archives of Disease in Childhood. Fetal and Neonatal Edition. 2017;**102**(1):F85-F87

[42] Espinoza ML, Cheung PY, Lee TF, O'Reilly M, Schmölzer GM. Heart rate changes during positive pressure ventilation after asphyxia-induced bradycardia in a porcine model of neonatal resuscitation. Archives of Disease in Childhood. Fetal and Neonatal Edition. 2019;**104**(1):F98-F101

[43] Wagner M, Cheung PY, Li ES, Lee TF, Lu M, O'Reilly M, et al. Effects of epinephrine on hemodynamic changes during cardiopulmonary resuscitation in a neonatal piglet model. Pediatric Research. 2018;**83**(4):897-903

Chapter 11

High-Resolution Ultrasound Imaging System for the Evaluation of the Vascular Response to Stent or Balloon Injuries in the Rabbit Iliac Arteries

Aurélien Frobert, Guillaume Ajalbert, Jérémy Valentin, Stéphane Cook and Marie-Noëlle Giraud

Abstract

For novel therapeutic approaches of cardiovascular diseases, the preclinical investigation is of paramount and required appropriate technologies. We investigated the use of high-resolution ultrasound imaging system to evaluate the progression of vascular lesions in a rabbit model. Animals underwent vascular injury using two standard procedures. A bare-metal stent was placed within the left iliac artery, and a balloon injury was induced in the contralateral artery. The animals were kept on a regular diet for 8 weeks. A Vevo3100© VisualSonic high-resolution ultrasound imaging system and the associated software VevoVasc were used for the longitudinal evaluation of the injured arteries and the distal abdominal aorta. The lumen size increased rapidly after the intervention in both iliac arteries. In the balloon-injured artery, the augmentation was transient and significantly reversed, inducing an alteration of the blow flow. In contrast, in the stented segment, the lumen size was maintained enlarged overtime. We demonstrated a significant correlation for the wall thickness and the lumen size between ultrasonic and histological quantification. High-resolution ultrasound imaging in rabbit iliac arteries and the distal abdominal aorta is feasible, reliable and of relevance to investigate novel strategies for the inhibition of hyperplasia induced with standard injury models.

Keywords: echography, iliac arteries, vascular stent, rabbit, longitudinal study

1. Introduction

Coronary heart diseases remain a prominent cause of morbidity and mortality [1]. Percutaneous coronary intervention (PCI) is the current standard treatment and aims to widen the lumen, restore the blood flow into the vessel and consequently re-perfuse the ischaemic myocardium. A catheter is fed through the femoral artery until the blocked coronary and the balloon inflated. A stent is then placed and maintains the artery opened to limit adverse vessel remodelling and elastic recoil.

Since 2003, the standard of care is the balloon-expandable, drug-eluting metallic stent. Steady improvement of stent technology promoted a rapid evolution from the first generation of the bare-metal stents (BMS, permanent metallic structure without drug release) to the last generation of drug-eluting stents (DES, permanent metallic structure with anti-proliferative drug release). The BMS and DES are mostly made of a cobalt-chromium alloy and remain lifelong in the artery of the patient. Several studies showed, however, that life-threatening complication, emerging several months or years after implantation, may occur, including restenosis due to neointimal hyperplasia and late in-stent thrombosis [2]. Novel therapeutic approaches to reduce persistent inflammation, stenosis and thrombosis are focused on anti-proliferative and anti-inflammatory processes such as drug-eluting stents [3], pharmaceutical [4, 5] or laser-based approaches [6, 7] as well as bioresorbable stents [8].

In this context, an appropriate animal model is paramount to foster the development of new therapies, to provide in vivo preclinical proof of concept, to evaluate the treatment performance and to promote translation to the clinic. The rabbit-injured iliac artery model has been well established to investigate the vascular response to hyperplasia and stenosis or thrombosis [3, 8, 9].

In the present study, we evaluated the vascular responses to bilateral iliac artery injuries performed by balloon denudation and stent overexpansion, using a high-resolution ultrasound imaging system. We explored the longitudinal evolution of the vessel morphometries and the blow flow.

2. Methods

Three male New Zealand white rabbits (3.5–4 kg) were obtained from the Charles River Laboratories, France. The animals were housed in the animal centre facility at the University of Fribourg (Switzerland). All animals received humane care in compliance with the European Convention on Animal Care and in accordance with the Swiss Animal Protection Law after obtaining permission from the State Veterinary Office, Fribourg approved by the Swiss Federal Veterinary Office, Switzerland (FR-2016/16).

Angioplasty: Under general anaesthesia induced with s.c. injection of Narketan (65 mg/ml) and Xylapan (4 mg/ml) and maintained by perfusion (Narketan 65 mg/50 ml, Xylapan 4 mg/50 ml, infusion 15–20 ml/h), heparin (100 UI/ml) was administered in the marginal vein using a 24 GA (BD Insyte) catheter. Body parameters, including temperature, heart rate and pO_2 were controlled by a veterinary monitor (Midmark Cardell touch). In clean condition, an arteriotomy of the left coronary artery was performed, and a 6-French introducer sheath (Glidesheath Slender, Terumo) positioned. Two ml of a contrast agent (Bracco, Iomeprol 35 g) was injected, and a 0.36-mm guidewire is advanced through an introducer sheath up to the right iliac artery. Under angiographic monitoring of the pelvic area (**Figure 1**), the stent from Baxter (Coroflex Blue Neo) was deployed and overexpanded. The balloon was then retracted and then directed within the contralateral iliac artery (left), positioned at the same distance from the aortic bifurcation and inflated at 10 atm for 30 seconds to induce endothelial injury. At the end of the procedure, s.c. injection of Temgésic (1 ml/kg), Trimethazol (Werner Stricer, composed of sulfadoxine 40 mg/kg and trimethoprim 8 mg/kg) and carprofen 2.2 mg/kg (Rimadyl, Zoetis) were performed. The rabbit awoke within 1 h. Animals were kept on a normal diet for up to 8 weeks.

High-resolution ultrasound image acquisitions were performed every week under general gaze anaesthesia induced by isoflurane 4–5% and O_2 2–3 L/h. The animals were placed on an in-house-made platform, on a heating pad. The animal temperature was monitored through the rectal probe provided with the Vevo3100. ECG stainless steel needle electrodes provided with the Vevo3100 were placed

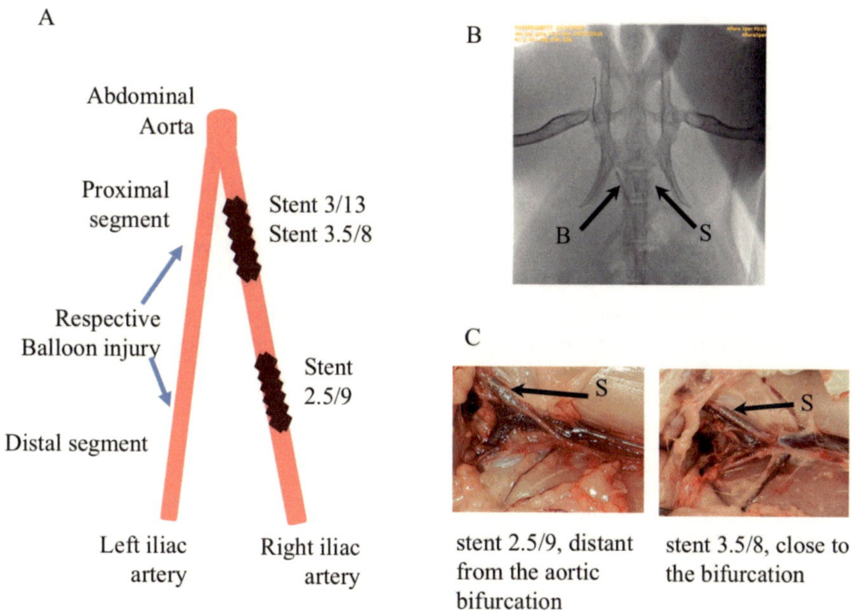

Figure 1.
Bilateral iliac artery injuries performed by an inflated balloon (left artery) and stent overexpansion (right artery). (A) Diagram showing the positioning of the stent and contralateral balloon placement. Each animal received a single stent that was placed either close to the aortic bifurcation or distally. (B) Angiography of both iliac arteries: The stent was placed first in the right artery and overexpanded, and the balloon was then retracted back to the aorta and introduced into the left artery. Balloon inflation was performed at a similar distance from the aortic bifurcation. S, stent; B, balloon. (C) Illustration of the positioning of the stent: 8 weeks post-intervention, the right iliac was exposed, and the transmural visualisation of the stent confirmed the site of the stent placement.

subcutaneously on the four limbs (right and left upper, right and left lower). The respiration rate was derived from the ECG signal. Longitudinal analyses were performed with a Vevo3100, VisualSonic high-resolution ultrasound system equipped with a transducer MX400 (20–46 Mhz) hold with the imaging station arm. The images were acquired at day 0, 14, 28, 42 and 56 or 63 in all animals. The image analyses were performed with the VevoVasc analysis module. The following parameters were quantified: wall thickness, lumen diameter and cross-sectional area (CSA). Using a Doppler pulse-wave mode, the velocity-time integral (VTI) was extracted. The following calculation was performed with the measured parameters: the blood flow was calculated as the product of the VTI and the CSA of the respective segments. The percentage changes in parameters relative to the values before the intervention were calculated as a ratio of the parameter at day 14, 28 or 42 to the same parameter at day 0 and multiplied by 100.

Artery harvesting was performed after euthanasia. Distal aortic segment and both iliac arteries were harvested according to the well-established procedure [10] and cut in short segments for histological analysis.

Histology characterisation: OCT embedding the vessel segments was frozen in the vapour of 2-methylbutane placed in liquid nitrogen. Sections of 5 μm are obtained using a cryocut and were processed for Movat Pentachrome staining. Briefly, sections were fixed 1 h in Bouin for 56°C, stained with Alcian Blue followed by Verhoeff's Elastic Stain, differentiated in ferric chloride solution, stained in brilliant crocein 1% and acid fuchsin 1%, placed in 5% phosphotungstic acid and stained with crocin. Sections were mounted with EUKITT®.

Stented segments were embedded in epoxy. 0.8-μm sections were cut with an ultramicrotome and stained with methyl blue.

The Bersoft Image Analysis software (Bersoft Technology and Software; Lunenburg, Canada) was used to quantify the vessel diameter and the wall thickness (including intimal, media and adventitia layers).

2.1 Statistics

Values are presented as mean ± SEM. The percentage changes in wall thickness, cross-sectional area and blood flow were analysed using two-way ANOVA; Fisher's LSD multi-comparisons were performed for the different segments and for the time effect. Linear regression and the parametric Pearson test were computed in a two-tailed manner. Analyses were performed using Prism software. Values were considered significantly different when $p < 0.05$.

3. Results

3.1 Stent placement

We report the results from three rabbits that received a Coroflex Blue Neo BMS. In concordance with the standard clinical procedure, each animal received a stent with an appropriate diameter defined under angiographic monitoring. The respective diameter (mm) and length (mm) of the stents were 2.5/9, 3/13 and 3.5/8. The stent positioning and contralateral balloon injury were proximal from the abdominal aortic bifurcation for the two animals and distal for one animal, as shown in **Figure 1**. The interventions were performed successfully without complication. All animals appeared healthy without significant weight loss. No infection, oedema or arterial thrombosis was encountered. The wound area was normal.

3.2 Morphometry: longitudinal quantification

The ultrasonographic vascular parameters were recorded before and after stenting or balloon injury every second week up to 8 weeks to evaluate the vessel structure. The imaging and measurements were performed at the injured and stented segments of the arteries as well as at the intact segments free from intervention situated distally and proximally of the lesion or stent, as controls (**Figure 2A**). The distal part of the abdominal aorta was also imaged and analysed (**Figure 2B**).

We report in **Figure 2A** the values of the diameters and wall thicknesses of the stented and balloon-injured iliac arteries obtained for each animal. We chose to present each animal data to visualise the individual variations. For instance, after balloon injury, the vessel diameter increased transiently up to 14 days at the site of the injury. However, for two animals the diameter returned to initial size while staying enlarged for the third animal. A transient vessel enlargement was also observed for the distal and proximal controls. Likewise, the aorta wall thickness followed three different evolutions from day 14 to day 60 when considering each animal separately, the aorta wall thickness increased consistently for one animal (that received 3.5/8 the stent), alternate phases of augmentation and diminution for the second one (stent 3/13), while it decreased for the last rabbit (stent 2.5/9) after a transient increase (**Figure 2B**).

For the iliac arteries, as presented in **Figure 2C**, we calculated the mean percentage changes both, the wall thickness and the CSA relative to the pre-intervention. We showed a change of the wall thickness at the site of the stent. At day 14, the wall

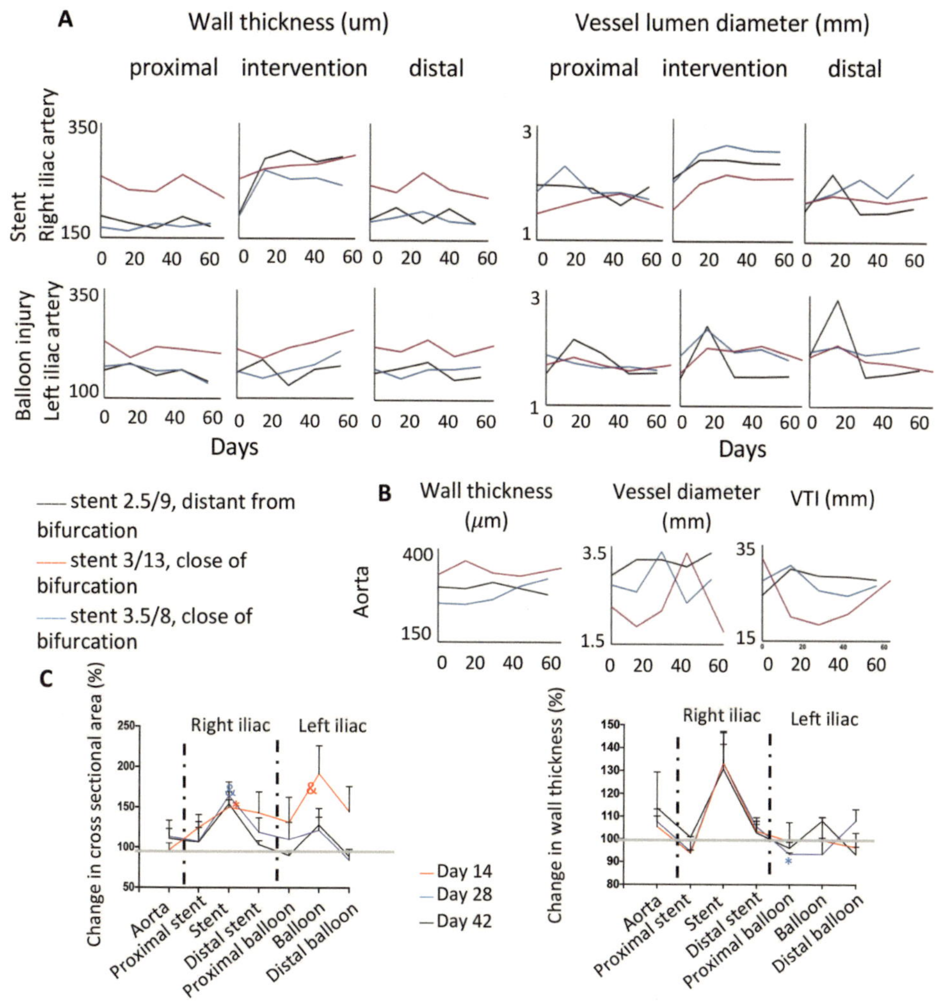

Figure 2.
*Longitudinal evaluation of the vessel structure. (A) The longitudinal measurements of the wall thickness and the lumen diameter of the right and left iliac arteries performed in different segments: at the site of the intervention (i.e. stent placement for the right artery and balloon inflation for the left artery) and at the proximal and distal uninjured segments as controls. Each animal and the corresponding stent are indicated and represented by a colour line. (B) Structure and VTI measured at the distal abdominal aorta. (C) Percentages of change in the crosssectional area of the vessel and the wall thickness relative to day 0 (preintervention, grey line) represented according to the vessel segment. Each line represents a different time post-intervention (day 14, 28 and 42). The results are shown as the mean of three animals and SEM. * $p < 0.05$ vs. day 0 and $p < 0.05$ vs. distal and proximal control segments.*

thickness showed a 133 ± 13% increase as compared with pre-intervention. For the proximal control segments, the wall thickness was not significantly altered over time.

In contrast, following balloon injury, the wall thickness remained constant, over time with a slight, but statistically significant, reduction (94 ± 1%, p = 0.01) observed after 42 days in the proximal segment.

In parallel, we report a significant increase (150 ± 9%, p = 0.03) in the CSA in the stented segment. The CSA remained elevated over time.

In contrast, the lumen CSA transiently increased in the left artery. The maximal significant change was 192 ± 35% in the injured segment as compared to the proximal (131 ± 31%, p = 0.02) and distal (145 ± 31%, p = 0.006) controls at day 14. The lumen area returned then to initial values.

3.3 Blood flow

Employing a pulsed wave (PW) Doppler mode, we recorded the velocity, extracted the velocity-time integral (VTI) and calculated the blood flow. As presented in **Figure 3**, for each animal, VTI gradually decreased over time in both, the balloon and stent-injured areas, and consequently in the distal control except for one animal.

Overall, in the right artery, there was no significant difference in the mean percentage changes of the blood flow in the stented segments (**Figure 3C**) although a maximal mean reduction of 72 ± 12% was observed at day 14. Looking at the individual data of the VTI, we observed a large variability. In contrast, in the left artery, the longitudinal analysis showed a significant change of the

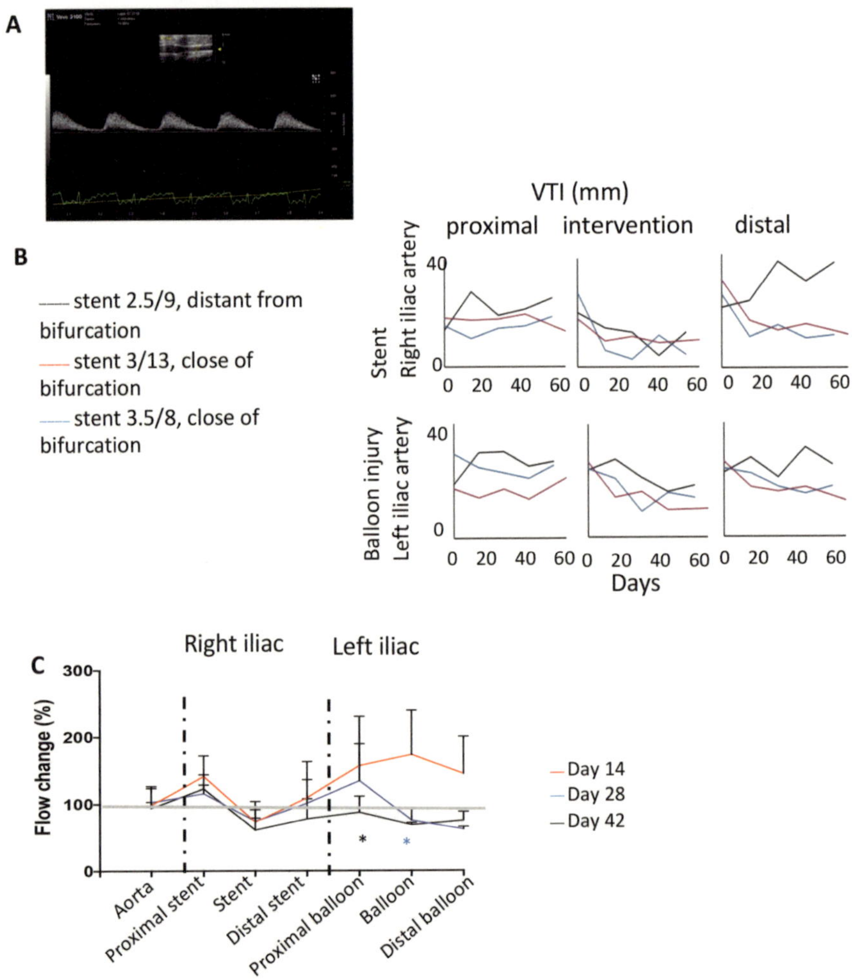

Figure 3.
*Height and weeks for longitudinal evaluation of the VTI and blood flow. (A) Illustration of the PW Doppler recording of left stented iliac artery velocity. The measurements were performed in the distal segment. (B) The measurements of the VTI in the right and left iliac arteries before (day 0) and post-intervention were performed in different segments: at the site of the intervention—Stent placement for the right artery and balloon inflation for the left artery and at the proximal and distal uninjured segments as controls. Each animal and the corresponding stents implanted are indicated and represented by a colour line. (C) Percentages of change in the blood flow relative to day 0 (pre-intervention, grey line) represented in the function of the vessel segment. Each line represents a different time post-intervention (day 14, 28 and 42). The results are shown as the mean of three animals and SEM * $p < 0.05$ vs. day 0.*

blood flow, respectively, 62 ± 4% ($p = 0.01$) at day 28 in the distal segment, and 68 ± 3% ($p = 0.01$) recorded at day 42 in the left iliac artery. The transient increase of the blood flow observed at day 14 (173 ± 66%) was not statistically significant.

3.4 Endpoint measurement: comparison between ultrasound measurement and histology

The lumen diameters at seven sites of the iliac arteries and aorta in the three rabbits subjected to balloon injury and stenting were quantified using both high-resolution ultrasound and histology. The correlations between the measurements performed by ultrasound and the histological analysis of the vessel diameter and the wall thickness, 8 weeks after the intervention, are presented in **Figure 4A**. We demonstrated a significant correlation between the two analytical procedures for both parameters, respectively, for the vessel diameter, $r^2 = 0.5$, $p = 0.006$ and for the wall thickness, $r^2 = 0.21$, $p = 0.04$. The thickening of the intima was observed prominently in the stented segment (**Figure 4B**) and for one animal in the distal area of the balloon-injured segment (**Figure 4C**). Alignments were performed on the different parts of the figures.

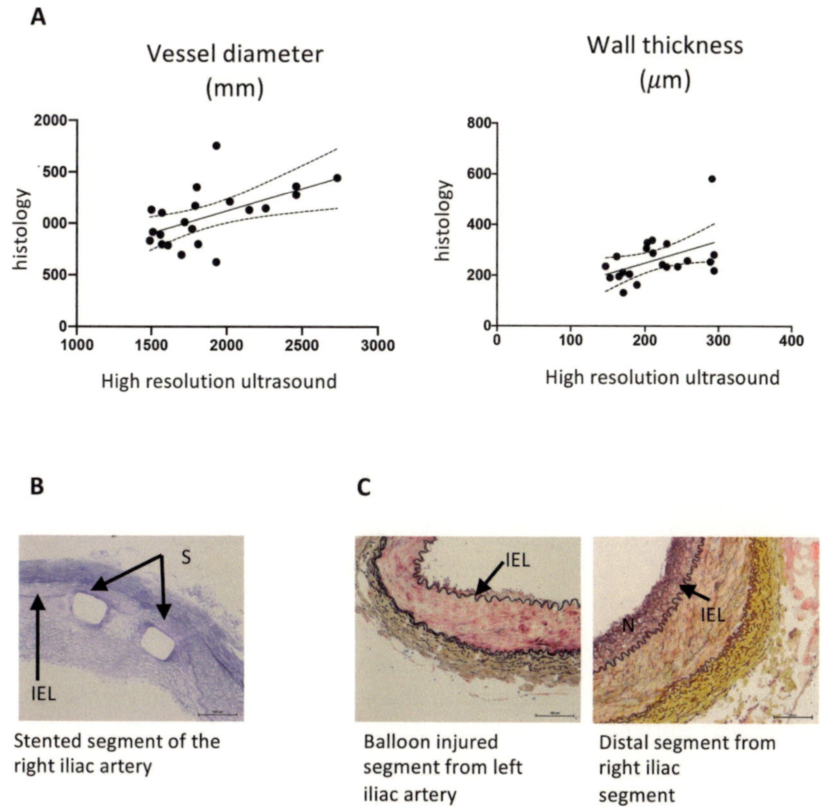

Figure 4.
Histology evaluation of rabbit iliac arteries. (A) Correlation between the high-resolution ultrasound and histological measurements of the vessel diameters and wall thicknesses. The individual dots represent each value obtained in all the segments assessed 8 weeks post-intervention. (B) Representative methyl blue stained histologic cross section of an 8-week rabbit-stented right iliac artery showing the intimal hyperplasia. The staining pattern shows prominent intimal thickening in the stented segment of the iliac artery. S indicates the stent struts. (C) Movat Pentachrome stained histologic cross section of a rabbit left iliac artery 8 weeks after the balloon injury. N indicates luminal neointimal formation. Arrow indicates internal elastic lamina (IEL).

4. Discussion

Mechanical endothelium injury is a standard strategy to induce hyperplasia in various animal models such as mice, rabbits or minipigs [9, 10, 17]. In the present study, the endothelial injury was induced by two different approaches in rabbits: first by overexpansion of a BMS in the right iliac artery and second with an inflated balloon in the contralateral left iliac artery. Stenting and balloon injury resulted in a respective permanent and transient overexpansion of the vessel inducing a well-recognised vascular response [11]. An acute inflammation rapidly follows the induced endothelium injury or denudation and peaks after a few days; then, the inflammation temporally declines during the resolution phase. Acute inflammation is accompanied by the proliferation of smooth muscle cells that results in hyperplasia. Consequent thickening of the wall (hyperplasia) may induce a narrowing of the vessel lumen (stenosis) and reduction of the blood flow.

In the present study, we showed that the rabbit iliac arteries and the distal abdominal aorta could be successfully monitored using high-resolution ultrasound for longitudinal and non-invasive investigation. The quality of the images acquired allowed rigorous measurements of the wall and vessel sizes. It is important to note that for some measurements, the possible presence of oedema could impair the quality of the image acquisition and may explain variations of the parameters in consecutive weekly measurements.

Besides, the metallic structures of the BMS were visualised as shadows. Although the presence of the metallic stent did not impair the wall thickness and vessel diameters' quantification, the blood flow velocity measurements were often challenging to perform.

Comparing the acute balloon injury with the chronic injury associated with stent implantation, we reported that the response of the vessel differed with the type of intervention. The primary function of the BMS stent is to provide mechanical support. As expected, the stent allowed the maintenance of increased vessel diameter in all animals. In contrast, balloon inflation induced a short-term increase in the lumen size followed by a reduction suggesting a weakening of the artery.

Notably, the overexpansion of the stent affected the wall thickness that increased rapidly and remained elevated. Histology analysis revealed the formation of neointimal hyperplasia. In contrast, balloon injury results in a transient wall thickening recorded 2 weeks post-intervention with a successive return to the initial dimensions. Transient inflammatory response to the balloon injury may explain the wall. Accordingly, Welt et al. [12] reported a transient inflammation following a balloon injury model as compared with a sustained accumulation of inflammatory cells such as monocytes in stented iliac arteries of rabbits up to 14 days.

Furthermore, Virmani et al. [13] reviewed the temporal vascular response to BMS implanted in the rabbit iliac artery. Between 7 and 14 days, the intimal thickness increased due to inflammation and cell proliferation and then peaked at 1 month with a return of the cell proliferation to the basal level. Further shrinkage has been reported from 3 to 6 months due to the extracellular matrix remodelling. In agreement, we also report the shrinkage of the wall thickness observed at day 28 in the left artery.

As far as the evaluation of the blood flow is concerned, the VTI and the calculated changes in flow revealed notable individual variations. Nevertheless, the longitudinal evaluation showed that in all animals, the VTI decreased overtime in the left and right segments but remained stable in the proximal segments. Meanwhile, the calculated blood flow was significantly reduced at the site of the balloon injury and the distal segment but maintained unchanged in the stented segment.

Despite the neointimal hyperplasia developed within the stent, the blood flow was maintained due to the diameter enlargement resulting from the stent overexpansion. Nevertheless, it is essential to note a maximal 72% reduction of the blow flow, although the statistical significance was not reached due to the small number of animal and large variability.

The aorta underwent twice the passage of the catheter, once with the crimped stent followed by its retraction and second with the successive introduction into the contralateral artery. The vessel structure remained similar in all the animals, suggesting the absence of injury due to the procedure. An exception was observed for one animal that showed a continuous increase in the wall thickness. These effects may be explained by the large stent diameter and the placement site, close to the aortic bifurcation. Notably, the variations observed in the right artery and the aorta might be dependent on the size and location of the stents that varied between animals.

Moreover, we provide evidence that the wall thickness and the vessel diameter significantly correlated with the histological evaluation. Besides, histology provides evidence that the wall thickening in the stented area resulted in intimal hyperplasia. Neointimal hyperplasia was also observed in the left and right proximal control segments of two animals. Our results corroborate the well-established proliferation of smooth muscle cells resulting in the wall artery thickening induced by the stent and balloon injury [14–16].

In agreement with the literature, under conventional diet and following injury, rabbits developed wall hyperplasia that is known to be associated with inflammation and smooth muscle proliferation rather than atherosclerotic plaques that can be observed in hyperlipidemic rabbits [9, 17].

Although histological analysis is essential to evaluate wall composition, inflammation and smooth muscle proliferation, longitudinal study provides a useful tool to record transient variation in the vessel dimensions.

5. Conclusion

The non-invasive, real-time imaging of the rabbit iliac arteries and the distal abdominal aorta for the quantification of lumen diameter and wall thickness using high-resolution ultrasound permit the monitoring of the progression of the wall and vessel following balloon angioplasty and endovascular stent implantation. Conveniently combined with the blood flow analysis, this methodological approach would be essential to evaluate novel therapeutic approaches to prevent hyperplasia.

Acknowledgements

The study was supported by the Swiss National Science Foundation (SNF 316030_157658), the University of Fribourg and the Fonds Scientifique Cardiovasculaire FSC, Fribourg Hospital.

Author details

Aurélien Frobert, Guillaume Ajalbert, Jérémy Valentin, Stéphane Cook and Marie-Noëlle Giraud*
Faculty of Sciences and Medicine, Cardiology, EMC Department, University of Fribourg, Switzerland

*Address all correspondence to: marie-noelle.giraud@unifr.ch

References

[1] Benjamin EJ, Muntner P, Alonso A, Bittencourt MS, Callaway CW, Carson AP, et al. Heart disease and stroke statistics-2019 update: A report from the American Heart Association. Circulation. 2019;**139**(10):e56-e528

[2] Secemsky EA, Matteau A, Yeh RW, Steg PG, Camenzind E, Wijns W, et al. Comparison of short- and long-term cardiac mortality in early versus late stent thrombosis (from pooled PROTECT trials). The American Journal of Cardiology. 2015;**115**(12):1678-1684

[3] Conway C, Desany GJ, Bailey LR, Keating JH, Baker BL, Edelman ER. Fracture in drug-eluting stents increases focal intimal hyperplasia in the atherosclerosed rabbit iliac artery. Catheterization and Cardiovascular Interventions. 2019;**93**(2):278-285

[4] Tolva V, Mazzola S, Zerbi P, Casana R, Albertini M, Calvillo L, et al. A successful experimental model for intimal hyperplasia prevention using a resveratrol-delivering balloon. Journal of Vascular Surgery. 2016;**63**(3):788-794

[5] Nishio H, Masumoto H, Sakamoto K, Yamazaki K, Ikeda T, Minatoya K. MicroRNA-145-loaded poly(lactic-co-glycolic acid) nanoparticles attenuate venous intimal hyperplasia in a rabbit model. The Journal of Thoracic and Cardiovascular Surgery. 2019;**157**(6):2242-2251

[6] Jain M, Zellweger M, Wagnieres G, van den Bergh H, Cook S, Giraud MN. Photodynamic therapy for the treatment of atherosclerotic plaque: Lost in translation? Cardiovascular Therapeutics. 2017;**35**(2):e12238

[7] Lanvin T, Conkey DB, Frobert A, Valentin J, Goy JJ, Cook S, et al. Subsurface ablation of atherosclerotic plaque using ultrafast laser pulses. Biomedical Optics Express. 2015;**6**(7):2552-2561

[8] Lin W, Qin L, Qi H, Zhang D, Zhang G, Gao R, et al. Long-term in vivo corrosion behavior, biocompatibility and bioresorption mechanism of a bioresorbable nitrided iron scaffold. Acta Biomaterialia. 2017;**54**:454-468

[9] Zaragoza C, Gomez-Guerrero C, Martin-Ventura JL, Blanco-Colio L, Lavin B, Mallavia B, et al. Animal models of cardiovascular diseases. Journal of Biomedicine and Biotechnology. 2011;**2011**:497841

[10] Jain M, Frobert A, Valentin J, Cook S, Giraud MN. The rabbit model of accelerated atherosclerosis: A methodological perspective of the iliac artery balloon injury. Journal of Visualized Experiments. 2017;(128)

[11] Newby AC. An overview of the vascular response to injury: A tribute to the late Russell Ross. Toxicology Letters. 2000;**112-113**:519-529

[12] Welt FG, Tso C, Edelman ER, Kjelsberg MA, Paolini JF, Seifert P, et al. Leukocyte recruitment and expression of chemokines following different forms of vascular injury. Vascular Medicine. 2003;**8**(1):1-7

[13] Virmani R, Kolodgie FD, Farb A, Lafont A. Drug eluting stents: Are human and animal studies comparable? Heart. 2003;**89**(2):133-138

[14] Chaabane C, Otsuka F, Virmani R, Bochaton-PiallatML.Biologicalresponses in stented arteries. Cardiovascular Research. 2013;**99**(2):353-363

[15] Carter AJ, Laird JR, Farb A, Kufs W, Wortham DC, Virmani R. Morphologic characteristics of lesion formation

and time course of smooth muscle cell proliferation in a porcine proliferative restenosis model. Journal of the American College of Cardiology. 1994;**24**(5):1398-1405

[16] Curcio A, Torella D, Indolfi C. Mechanisms of smooth muscle cell proliferation and endothelial regeneration after vascular injury and stenting: Approach to therapy. Circulation Journal. 2011;**75**(6):1287-1296

[17] Yamashita A, Asada Y. A rabbit model of thrombosis on atherosclerotic lesions. Journal of Biomedicine and Biotechnology. 2011;**2011**:424929

Chapter 12

Impact of Oxidative Stress on Inflammation in Rheumatoid and Adjuvant Arthritis: Damage to Lipids, Proteins, and Enzymatic Antioxidant Defense in Plasma and Different Tissues

Silvester Ponist, Miloslav Zloh and Katarina Bauerova

Abstract

Animal models of rheumatoid arthritis (RA) are widely used for testing potential new therapies for RA. The most commonly used models of human RA are adjuvant-induced arthritis (AIA) and collagen-induced arthritis in rats and mice. In this chapter, we will focus on inflammatory and oxidative stress (OS) processes during the development of AIA. OS is a result of increased production of reactive oxygen species (ROS) or a reduction in the body's endogenous antioxidant defense system. ROS and reactive nitrogen species (RNS) can contribute to the pathogenesis of RA by the induction of membrane oxidation, irreversible damage to proteins and DNA, cartilage damage, and induction of bone resorption. ROS/RNS can also modulate a variety of signaling events that control gene expression and affect cellular processes that participate in chronic inflammation. Our research team has been studying the course of OS during the development of rat AIA for more than a decade. We have analyzed the course of OS using markers of lipid peroxidation (malondialdehyde, 4-hydroxy-2-nonenal, and F-2 isoprostanes), protein carbonyls, antioxidant enzymes (heme oxygenase and gamma-glutamyl transferase), and levels of endogenous antioxidants (coenzyme Q_{10} and Q_9, gamma-tocopherol) in plasma and different tissues (joint, liver, lung, skeletal muscle, and spleen).

Keywords: animal models, arthritis, redox signaling, cachexia, antioxidants

1. Introduction

Research on animal models is necessary to better understand the etiopathology of rheumatoid arthritis (RA) and has enabled successful new strategies for innovative drug research. Recently the discovery of novel biomarkers of presymptomatic and emerging stages of human RA focused the attention on interventions that underlie different disease variants. This development in the field underlying RA pathogenesis has also led to the increased need of new animal models. Integration of the knowledge on human and animal models will allow to create a comprehensive "pathogenesis map" to the subset of disease they mimic [1].

Rheumatoid arthritis occurs due to the continuous deterioration of cells and tissues that ultimately affects major organs. Both oxidative stress (OS) and inflammation are considered major role players in the pathogenesis of RA [2]. Even if there is a lot of evidence from animal models of RA and human RA, about that OS plays an important role in tissue damage and also promotes cardiovascular diseases in patients with RA [3]; until now, a therapeutic strategy to reduce OS in RA has not yet been established. Thus, understanding how the OS is influencing the development of animal and human RA is of great importance.

In this chapter, we will discuss the importance of OS in the pathogenesis of human RA and its experimental model, rat adjuvant arthritis (AIA).

2. Pathogenesis of rheumatoid arthritis

RA is a chronic, progressive, inflammatory autoimmune disease associated with articular, extra-articular, and systemic effects. It has been reported that RA affects multiple comorbidities [4]. Mortality rates are more than twice as high in patients with RA as in the general population (Wolfe et al. [5]). Although the exact cause of RA remains unknown [5, 6], several findings suggest a genetic basis for disease development. More than 80% of patients carry the epitope of the HLA-DRB1*04 cluster [7], and patients expressing two HLA-DRB1*04 alleles are at elevated risk for major organ involvement and surgery related to joint destruction [8]. Environmental factors, such as smoking and infection, may also influence the development, rate of progression, and severity of RA [9, 10]. In addition to joint symptoms, many patients experience extra-articular or systemic manifestations or both. Extra-articular manifestations include rheumatoid nodules, vasculitis, pericarditis, uveitis, and rheumatoid lung [11]. Systemic manifestations include often anemia, cardiovascular disease, osteoporosis, fatigue, and depression [12, 13]. The earliest event in RA pathogenesis is the activation of the innate immune response that includes the activation of dendritic cells by exogenous material and autologous antigens. Antigen-presenting cells, including dendritic cells, macrophages, and activated B cells, present arthritis-associated antigens to T cells. T-cell activation and B-cell activation result in increased production of cytokines and chemokines. In addition to antigen presentation, macrophages are involved in osteoclastogenesis and are a major source of cytokines, including TNF-α, IL-1, and IL-6 [6, 7]. Within the synovial membrane, there is a great increase in activated fibroblast-like synoviocytes, which also produce inflammatory cytokines, prostaglandins, and matrix metalloproteinases (MMPs). Synoviocytes contribute to the destruction of cartilage and bone by secreting MMPs into the synovial fluid (SF) and by direct invasion into these tissues [7]. Pro-inflammatory cytokines are involved in the pathogenesis of RA [2, 14]. TNF-α and IL-6 play dominant roles in the pathobiology of RA; however, IL-1, vascular endothelial growth factor (VEGF), and IL-17 have also a significant impact on the disease process. These cytokines activate genes associated with inflammatory responses, including additional cytokines and MMPs involved in tissue degradation [6]. Th-17 lymphocytes have a critical role in synovitis in RA patients [15]. TNF-α, IL-6, and IL-1 are key mediators of cell migration and inflammation in RA [7]. IL-6 acts directly on neutrophils through membrane IL-6 receptors that contribute to inflammation and joint destruction by secreting proteolytic enzymes and reactive oxygen intermediates [12]. Furthermore, an in vitro study with fibroblasts from patients with RA demonstrated the role of IL-6 in promoting neutrophil recruitment by activated fibroblasts [16]. The principal cause of bone erosion is the pannus that is found at the interface with the cartilage and bone. Angiogenesis is a key process in the formation and maintenance of pannus because invasion of cartilage and bone requires increased blood supply. In patients with RA, many pro-angiogenic factors are expressed in synovium, among them, VEGF plays the central role in new blood vessel development [17]. Cartilage degradation in RA occurs when TNF-α, IL-1, and IL-6 activate synoviocytes, resulting in the secretion of MMPs into the SF [6, 7]. Cytokines also activate chondrocytes (**Figure 1**), leading to the direct release of additional MMPs into the cartilage [7]. ROS have been produced mainly during oxidative phosphorylation and by activated phagocytic cells during oxidative burst. It has been known that ROS can function as a second messenger to activate nuclear factor kappa-B (NF-κB) which orchestrates the expression of a spectrum of genes involved in the inflammatory response. Several cytokines, including TNF-α and IL-1β, are known initiators of NF-κB activation cascade [18] and are under its transcriptional control.

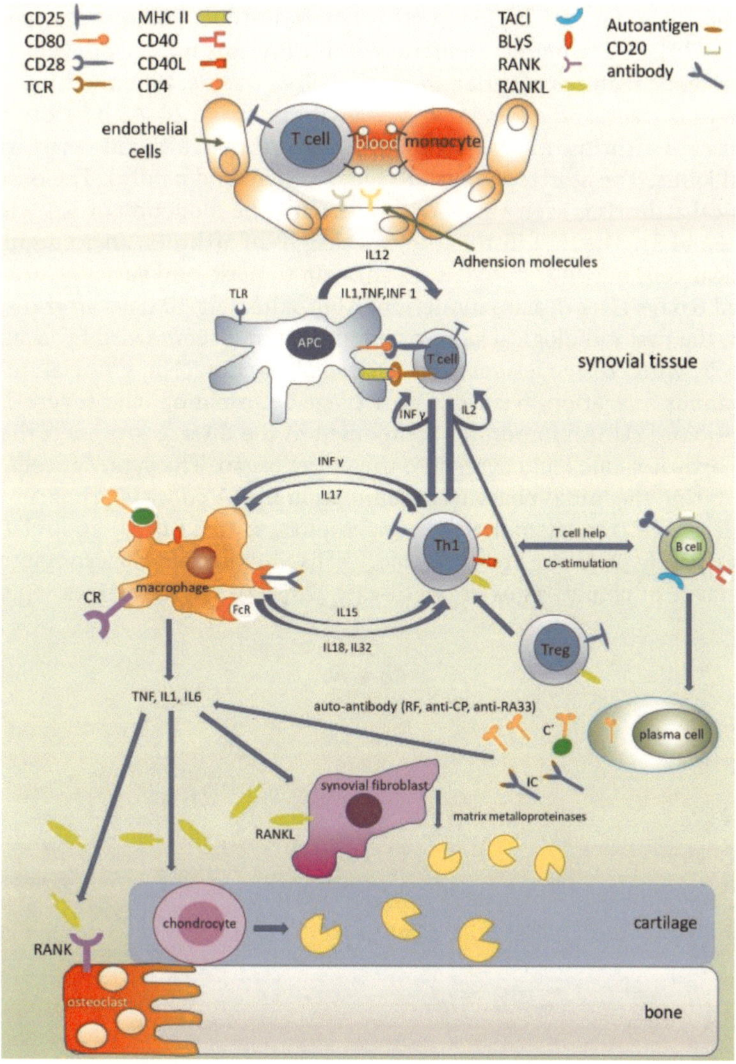

Figure 1.
Pathogenesis of cartilage and bone damage in rheumatoid arthritis. MHC, major histocompatibility complex; TCR, T-cell receptor; TACI, transmembrane activator and CAML interactor; BLyS, B-lymphocyte stimulator; RANK, receptor activator of nuclear factor κ B; RANKL, receptor activator of nuclear factor κ B ligand; TNF, tumor necrosis factor; INF, interferon; IL, interleukin; CR, complement receptor; RF, rheumatoid factor.

TNF-α participates positively in the phosphorylation of kinase kappa inhibitor, allowing NF-κB dimers (p50 and p65 portions) to migrate to the nucleus and then bind to promoters of pro-inflammatory genes [19] and stimulate the NADPH oxidase activation. Increased cytokine production driven by NF-κB can enhance expression of vascular adhesion molecules that attract leucocytes into the joint as well as MMPs.

3. Rat adjuvant arthritis

Animal models of arthritis play an important role in unraveling mechanisms of chronic inflammation in rheumatoid synovial tissue. They are used extensively to study new treatment strategies for RA. AA can be induced by intradermal or footpad injection of heat-killed mycobacterial species, preferably in a fine suspension in a mineral or vegetable oil (CFA). The disease is restricted to susceptible rodents, mostly certain rat strains, such as Lewis, Buffalo, Sprague-Dawley, and Wistar rats [20]. Following AA induction with CFA, rats not only develop arthritis but also systemic features of inflammation, such as uveitis, inflammation of the gastrointestinal tract, and a loss in body weight that starts 24–48 h before the clinical onset of arthritis. AA is a symmetric polyarthritis, affecting primarily the peripheral joints. The affected joints are red, swollen, and painful. The onset of overt clinical arthritis is seen 10–14 days following the induction of AA with CFA (**Figures 2** and **3**). The first histopathological signs of arthritis, an accumulation of mononuclear cells in connective tissues adjacent to periosteal surfaces, are already manifested 6 days after disease induction. Approximately 10 days after disease induction, the first radiological signs of inflammation become visible: localized osteoporosis, with erosive lesions, and periosteal reaction. The synovial infiltrate leads to pannus formation, resulting in cartilage deformation, and severe destruction of the joint [21]. An important component of the disease process is the trafficking of arthritogenic leukocytes into the target organ. The synovial cellular infiltrate during the initial phase of inflammation in AA consists primarily of mononuclear cells (mostly monocytes, macrophages, and T cells) and relatively fewer neutrophils [22]. The arthritogenic T cells migrate into the synovium before the appearance of clinical signs of the disease [23]. Data in AA suggesting that

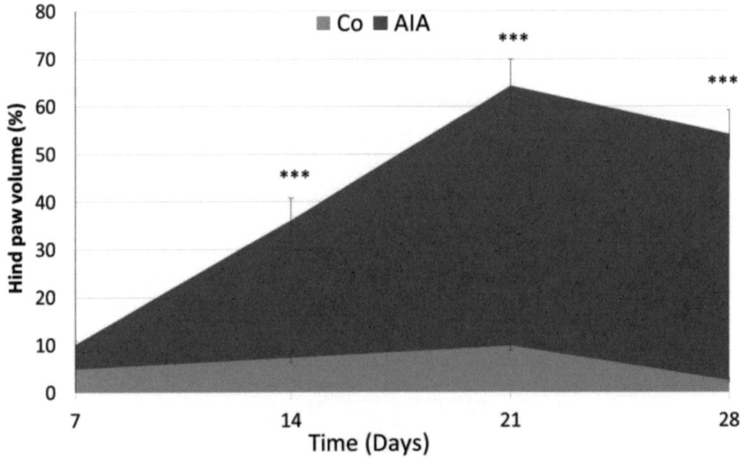

Figure 2.
Changes in hind paw volume during development of adjuvant-induced arthritis. Co, control healthy rats; AIA, adjuvant-induced arthritic rats.

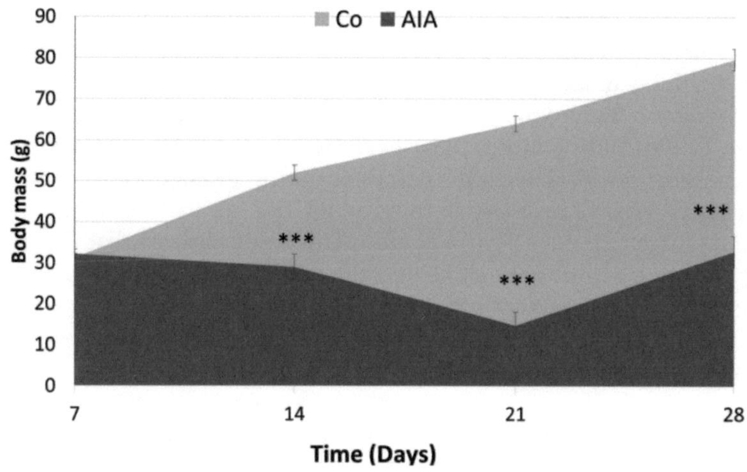

Figure 3.
Changes in body mass during development of adjuvant-induced arthritis. Co, control healthy rats; AIA, adjuvant-induced arthritic rats.

immune-stimulatory DNA sequences (ISS) may be a critical factor contributing to the chronicity of inflammation in chronic autoimmune arthritis. ISS can stimulate the expression of co-stimulatory molecules and the production of cytokines such as IL-12, TNF-α, and interferons by macrophages, dendritic cells, B cells, and NK cells [24] and are capable of skewing an immune response toward a strong and prolonged Th1 type of immunity [25]. AA has been used in the evaluation of nonsteroidal inflammatory drugs, such as phenylbutazone and aspirin during the early 1960s, and later in cyclooxygenase-2 inhibitors such as celecoxib. AA in rats shares many features with human arthritis, including genetic linkage, synovial CD4 + cells, and T-cell dependence [26].

4. Oxidative stress and inflammation

Inflammation is a natural defense mechanism against pathogens. It occurs in many pathogenic diseases (microbial and viral infections, exposure to allergens, radiation and toxic chemicals, autoimmune diseases, etc.). Chronic diseases linked with higher production of ROS result in OS and variety of protein oxidations [27]. Furthermore, some oxidized proteins trigger a release of inflammatory signal molecules, and peroxiredoxin 2 (PRDX2), which has been recognized as an inflammatory signal [28]. Relationship between OS and inflammation has been documented by many authors. Evidences indicated that OS plays a pathogenic role in chronic inflammatory diseases. Damage of OS such as oxidized proteins, glycated products, and lipid peroxidation results in neuron degenerations mostly reported in brain disorders [29]. ROS generated in brain tissues can modulate synaptic and non-synaptic communication between neurons that result in neuro-inflammation and cell death and then in neurodegeneration and memory loss [29]. Tripeptide glutathione (GSH) is an intracellular thiol antioxidant; lower level of this GSH causes higher ROS production, which results in imbalanced immune response, inflammation, and susceptibility to infection [30]. A study was conducted on the role of GSH and its oxidized form and their regulatory function and gene expressions beyond free radical scavenging activities linked with GSH. GSH is involved in the redox regulation of immune system [31] through disulfide bounds between protein cysteines and GSH. This process is called as glutathionylation, which regulates signaling proteins and transcription factors [32]. Inflammatory stimuli induce the release of PRDX2, a ubiquitous redox-active intracellular enzyme. PRDX2 is a redox-dependent inflammatory mediator, which activates macrophages to produce and release TNF-α. During intracellular oxidative stress GSH binds with PRDX2 and this protein glutathionylation occurs before or during PRDX2 release, and glutathionylated PRDX2 regulates immunity. PRDX2 is a part of inflammatory cascade and is able to induce TNF-α release. This study showed that PRDX2 and thioredoxin from macrophages could alter the redox balance of cell surface receptors and enable the induction of inflammatory process [28].

4.1 Oxidative stress in rheumatoid arthritis

RA is one of the conditions that induces OS. A fivefold increase in mitochondrial ROS production in whole blood and monocytes of RA patients–compared with healthy subjects–suggests that OS is a pathogenic hallmark in RA. Free radicals are indirectly implicated in joint damage because they also play a role as secondary messengers in inflammatory and immune cellular response in RA. T-cell exposure to increased OS becomes refractory to several stimuli including those for growth and death and may perpetuate the abnormal immune response [33]. On the other hand, free radicals can degrade directly the joint cartilage, attacking its proteoglycan and inhibiting its synthesis [34]. Oxidative damage of hyaluronic acid and lipoperoxidation products and oxidation of low-density lipoproteins and carbonyl increment resulting from protein oxidation have been demonstrated in RA. Increased levels of 4-HNE have been assessed in serum (or plasma) and synovial fluid of patients with RA [35, 36]. Peroxidative damage induced by free radicals has been demonstrated to play a role in the pathogenesis not only of RA but also of systemic lupus erythematosus, progressive systemic sclerosis, diabetes mellitus type 1, and myasthenia gravis. Increased OS has been associated with increased lipid peroxidation in these patients. Lipid peroxidation occurs as a result of increased OS stemming from deranged pro-oxidant/antioxidant balance and represents an important pathogenic process in the oxygen toxicity. As a result of lipid peroxidation increases in the levels of conjugated dienes, isoprostanes, 4-HNE, and malondialdehyde have been demonstrated [37]. Study of Basu et al. [38] has shown that blood and synovial fluid from patients with various rheumatic diseases have high levels of both free radical-mediated F2-isoprostanes and the cyclooxygenase-derived PGF2 metabolite. This suggests that both oxidative injury and inflammation play a part to various degrees in these chronic inflammatory diseases. The measuring of arachidonic acid metabolites in body fluids opens unique opportunities for studying the role of lipid peroxidation [38]. ROS-induced genotoxic events have also been linked to mutation of p53 in RA-derived fibroblast-like synoviocytes [39]. Furthermore, it has been suggested that antioxidants systems, either enzymatic or not, are impaired in RA. Low levels of glutathione [40], tocopherols, β-carotene, and retinols and low activities of glutathione reductase and superoxide dismutase have been observed in patients with RA [41]. In a recent study, RA patients were, as usually, sub-grouped according to the presence or absence of rheumatoid factor, disease activity score, and disease duration. In addition, RA patients and healthy controls were evaluated for the oxidant-antioxidant status by monitoring ROS production, biomarkers of lipid peroxidation, protein oxidation, and DNA damage. The endogenous levels of enzymatic and nonenzymatic antioxidants were also measured. RA patients showed a marked increase in ROS formation, lipid peroxidation, protein oxidation, DNA damage, and decrease in the activity of antioxidant defense system leading to OS, which obviously contributes to tissue damage and to the chronicity of the disease [42]. Oxidative modification of proteins has been shown to elicit antibodies in a variety of diseases including systemic lupus erythematosus (SLE), alcoholic liver disease, diabetes mellitus, and finally RA. Oxidative stress processes enhance the reactivity of the adaptive response. Oxidation of carbohydrates increased the antibody response to coadministered coantigens. In addition, the use of the Schiff base-forming agent Tucaresol during immunization with protein antigen increased T-cell-dependent immune response. Direct modification of protein antigen has been shown to be required for the enhancement of the immune response [43]. In SLE, oxidatively modified DNA and low-density lipoproteins (LDL) are present and induce a premature atherosclerosis. In an animal model of SLE, immunization with 4-hydroxy-2-nonenal (HNE)-modified autoantigens accelerated epitope spreading. Pentosidine, an advanced glycation end product (AGE), and AGE-modified IgG have correlated with RA disease activity. Oxidatively modified glutamic acid decarboxylase is important in type 1 diabetes mellitus. Oxidative modification induced fragmentation of scleroderma-specific autoantigens and seems to be responsible for the production of autoantibodies. Growing evidence for the involvement of oxidative damage in autoimmunity is pointing to the administration of antioxidants could be a viable untried alternative for preventing or ameliorating autoimmune disease [37]. OS occurring during inflammation can cause proteins to become nonenzymatically damaged by glyoxidation. This process results in the generation of AGE. The immunoglobulin molecule can also undergo similar glyoxidation to generate AGE-IgG. In inflammatory arthritis, they have shown that antibodies to AGE-IgG are specifically associated with RA, whereas the actual formation of AGE-IgG is related to the intensity of the systemic inflammatory response [44].

Studies focusing on direct detection of ROS and RNS found all these biomarkers elevated in RA patients suggesting an active OS. The redox status of neutrophils sourced from SF was measured by flow cytometry in terms of total ROS and hydroxyl radicals. Neutrophils a major cellular component of the SF of RA patients and their levels of ROS correlated strongly with protein carbonylation and lipid peroxidation. In patients with RA, the strong correlation between DAS28 score, levels of ROS, and markers of oxidative damage suggests that measurement of OS could serve as a marker for monitoring disease severity [45]. In another study, RA patients had significantly higher levels of ROS (O_2^-, H_2O_2) than controls. Significant differences where monitored in serum levels of NO in patients with high activity of disease. More intensive response in samples with higher disease activity suggests that oxidative/nitrosative stress markers may be valuable in evaluating the RA progression and helpful in elucidating the mechanisms of disease pathogenesis [46]. The chronic OS in the RA synovium increases ROS production in the cellular oxidative phosphorylation and induces repetitive cycles of hypoxia/reoxygenation. The hypoxia in RA joints whose origin is a consequence of the rapid cellular proliferation induced by the inflammatory response, however, precedes inflammation at least in an animal arthritis model [47]. From the "danger model," in which the synoviocyte is an impaired cell, this sequence of events could be happening in the human disease [48]. Activated phagocytic cells can also enhance this OS during oxidative burst. Kundu et al. [49] showed neutrophils as most important phagocytes responsible for elevating OS in synovial infiltrates and peripheral blood of RA patients: The basal levels of total ROS, superoxide, and hydroxyl radicals were significantly increased in neutrophils from peripheral blood and synovial infiltrate. Furthermore, raised levels of superoxide in neutrophils of synovial infiltrate showed a positive correlation with NADPH oxidase activity in synovial fluid. However, there was no major increase in the RNS generated in monocytes from both sources.

4.2 Oxidative stress in adjuvant arthritis

In the development of AIA, not only immunological and inflammatory pathological changes are involved, but also the redox homeostasis is shifted toward increased production of ROS and RNS. Overproduction of ROS and RNS damages lipids, proteins, and DNA (also exhausts the natural enzymatic and nonenzymatic antioxidant defense), which is possible to detect with different markers of oxidation in biological structures. In human RA OS-mediated damage to lipids, proteins, and DNA and changes in enzymatic and nonenzymatic antioxidant defense are extensively studied. AIA in animals resembles the OS caused damage in human rheumatic diseases; therefore, it is a very useful tool to study process of OS during autoimmune diseases. Since there has been no standard therapy to reduce OS damage in diseases established yet, AIA could be a promising candidate for developing this type of therapy.

4.2.1 Peroxidation of lipids

4.2.1.1 4-HNE and MDA

The 4-HNE is one of the aldehydes specific to lipid peroxidation. 4-HNE is believed to be predominantly responsible from the cytopathologic effects seen during OS. Any factor compatible with stress or activity of antioxidant enzymes may trigger potentially dangerous metabolic pathway of peroxidative damage [50]. Our results showed that the level of HNE protein adducts was significantly increased on day 14 in rat AA [51]. The level of malondialdehyde (MDA) in the plasma of arthritic animals was also elevated [52–54] (**Table 1**). He et al. demonstrated an increased level of MDA in serum of AIA rats, which was significantly decreased by the administration of anthocyanins from cherries [53]. AA induced in male Sprague-Dawley rats increased plasma MDA levels, levels of glutathione, enzyme activities of SOD and GPx were decreased [55]. Also, Wang et al. demonstrated a significant increase of MDA and moreover nitrites in plasma of AIA rats [56]. Levels of anti-type II collagen antibody, nitrite/nitrate, and lipid peroxidation (levels of 4-HNE and MDA) were determined in the serum, joints, and brain. CIA elevated levels of nitrite/nitrate and 4-HNE and MDA levels in serum and the brain [57]. We also measured an increased levels of 4-HNE and MDA in plasma and the brain of AIA rats (**Tables 1** and **2**) [58].

4.2.1.2 Isoprostanes

Isoprostanes are a complex family of compounds produced from arachidonic acid via a free radical-catalyzed mechanism. They are reliable markers of lipid peroxidation. A strong link between lipid peroxidation and diseases associated with ischemia-reperfusion, atherosclerosis, and inflammation has been suggested by

Oxidative stress in plasma	MDA (μg/mL)	HNE (ng/mL)	Protein carbonyls (nmol/mL)
CO	2.4 ± 0.39	1.54 ± 0.16	391.2 ± 14.34
AIA	$5.79 \pm 0.44^{***}$	$2.5 \pm 0.19^{***}$	$457.72 \pm 11.09^{**}$

Values are expressed as average ± standard error of mean, statistical significance (ANOVA-Tukey-Kramer post hoc test): $^{**}p < 0.01$, $^{***}p < 0.01$ vs. CO.

Table 1.
Markers of oxidative stress (malondialdehyde (MDA), 4-hydroxynonenal (HNE), and protein carbonyls) in plasma of arthritic animals measured on day 28.

Oxidative stress in brain	MDA (μg/g tissue)	HNE (ng/g tissue)
CO	5.38 ± 0.73	3.26 ± 0.17
AIA	10.12 ± 1.01***	4.78 ± 0.5**

*Values are expressed as average ± standard error of mean, statistical significance (ANOVA-Tukey-Kramer post hoc test): **p < 0.01, ***p < 0.01 vs. CO.*

Table 2.
Markers of oxidative stress (malondialdehyde (MDA) and 4-hydroxynonenal (HNE) in the brain of arthritic animals measured on day 28).

elevated levels of F2-isoprostanes observed in such diseases. Quantification of F2-isoprostanes as pathophysiological markers is suitable for the investigation of lipid peroxidation in human diseases and provides an interesting biomarker of antioxidant efficacy in diseases where OS might be involved [59]. There are only few evidences about F2-isoprostanes in animal models of RA. In one of our previous experiments, we have measured an elevated level of F2-isoprostanes in plasma of AIA rats, which were significantly increased compared to control healthy animals [60]. In a CIA model, authors investigated the ability of grape seed proanthocyanidin extract (GSPE) to reduce the development of mice arthritis. They have found that CIA significantly increased the level of 8-isoprostane in plasma. Plasma levels of 8-isoprostane and serum level of collagen type II-specific IgG2a in GSPE-treated mice were significantly decreased than those in the control mice [61]. Authors demonstrated that F2-isoprostanes are increased also in the urine of CIA mice [62]. F2-isoprostanes as an important marker of lipid peroxidation should be more extensively studied in AIA animal models, to obtain a better picture about the similarity with human RA.

4.2.2 Oxidation of proteins

Protein carbonyls (aldehydes and ketones) are produced directly by oxidation or via reactions with other molecules generated by the oxidation process. Autoimmune attack, resulting from abrogation of self-tolerance, is involved in many human diseases. Autoimmune disease may be either organ specific (type 1 diabetes, thyroiditis, myasthenia gravis, and primary biliary cirrhosis) or systemic (RA, progressive systemic sclerosis, and systemic lupus erythematosus). Nearly all these diseases have autoantibodies. Autoantibodies are typically present several years prior to diagnosis of SLE and serve as markers for future disease. Inflammation, infection, drugs, ROS, and environmental factors induce formation of neo-antigens [63]. The protein thiol groups were 59% diminished by AIA. The protein carbonyls content, an indicative of protein damage, was increased by arthritis (41%). Protein damage in both liver and brain was estimated as the tissue content of protein carbonyl groups. Corroborating previous results, arthritis increased protein damage in both tissues, 55% in the liver and 51% in the brain [64]. Authors Hemshekhar et al. [65] also showed a significant decrease in total protein thiol content with reference to saline-fed rats up to 51 and 36.05%, respectively, in liver and spleen homogenates of arthritic rats [65]. In a study about protective effects of green tea extract in AIA rats, authors detected a significant OS-caused damage to proteins and lipids in the liver, brain, and plasma [66]. The antioxidant defense, reduced in arthritis, is improved by the green tea treatment, as shown in the restoration of the GSH and protein thiol levels and by the tendency for normalizing the activities of the antioxidant enzymes. In arthritis rats, we found a significant increase of protein carbonyls in plasma [66–69] (**Table 1**). This finding emphasizes the role of OS in inflammatory diseases such as AIA, not only in tissues directly affected by the disease (cartilage, bone, and skeletal muscle) (**Table 2**).

4.2.3 Production of reactive oxygen species by neutrophils

Recent evidence from animal models of RA emphasized the importance of neutrophils in the initiation and progression of AIA [70]. Progressive erosion of articular cartilage is a prominent feature of this disease. Not surprisingly, immunosuppressive approaches such as blockade of CD4+ lymphocytes effectively reduce the intensity of damage and the progression of AIA. The report of Santos et al. [71] convincingly demonstrates a requirement not only for CD4+ lymphocytes but also for neutrophils, the latter determined by the protective effects of neutrophil depletion. The sequence of events showed that CD4+ cells are necessary for the establishment of the immune response, which leads to the recruitment of neutrophils, with the involvement of cytokines (TNF-α, IL-1) and the IL-8 family of chemokines. The combination of products (oxidants, proteinases, and cytokines) from stimulated neutrophils, synovial macrophages, and lymphocytes is important to set the stage for acute and progressive polyarthritis [72]. We assessed ROS production in stimulated neutrophils of arthritic rats, and it was found to be increased, with a maximum on day 14 and 21 of AIA. Neutrophils in the whole blood of AIA animals reacted excessively to stimulation and produced 6–9 times more ROS [73]. We also demonstrated oxidative damage of tissues in AIA: ROS levels in the joint and the spleen were significantly elevated [74] (**Table 3**).

4.2.4 Levels of endogenous antioxidants

The mammal organism has several mechanisms to counteract with OS by producing antioxidants, which are either produced in situ or externally supplied with foods or supplements. The nonenzymatic antioxidants are distinguished as metabolic antioxidants and nutrient antioxidants. Metabolic antioxidants referred also as endogenous antioxidants such as glutathione, lipoid acid, L-arginine, melatonin, coenzyme Q_{10}, uric acid, bilirubin, metal-chelating proteins, and transferrin are produced by metabolic processes, while nutrient antioxidants are compounds that cannot be produced in the body and must be provided through foods or supplements, such as vitamin E, vitamin C, carotenoids, trace metals (selenium, manganese, zinc), flavonoids, and omega-3 and omega-6 fatty acids [75]. Decreased levels of nonenzymatic antioxidant glutathione and vitamin C were observed in the liver of AIA rats compared to the normal rats [76]. Antioxidant state showed that plasma vitamin E, vitamin C, vitamin A, and β-carotene were significantly lower in arthritic control rats than normal rats [77]. Reduction of plasmatic antioxidants is indicating reduced antioxidant capacity and elevation of oxidative stress during adjuvant arthritis which is similar to rheumatoid arthritis in human [78].

Chemiluminescence (RLU*s)	Spontaneous	PMA stimulated	Neutrophil count in 1 μL of blood
CO	41,802 ± 2452	150,789 ± 9159	12,174 ± 747
AIA	168,203 ± 12815***	1,165,603 ± 94470***	40,260 ± 3325***

*RLU*s, relative light units; PMA, phorbol-12-myristate-13-acetate; values are expressed as average ± standard error of mean, statistical significance (ANOVA-Tukey-Kramer post hoc test): ***$p < 0.001$ vs. CO.*

Table 3.
Spontaneous and stimulated chemiluminescence and neutrophil count in whole blood of arthritic rats.

CoQ_{10} plays a central role in the electron transport chain and as a radical-scavenging antioxidant; therefore we studied its level in plasma during AA. In our experiments the arthritis process increased significantly the level of CoQ_{10} in comparison with healthy control rats. The arthritic processes also stimulated the synthesis of CoQ_9 (dominant form of CoQ in rats) and its transport to plasma [79] (**Table 4**). In the skeletal muscle mitochondria, we have measured significant changes in levels of α- and γ-tocopherol (**Table 5**).

Similarly in AIA, also in patients with RA, a depletion of endogenous antioxidants was measured. The plasma concentration of beta-carotene and vitamin E, hemoglobin, and hematocrit were significantly lower in patients with RA than in controls. These results provide evidence for a potential role of raised lipid peroxidation and lowered enzymic and nonenzymic antioxidants in RA because of its inflammatory character. These results suggested that OS plays a very important role in the pathogenesis of RA [80, 78].

4.2.5 Changes in antioxidant enzymes

In order to protect tissues from oxidative injuries, the body possesses enzymatic antioxidant enzymatic systems such as superoxide dismutases and catalase enzymes. It has been reported that AA decreases serum or synovial SOD and catalase activities together with other endogenous antioxidant systems [81]. Ramos-Romero et al. [82] showed a decrease in splenic catalase activity and, paradoxically, an increase in splenic total and mitochondrial SOD in AIA. The decreased catalase activity could be associated with the consumption of catalase in neutralizing the H_2O_2. On the other hand, increased splenic SOD activities could reflect the response of the body to increased ROS concentrations, and/or it could be due to the fact that arthritis was in its recovery phase 1 month after its induction. Moreover, SOD increase could also be explained by the increase in the oxidative stress found in arthritic rats and by the increased TNF-α secretion present in arthritis [82]. Both OS and TNF-α are shown to induce SOD synthesis [83]. It should be added that a similar increase in SOD activity was found in the plasma of RA patients [84] and in

Plasma	CoQ9TOT (μmol/L)	CoQ10TOT (μmol/L)	αT (μmol/L)	γT (μmol/L)
CO	0.328 ± 0.023	0.031 ± 0.004	19.9 ± 1.13	0.643 ± 0.051
AIA	0.468 ± 0.044**	0.027 ± 0.003	21.6 ± 0.72	0.834 ± 0.060*

*Values are expressed as average ± standard error of mean, statistical significance (ANOVA-Tukey-Kramer post hoc test): *$p < 0.05$, **$p < 0.01$ vs. CO.*

Table 4.
Concentrations of total coenzyme Q_9 ($CoQ_{9\text{-}TOT}$), total coenzyme Q_{10} ($CoQ_{10\text{-}TOT}$), α-tocopherol (αT), and γ-tocopherol (γT) in plasma.

Skeletal muscle mitochondria	CoQ9TOT (μmol/L)	CoQ10TOT (μmol/L)	αT (μmol/L)	γT (μmol/L)
CO	43.1 ± 3.01	1.90 ± 0.160	23.0 ± 1.23	0.98 ± 0.042
AIA	32.7 ± 2.49*	1.63 ± 0.187	18.7 ± 0.829*	1.39 ± 0.155*

*Values are expressed as average ± standard error of mean, statistical significance (ANOVA-Tukey–Kramer post hoc test): *$p < 0.05$ vs. CO.*

Table 5.
Concentrations of total coenzyme Q_9 ($CoQ_{9\text{-}TOT}$), total coenzyme Q_{10} ($CoQ_{10\text{-}TOT}$), α-tocopherol (αT), and γ-tocopherol (γT) in skeletal muscle mitochondria.

the synovial membrane of mice with collagen-induced arthritis [85]. Catalase catalyzes the decomposition of hydrogen peroxide to water and oxygen, thus preventing the oxidation of biological structures by hydrogen peroxide. Authors demonstrated the elevated and LPO activity and NO level and decreased GSH, SOD, and catalase activities in AIA rats [86]. OS in AIA model is depleting antioxidant enzymes, which is in good agreement with human RA studies.

Activity of glutathione peroxidase (GPx) in blood serum and muscles of rats with AIA increased and activity of glutathione reductase (GR) in these tissues increased in comparison with the control. Probably, changes in enzyme activity are a defense response of the body to ROS generation in RA and can be a result of ROS activation or stimulation of their synthesis [87]. Similarly in the study of Sahu et al. [88], CIA increased antioxidant enzyme GPx and GR activities in joints, liver, kidney, and spleen tissues of rats.

Several pathologic factors have been suggested to be involved in the overexpression of heme oxygenase-1 HO-1 in RA lesions. In addition to superoxides and pro-inflammatory cytokines, hypoxia may play an important role in HO-1 expression in the lesions [89, 90]. AIA is an experimental model widely used to evaluate etiopathogenetic mechanisms in chronic inflammation. Devesa et al. [91] have examined the participation of HO-1 in AIA. They have found an increased nitric oxide (NO) production in the paw preceded the upregulation of HO-1, whereas selective inhibition of inducible NO synthase (iNOS) after the onset of arthritis lowered HO-1 expression, suggesting that this enzyme may depend on NO produced by iNOS. Administration of the HO-1 inhibitor protoporphyrin IX ameliorated the symptoms of arthritis. This compound significantly decreased leukocyte infiltration, erosion of articular cartilage, and osteolysis, as well as the production of inflammatory mediators. In this model, HO-1 can be involved in vascular endothelial growth factor production and angiogenesis. These results support a role for HO-1 in mediating the progression of the disease in this model of chronic arthritis [91]. Our research group showed that extra-articular manifestations of AIA are present also in lung, where the expression of heme oxygenase-1 was reduced during AIA [60].

5. Potential role of free radicals in rheumatoid cachexia

Cachexia is one of the major causes of progressive weight loss and affects up to 20% of RA patients [92]. Unlike sarcopenia, which is a normal physiological process of body mass reduction affected by aging, cachexia appears to be a secondary manifestation of an already ongoing disease [93]. Cachexia associated with RA can occur in two forms. The first form is cachectic RA or rheumatoid cachectic obesity, which is manifested by severe muscle wasting, with little or no fat mass loss. It is a less threatening form of cachexia mainly because the energy demands of muscles can be compensated by lipid metabolism [94]. The second form is rheumatoid cachexia, which is manifested by severe muscle wasting as well as fat loss.

Rheumatoid cachexia (RC) is a progressive form of RA, which is primarily thought to be caused by the abnormal production of the pro-inflammatory cytokines produced by the immune cells localized in the synovial tissue of the affected joints. Excessive concentrations of several cytokines, especially TNF-α, IL-1β, IL-6, and INF-γ, could potentially affect the intracellular mechanisms of muscle fibers, leading to severe muscle atrophy and weakness [95]. The most dominant cytokine in RA and RC pathogenesis appears to be TNF-α which acts synergically with IL-1β. When bound to their specific receptors, these cytokines cause activation of NF-κB signaling cascade. A study by Cai et al. [96] suggests that muscle atrophy is predominantly promoted by the NF-κB pathway via the activation of MuRF1 transcription factor which ultimately induces immoderate proteolysis of muscle proteins by activating the ubiquitin-proteasome system. Moreover, Castillero et al. [97] observed overexpression of MuRF1 as well as several other myogenic factors, such as atrogin-1/MAFbx ubiquitin ligases in adjuvant arthritis.

Another important pathogenic factor of RC is reduced physical activity, which appears to be the result of either poor pain management of inflamed and swollen joints, metabolic changes, or merely general caution for physical activity. Lower physical activity leads to reduced muscle fiber stimulation, which significantly disrupts the cycle of muscle proteolysis and proteosynthesis in favor of proteolysis [98]. One of the possible triggers of RC could also be increased free radical concentration and onset of OS.

As mentioned in the previous text, ROS and RNS concentrations have been reported to be elevated in the joint area as well as plasma. This may indicate that an increase in free radicals levels could also be found in skeletal muscle tissue. There are several sites of free radical production in muscles including mitochondria, sarcoplasmic reticulum, and sarcolemma [99]. As metabolically highly active organs, muscles dramatically increase their oxygen consumption during physical activity in order to compensate various energy-dependent processes. Concurrently excessive amounts of oxidants are produced, which then serve as messenger molecules in multiple intracellular cascades. The main site of free radical generation is mitochondria during aerobic metabolism and oxidative phosphorylation. It has been shown that complexes I and III and more recently complex II of mitochondrial electron transport chain are key producers of ROS in muscle fibers [100]. Several authors suggest that the major ROS produced in muscle cells is superoxide anion ($O2^{\bullet -}$), which is a very unstable radical and rapidly undergoes reduction resulting in dismutation into hydrogen peroxide (H_2O_2) [101]. Even though H2O2 is quite a stable nonradical molecule, excessive concentrations of H_2O_2 could ultimately result in increased generation of hydroxyl radical (•OH)–a highly reactive ROS which could potentially damage various cellular molecules and disrupt many intracellular mechanisms. Free radicals are also regularly produced by several enzymes such as nicotinamide adenine dinucleotide phosphate (NADPH) oxidase (Nox) family as well as xanthine oxidase (XO) [99]. In skeletal muscles only Nox 2 and Nox 4 isoforms have been found, and it is believed that both of these isoforms are localized primarily in mitochondria [102]. However, the precise mechanism by which the increased activity of these enzymes is promoted is to date poorly understood. Under physiological conditions, excessive concentrations of free radicals are regularly scavenged and converted into non-radical molecules by antioxidant defense system molecules. However, several studies have observed low concentrations of some nonenzymatic antioxidants such as GSH [41] as well as low activity of enzymatic antioxidants such as SOD and glutathione peroxidase (GPX) in RA, which could potentially affect muscle tissue [103]. It has been proposed that decreased physical activity in RC patients could play a major role in oxidative damage of muscle cells since lower muscle stimulation reduces antioxidant capacity thus causing impaired balance in oxidant-antioxidant ratio [104].

Several long-term studies have reported a number of negative effects of free radicals in muscles at the molecular level. Oxidative damage of lipids, particularly in cell membranes [105], as well as nucleic acids in the DNA [106] is of great importance to normal cellular functioning, lately there has been a great deal of emphasis on protein modifications caused by ROS in multiple diseases.

Proteins as functional units of the cell can cause great damage to the cell itself if its space conformation is disrupted. Perhaps the most common protein modification caused by imbalance of oxidative status is carbonylation of side chains of multiple amino acids such as arginine, lysine, threonine, and proline [107]. Moreover, carbonylation of proteins that are part of the contractile apparatus could be crucial in RC muscle dysfunction. Fedorova et al. [108] showed that carbonylation of actin could very much affect actomyosin ATPase activity, thus promoting subsequent muscle atrophy. Taken together, action of pro-inflammatory cytokines, mitochondrial dysfunction, and enhanced activity NADPH oxidase and xanthine oxidase contribute to the overproduction of ROS/RNS.

Decreased physical activity results in downregulation of antioxidants. Combination of these factors consequently leads to imbalance in protein synthesis and degradation resulting in muscle wasting (**Figure 4**).

5.1 Animal models in rheumatoid cachexia

Rheumatoid cachexia still remains a poorly investigated disease, and many scientists are trying to understand the exact mechanism by which the disease takes place. Several animal models of RA are used in the study of this condition. The best studied animal model to date has been CIA, which, by its characteristics, offers the most accurate comparison with humans, as the onset of this affection is relatively slow and the immune mechanisms driving the onset of cachexia are closest to rheumatoid cachexia in people [109].

Recently, Albarse et al. [110] have been investigating the development of cachexia in CIA in DBA1/J mice. In their study, they have observed significant increase in free exploratory locomotion as well as grip strength and endurance exercise performance. Additionally, they registered reduction of muscle weight in several muscles, which could indicate that mechanisms, which led to the onset of arthritis, could subsequently promote muscle atrophy and weakness.

Another model of rheumatoid arthritis and adjuvant arthritis was also used to investigate muscle wasting in male and female Lewis rats in the study of Roubenoff et al. [111]. It was shown that adjuvant-induced rats also manifested severe muscle loss when compared to control as well as pair-fed groups. This makes adjuvant arthritis suitable model for study of cachexia in chronic inflammatory diseases. Even though there have been multiple authors dedicated to unraveling the true cause of rheumatoid-induced cachexia, much more study is needed in order to

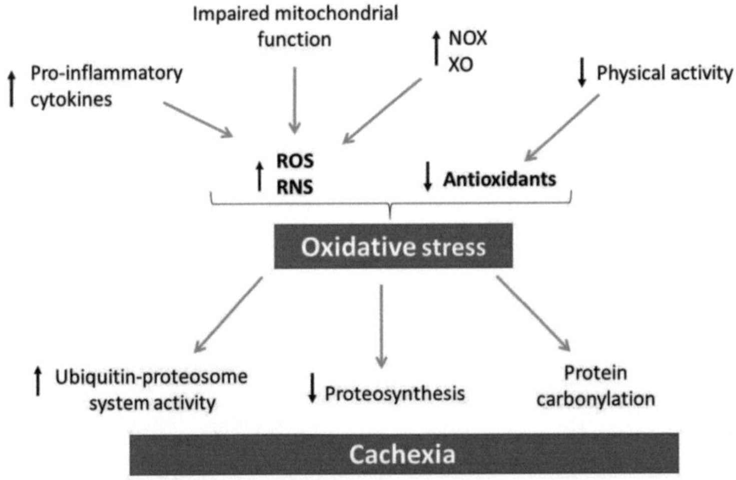

Figure 4.
Mechanism of the effect of oxidative stress on the onset of cachexia in rheumatoid arthritis. NOX, nicotinamide adenine dinucleotide phosphate (NADPH) oxidase; XO, xanthine oxidase; ROS, reactive oxygen species; RNS, reactive nitrogen species.

sufficiently understand precise mechanism by which this serious condition occurs. This could greatly improve quality of RA patients and thus contribute to modern medicine.

6. Conclusion

The animal model adjuvant arthritis gives a broad spectrum of possibilities to study different pathological mechanism of rheumatoid arthritis. One important pathological pathway is the connection between inflammation and oxidative stress, which is studied on both systemic and local levels. From our original results as well as from results reported by other authors, it is evident that treatment with compounds possessing redox balance modulating properties might be of great relevance for new strategies for therapy of rheumatoid arthritis. For this purpose, adjuvant arthritis seems to be an ideal animal model. Moreover, this animal model has also a good potential in the research of inflammatory cachexia and its pharmacological intervention.

Acknowledgements

Our experimental studies were supported by grants: APVV-15-0308, APVV SK-PT-18-0022, and VEGA 2/0115/19. We thank Martin Chrastina, MSc, for technical assistance.

Conflict of interest

The authors declare that they do not have any conflict of interest.

Author details

Silvester Ponist*, Miloslav Zloh and Katarina Bauerova
Center of Experimental Medicine, Institute of Experimental Pharmacology and Toxicology, Slovak Academy of Sciences, Bratislava, Slovakia

*Address all correspondence to: silvester.ponist@savba.sk

References

[1] Kollias G, Papadaki P, Apparailly F, et al. Animal models for arthritis: Innovative tools for prevention and treatment. Annals of the Rheumatic Diseases. 2011;**70**(8):1357-1362

[2] McInnes IB, Schett G. The pathogenesis of rheumatoid arthritis. The New England Journal of Medicine. 2011;**365**(23):2205-2219

[3] Sattar N, McCarey DW, Capell H, McInnes IB. Explaining how "high-grade" systemic inflammation accelerates vascular risk in rheumatoid arthritis. Circulation. 2003;**108**(24): 2957-2963

[4] Plenge RM. Rheumatoid arthritis genetics: 2009 update. Current Rheumatology Reports. 2009;**11**:351-356

[5] Wolfe F, Mitchell DM, Sibley JT, et al. The mortality of rheumatoid arthritis. Arthritis and Rheumatism. 1994;**37**:481-494

[6] Smolen JS, Steiner G. Therapeutic strategies for rheumatoid arthritis. Nature Reviews. Drug Discovery. 2003; **2**:473-488

[7] Smolen JS, Aletaha D, Koeller M, Weisman MH, Emery P. New therapies for treatment of rheumatoid arthritis. Lancet. 2007;**370**:1861-1874

[8] Weyand CM, Hicok KC, Conn DL, Goronzy JJ. The influence of HLA-DRB1 genes on disease severity in rheumatoid arthritis. Annals of Internal Medicine. 1992;**117**:801

[9] Klareskog L, Padyukov L, Alfredsson L. Smoking as a trigger for inflammatory rheumatic diseases. Current Opinion in Rheumatology. 2007;**19**:49-54

[10] Getts MT, Miller SD. 99th Dahlem conference on infection, inflammation and chronic inflammatory disorders: Triggering of autoimmune diseases by infections. Clinical and Experimental Immunology. 2010;**160**:15-21

[11] Hochberg MC, Johnston SS, John AK. The incidence and prevalence of extra-articular and systemic manifestations in a cohort of newly-diagnosed patients with rheumatoid arthritis between 1999 and 2006. Current Medical Research and Opinion. 2008;**24**:469-480

[12] Dayer JM, Choy E. Therapeutic targets in rheumatoid arthritis: The interleukin-6 receptor. Rheumatology. 2010;**49**:15-24

[13] Pollard L, Choy EH, Scott DL. The consequences of rheumatoid arthritis: Quality of life measures in the individual patient. Clinical and Experimental Rheumatology. 2005;**23**: S43-S52

[14] Firestein GS. Evolving concepts of rheumatoid arthritis. Nature. 2003;**423**: 356

[15] Nalbandian A, Crispin JC, Tsokos GC. Interleukin-17 and systemic lupus erythematosus: Current concepts. Clinical and Experimental Immunology. 2009;**157**:209

[16] Lally F, Smith E, Filer A, et al. A novel mechanism of neutrophil recruitment in a coculture model of the rheumatoid synovium. Arthritis and Rheumatism. 2005;**52**:3460

[17] Paleolog EM. Angiogenesis in rheumatoid arthritis. Arthritis Research. 2002;**4**:S81-S90

[18] Okamoto T. Oxidative stress in rheumatoid arthritis. In: Surh YJ, Packer L, editors. Oxidative Stress, Inflammation and Health. California: Taylor & Francis; 2005. pp. 245-270

[19] Moynagh PN. The NF-kappaB pathway. Journal of Cell Science. 2005; **118**:4589-4592

[20] Waksman BH. Immune regulation in adjuvant disease and other arthritis models: Relevance to pathogenesis of chronic arthritis. Scandinavian Journal of Immunology. 2002;**56**:12

[21] Prakken BJ, Roord S, Ronaghy A, Wauben M, Albani S, van Eden W. Heat shock protein 60 and adjuvant arthritis: A model for T cell regulation in human arthritis. Springer Seminars in Immunopathology. 2003;**25**:47-63

[22] Pearson CM, Wood FD. Studies of arthritis and other lesions induced in rats by the injection of mycobacterial adjuvant. VII. Pathologic details of the arthritis and spondylitis. The American Journal of Pathology. 1963;**42**:73-95

[23] Bush KA, Walker JS, Lee CS, Kirkham BW. Cytokine expression and synovial pathology in the initiation and spontaneous resolution phases of adjuvant arthritis: Interleukin-17 expression is upregulated in early disease. Clinical and Experimental Immunology. 2001;**123**:487-495

[24] Krieg AM. The role of CpG motifs in innate immunity. Current Opinion in Immunology. 2000;**12**:35

[25] Chu RS, Targoni OS, Krieg AM, et al. CpG oligo-deoxynucleotides act as adjuvants that switch on T helper 1 (Th1) immunity. The Journal of Experimental Medicine. 1997;**186**:1623

[26] Bendele A, McComb J, Gould T, McAbee T, Sennello G, Chlipala E, et al. Animal models of arthritis: Relevance to human disease. Toxicologic Pathology. 1999;**27**(1):134-142

[27] Berlett BS, Stadtman ER. Protein oxidation in aging, disease, and oxidative stress. The Journal of Biological Chemistry. 1997;**272**(33): 20313-20316

[28] Salzanoa S, Checconia P, Hanschmannc EM, et al. Linkage of inflammation and oxidative stress via release of glutathionylated peroxiredoxin-2, which acts as a danger signal. Proceedings of the National Academy of Sciences of the United States of America. 2014;**111**(33): 12157-12162

[29] Popa-Wagner A, Mitran S, Sivanesan S, Chang E, Buga AM. ROS and brain diseases: The good, the bad, and the ugly. Oxidative Medicine and Cellular Longevity. 2013;**2013**:963520, 14 pages

[30] Ghezzi P. Role of glutathione in immunity and inflammation in the lung. International Journal of General Medicine. 2011;**4**:105-113

[31] Ghezzi P. Protein glutathionylation in health and disease. Biochimica et Biophysica Acta, General Subjects. 2013; **1830**(5):3165-3172

[32] Fratelli M, Demol H, Puype M, et al. Identification by redox proteomics of glutathionylated proteins in oxidatively stressed human T lymphocytes. Proceedings of the National Academy of Sciences of the United States of America. 2002;**99**(6):3505-3510

[33] Hassan SZ, Gheita TA, Kenawy SA, Fahim AT, El-Sorougy IM, Abdou MS. Oxidative stress in systemic lupus erythematosus and rheumatoid arthritis patients: Relationship to disease manifestations and activity. International Journal of Rheumatic Diseases. 2011;**14**:325-331

[34] Hadjigogos K. The role of free radicals in the pathogenesis of rheumatoid arthritis. Panminerva Medica. 2003;**45**:7-13

[35] Uchida K. A lipid-derived endogenous inducer of COX-2: A bridge between inflammation and oxidative stress. Molecules and Cells. 2008;**25**: 347-351

[36] Selley ML, Bourne DJ, Bartlett MR, Tymms KE, Brook AS, Duffield AM, et al. Occurrence of (E)-4- hydroxy-2-nonenal in plasma and synovial fluid of patients with rheumatoid arthritis and osteoarthritis. Annals of the Rheumatic Diseases. 1992;**51**:481-484

[37] Kurien BT, Scofield RH. Autoimmunity and oxidatively modified autoantigens. Autoimmunity Reviews. 2008;**7**:567-573

[38] Basu S, Whiteman M, Mattey DL, Halliwell B. Raised levels of F(2)-isoprostanes and prostaglandin F (2alpha) in different rheumatic diseases. Annals of the Rheumatic Diseases. 2001; **60**(6):627-631

[39] Hitchon CA, El-Gabalawy HS. Oxidation in rheumatoid arthritis. Arthritis Research & Therapy. 2004;**6**: 265-278

[40] Kalpakcioglu B, Senel K. The interrelation of glutathione reductase, catalase, glutathione peroxidase, superoxide dismutase, and glucose-6-phosphate in the pathogenesis of rheumatoid arthritis. Clinical Rheumatology. 2008;**27**:141-145

[41] Hassan MQ, Hadi RA, Al-Rawi ZS, Padron VA, Stohs SJ. The glutathione defense system in the pathogenesis of rheumatoid arthritis. Journal of Applied Toxicology. 2001;**21**:69-73

[42] Mateen S, Moin S, Khan AQ, Zafar A, Fatima N. Increased reactive oxygen species formation and oxidative stress in rheumatoid arthritis. PLoS One. 2016;**11**(4):e0152925

[43] Allison ME, Fearon DT. Enhanced immunogenicity of aldehyde-bearing antigens: A possible link between innate and adaptive immunity. European Journal of Immunology. 2000;**30**: 2881-2887

[44] Newkirk MM, Goldbach-Mansky R, Lee J, Hoxworth J, McCoy A, Yarboro C, et al. Advanced glycation endproduct (AGE)-damaged IgG and IgM autoantibodies to IgGAGE in patients with early synovitis. Arthritis Research & Therapy. 2003;**5**:R82-R90

[45] Datta S, Kundu S, Ghosh P, De S, Ghosh A, Chatterjee M. Correlation of oxidant status with oxidative tissue damage in patients with rheumatoid arthritis. Clinical Rheumatology. 2014; **33**(11):1557-1564

[46] Veselinovic M, Barudzic N, Vuletic M, Zivkovic V, Tomic-Lucic A, Djuric D, et al. Oxidative stress in rheumatoid arthritis patients: Relationship to diseases activity. Molecular and Cellular Biochemistry. 2014;**391**:225-232

[47] Jeon CH, Ahn JK, Chai JY, et al. Hypoxia appears at pre-arthritic stage and shows co-localization with early synovial inflammation in collagen induced arthritis. Clinical and Experimental Rheumatology. 2008;**26**: 646-648

[48] Pacheco-Tena C, Gonzalez-Chavez SA. The danger model approach to the pathogenesis of the rheumatic diseases. Journal of Immunology Research. 2015; **2015**:23

[49] Kundu S, Ghosh P, Datta S, Ghosh A, Chattopadhyay S, Chatterjee M. Oxidative stress as a potential biomarker for determining disease activity in patients with rheumatoid arthritis. Free Radical Research. 2012;**46**(12):1482-1489

[50] Esterbauer H, Schaur RJ, Zollner H. Chemistry and biochemistry of 4-hydroxynonenal, malonaldehyde and related aldehydes. Free Radical Biology & Medicine. 1991;**11**:81-128

[51] Ponist S, Mihalova D, Jancinova V, Snirc V, Ondrejickova O, Mascia C, et al. Reduction of oxidative stress in adjuvant arthritis. Comparison of

efficacy of two pyridoindoles: Stobadine dipalmitate and SMe1.2HCl. Acta Biochimica Polonica. 2010;**57**(2): 223-228

[52] Bauerova K, Paulovicova E, Mihalova D, Svik K, Ponist S. Study of new ways of supplementary and combinatory therapy of rheumatoid arthritis with immunomodulators. Glucomannan and Imunoglukán in adjuvant arthritis. Toxicology and Industrial Health. 2009;**25**(4-5):329-335

[53] He YH, Zhou J, Wang YS, Xiao C, et al. Anti-inflammatory and anti-oxidative effects of cherries on Freund's adjuvant-induced arthritis in rats. Scandinavian Journal of Rheumatology. 2006;**35**(5):356-358

[54] Sotnikova R, Ponist S, Navarova J, Mihalova D, Tomekova V, Strosova M, et al. Effects of sesame oil in the model of adjuvant arthritis. Neuroendocrinology Letters. 2009; **30**(1):22-24

[55] Tastekin N, Aydogdu N, Dokmeci D, Usta U, Birtane M, Erbas H, et al. Protective effects of L-carnitine and alpha-lipoic acid in rats with adjuvant arthritis. Pharmacological Research. 2007;**56**(4):303-310

[56] Wang B, Yao YY, Zhou AW, Ge ZD, Chen MZ, Xu SY. Protective effect of total glucosides of paeony on joint damage in adjuvant arthritis rats. Chinese Journal of Pharmacology and Toxicology. 1996;**10**(3):211-214

[57] Jiménez-Caliani AJ, Jiménez-Jorge S, Molinero P, Guerrero JM, Fernández-Santos JM, Martín-Lacave I, et al. Dual effect of melatonin as proinflammatory and antioxidant in collagen-induced arthritis in rats. Journal of Pineal Research. 2005;**38**(2):93-99

[58] Poništ S, Slovák L, Kuncírová V, Fedorova T, Logvinenko A, Muzychuk O, et al. Inhibition of oxidative stress in brain during rat adjuvant arthritis by carnosine, trolox and novel trolox-carnosine. Physiological Research. 2015;**64**(4): S489-S496

[59] Cracowski JL, Durand T, Bessard G. Isoprostanes as a biomarker of lipid peroxidation in humans: Physiology, pharmacology and clinical implications. Trends in Pharmacological Sciences. 2002;**23**(8):360-366

[60] Bauerova K, Acquaviva A, Ponist S, Gardi C, Vecchio D, Drafi F, et al. Markers of inflammation and oxidative stress studied in adjuvant-induced arthritis in the rat on systemic and local level affected by pinosylvin and methotrexate and their combination. Autoimmunity. 2015;**48**(1):46-56

[61] Cho ML, Heo YJ, Park MK, Oh HJ, Park JS, Woo YJ, et al. Grape seed proanthocyanidin extract (GSPE) attenuates collagen-induced arthritis. Immunology Letters. 2009;**124**(2): 102-110

[62] McCubbin MD, Hou G, Abrams GD, Dick R, Zhang Z, Brewer GJ. Tetrathiomolybdate is effective in a mouse model of arthritis. The Journal of Rheumatology. 2006;**33**(12):2501-2506

[63] Kurien BT, Hensley K, Bachmann M, Scofield RH. Oxidatively modified autoantigens in autoimmune diseases. Free Radical Biology & Medicine. 2006;**41**(4):549-556

[64] Gonçalves GA, Soares AA, Correa RCG, Barros L, Haminiuk CWI, Peralta RM, et al. Merlot grape pomace hydroalcoholic extract improves the oxidative and inflammatory states of rats with adjuvant-induced arthritis. Journal of Functional Foods. 2017;**33**: 408-418

[65] Hemshekhar M, Thushara RM, Jnaneshwari S, Devaraja S, Kemparaju K, Girish KS. Attenuation of

adjuvant-induced arthritis by dietary sesamol via modulation of inflammatory mediators, extracellular matrix degrading enzymes and antioxidant status. European Journal of Nutrition. 2013;**52**(7):1787-1799

[66] Gonçalves GA, de Sá-Nakanishi AB, Wendt MM, Comar JF, Bersani Amado CA, Bracht A, et al. Green tea extract improves the oxidative state of the liver and brain in rats with adjuvant-induced arthritis. Food & Function. 2015;**6**(8):2701-2711

[67] Bauerova K, Valentova J, Ponist S, Navarova J, Komendova D, Mihalova D. Effect of copper complexes on the development of adjuvant arthritis: Therapeutic and toxicological aspects. Biologia. 2005;**60**(17):65-68

[68] Strosova M, Karlovska J, Spickett CM, Orszaghova Z, Ponist S, Bauerova K, et al. Modulation of SERCA in the chronic phase of adjuvant arthritis as a possible adaptation mechanism of redox imbalance. Free Radical Research. 2009;**43**(9):852-864

[69] Ponist S, Drafi F, Kuncirova V, Mihalova D, Ondrejickova O, Trunova O, et al. Anti-inflammatory effect of carnosine on rat adjuvant arthritis. International congress on carnosine in exercise and disease, July 10-12, 2011 Ghent (Belgium). In: Programme and Abstract Book. 2011. p. 41

[70] Cross A, Barnes T, Bucknall RC, Edwards SW, Moots RJ. Neutrophil apoptosis in rheumatoid arthritis is regulated by local oxygen tensions within joints. Journal of Leukocyte Biology. 2006;**80**(3):521-528

[71] Santos LL, Morand EF, Hutchinson P, Boyce NW, Holdsworth SR. Anti-neutrophil monoclonal antibody therapy inhibits the development of adjuvant arthritis. Clinical and Experimental Immunology. 1997;**107**:248-253

[72] Ward PA. Neutrophils and adjuvant arthritis. Clinical and Experimental Immunology. 1997;**107**:225-226

[73] Nosal R, Jancinova V, Petríkova M, Ponist S, Bauerova K. Suppression of oxidative burst of neutrophils with methotrexate in rat adjuvant arthritis. Chemicke Listy. 2007;**101**:243-244

[74] Drabikova K, Perecko T, Nosal R, Bauerova K, Ponist S, Mihalova D, et al. Glucomannan reduces neutrophil free radical production in vitro and in rats with adjuvant arthritis. Pharmacological Research. 2009;**59**(6):399-403

[75] Pham-Huy LA, He H, Pham-Huy C. Free radicals, antioxidants in disease and health. International Journal of Biomedical Sciences. 2008;**4**(2):89-96

[76] Devi PR, Kumari SK, Kokilavani C. Effect of *Vitex negundo* leaf extract on the free radicals scavengers in complete Freund's adjuvant induced arthritic rats. Indian Journal of Clinical Biochemistry. 2007;**22**(1):143-147

[77] Doha AM, Sahar YAO. In vivo evaluation of antioxidant and anti-inflammatory activity of different extracts of date fruits in adjuvant arthritis. Polish Journal of Food and Nutrition Sciences. 2004;**54**(4):397-402

[78] Sarban S, Kocyigit A, Yazar M, Isikan UE. Plasma total antioxidant capacity, lipid peroxidation, and erythrocyte antioxidant enzyme activities in patients with rheumatoid arthritis and osteoarthritis. Clinical Biochemistry. 2005;**38**(11):981-986

[79] Bauerova K, Kucharska J, Ponist S, Gvozdjakova A. Coenzyme Q10 supplementation in adjuvant arthritis (pre-clinical study). In: Gvozdjakova A, editor. Mitochondrial Medicine: Mitochodrial Metabolism, Diseases,

Diagnosis and Therapy. Netherlands: Springer; 2008. pp. 340-342

[80] Kamanli A, Naziroğlu M, Aydilek N, Hacievliyagil C. Plasma lipid peroxidation and antioxidant levels in patients with rheumatoid arthritis. Cell Biochemistry and Function. 2004;**22**(1): 53-57

[81] Kripa KG, Chamundeeswari D, Thanka J, Uma Maheswara Reddy C. Modulation of inflammatory markers by the ethanolic extract of *Leucas aspera* in adjuvant arthritis. Journal of Ethnopharmacology. 2011;**134**: 1024-1027

[82] Ramos-Romero S, Pérez-Cano FJ, Pérez-Berezo T, Castellote C, Franch A, Castell C. Effect of a cocoa flavonoid-enriched diet on experimental autoimmune arthritis. British Journal of Nutrition. 2012;**107**:523-532

[83] Xu Y, Kiningham KK, Devalaraja MN, Yeh CC, Majima H, Kasarskis EJ, et al. An intronic NF-kappaB element is essential for induction of the human manganese superoxide dismutase gene by tumor necrosis factor-alpha and interleukin-1. DNA and Cell Biology. 1999;**18**:709-722

[84] Bhowmick K, Chakraborti G, Gudi NS, Kutty Moideen AV, Shetty HV. Free radical and antioxidant status in rheumatoid arthritis. Indian Journal of Rheumatology. 2008;**3**:8-12

[85] Kasama T, Kobayashi K, Sekine F, Negishi M, Ide H, Takahashi T, et al. Follow-up study of lipid peroxides, superoxide dismutase and glutathione peroxidase in the synovial membrane, serum and liver of young and old mice with collagen induced arthritis. Life Sciences. 1988;**43**:1887-1896

[86] Liu JY, Hou YL, Cao R, Qiu HX, Cheng GH, Tu R, et al. Protodioscin ameliorates oxidative stress, inflammation and histology outcome in complete Freund's adjuvant induced arthritis rats. Apoptosis. 2017;**22**(11): 1454-1460

[87] Kryl'skii ED, Popova TN, Kirilova EM. Activity of glutathione antioxidant system and NADPH-generating enzymes in rats with experimental rheumatoid arthritis. Bulletin of Experimental Biology and Medicine. 2015;**160**(1):24-27

[88] Sahu D, Saroha A, Roy S, Das S, Srivastava PS, Das HR. Suramin ameliorates collagen induced arthritis. International Immunopharmacology. 2012;**12**(1):288-293

[89] Kitamura A, Nishida K, Komiyama T, Doi H, Kadota Y, Yoshida A, et al. Increased level of heme oxygenase-1 in rheumatoid arthritis synovial fluid. Modern Rheumatology. 2011;**21**(2):150-157

[90] Kobayashi H, Takeno M, Saito T, Takeda Y, Kirino Y, Noyori K, et al. Regulatory role of heme oxygenase 1 in inflammation of rheumatoid arthritis. Arthritis and Rheumatism. 2006;**54**(4): 1132-1142

[91] Devesa I, Ferrándiz ML, Guillén I, Cerdá JM, Alcaraz MJ. Potential role of heme oxygenase-1 in the progression of rat adjuvant arthritis. Laboratory Investigation. 2005;**85**(1):34-44

[92] van Bokhorst de van der Schueren MA, Nicole PC, Konijn NPC, Irene EM, Bultink IEM, Lems WF, Earthman CP, Van Tuyl LH. Relevance of the new pre-cachexia and cachexia definitions for patients with rheumatoid arthritis. Clinical Nutrition 2012;31: 1008-1010

[93] Morley JE, Thomas DR, Wilson MMG. Cachexia: Pathophysiology and clinical relevance. The American Journal of Clinical Nutrition. 2006;**83**:735-743

[94] Lemmey AB, Williams SL, Marcora SM, Jones J, Maddison PJ. Are the benefits of a high-intensity progressive resistance training program sustained in rheumatoid arthritis patients? A 3-year followup study. Arthritis Care and Research. 2012;**64**(1): 71-75

[95] Roubenoff R, Roubenoff RA, Cannon JG, Kehayias JJ, Zhuand H, Dawson-Hughes B, et al. Rheumatoid cachexia: Cytokine-driven hypermetabolism accompanying reduced body cell mass in chronic inflammation. The Journal of Clinical Investigation. 1994;**93**(6):2379-2386

[96] Cai D, Frantz JD, Tawa NE Jr, Melendez PA, Oh BC, Lidov HG, et al. IKKbeta/NF-kappaB activation causes severe muscle wasting in mice. Cell. 2004;**119**(2):285-298

[97] Castillero E, Martín AI, López-Menduiña M, Granado M, Villanúa MA, López-Calderón A. IGF-1 system, atrogenes and myogenic regulatory factors in arthritis induced muscle wasting. Molecular and Cellular Endocrinology. 2009;**309**:8-16

[98] Roubenoff R. Rheumatoid cachexia: A complication of rheumatoid arthritis moves into the 21st century. Arthritis Research & Therapy. 2009;**11**(2):108

[99] Ábrigo J, Elorza AA, Riedel CA, Vilos C, Simon F, Cabrera D, et al. Role of oxidative stress as key regulator of muscle wasting during cachexia. Oxidative Medicine and Cellular Longevity. 2018;**2018**:2063179

[100] Murphy MP. How mitochondria produce reactive oxygen species. The Biochemical Journal. 2009;**417**:1-13

[101] Reid MB, Moylan JS. Beyond atrophy: Redox mechanisms of muscle dysfunction in chronic inflammatory disease. The Journal of Physiology. 2011; **589**:2171-2179

[102] Sun QA, Hess DT, Nogueira L, Yong S, Bowles DE, Eu J, et al. Oxygen-coupled redox regulation of the skeletal muscle ryanodine receptor – Ca2+ release channel by NADPH oxidase 4. Proceedings of the National Academy of Sciences of the United States of America. 2011;**108**(38):16098-16103

[103] Bae SC, Kim SJ, Sung MK. Inadequate antioxidant nutrient intake and altered plasma antioxidant status of rheumatoid arthritis patients. Journal of the American College of Nutrition. 2003;**22**(4):311-315

[104] Bouzid MA, Filaire E, Matran R, Robin S, Fabre C. Lifelong voluntary exercise modulates age-related changes in oxidative stress. International Journal of Sports Medicine. 2018;**39**(1):21-28

[105] Catalá A. Five decades with polyunsaturated fatty acids: Chemical synthesis, enzymatic formation, lipid peroxidation and its biological effect. Journal of Lipids. 2013;**2013**:710290

[106] Meunier B, Pratviel G, Bernadou J. Active species involved in oxidative DNA cleavage. Bulletin de la Société Chimique de France. 1994;**131**:933-943

[107] Stadtman ER, Levine RL. Free radical-mediated oxidation of free amino acids and amino acid residues in proteins. Amino Acids. 2003;**25**(3-4): 207-218

[108] Fedorova M, Kuleva N, Hoffmann R. Identification of cysteine, methionine and tryptophan residues of actin oxidized in vivo during oxidative stress. Journal of Proteome Research. 2010;**9**(3):1598-1609

[109] Williams RO. Collagen-induced arthritis as a model for rheumatoid arthritis. Methods in Molecular Medicine. 2004;**98**:207-216

[110] Albarse PVG, Lora PS, Silva JMS, Santo RCE, Freitas EC, de Oliveira MS,

et al. Collagen-induced arthritis as animal model of rheumatoid cachexia. Journal of Cachexia, Sarcopenia and Muscle. 2018;**9**(3):603-612

[111] Roubenoff R, Freeman LM, Smith DE, Abad LW, Dinarello CA, Kehayias JJ. Adjuvant arthritis as a model of inflammatory cachexia. Arthritis and Rheumatism. 1997;**40**(3): 534-539

Chapter 13

Effects of Morphometric Indicators on Incubation Values of Eggs and Sex of the Chicks of the Light Hen Hybrids

Milena Milojević, Živan Jokić and Sreten Mitrović

Abstract

The aim of this study was to establish incubation values of eggs (egg fertilization, absolute and relative embryo mortality, and hatchability of male and female chicks), morphometric indicators (preincubation egg mass; length, width, and egg shape index; hatched female and male chicken mass and their relative share in the egg mass before incubation), and the phenotype correlation between some traits in the younger parent flock (YF_{33}—33 weeks) and the older flock (OF_{49}—49 weeks) of the light Institut de Sélection Animale (ISA) Brown hybrid. With regard to incubation values, the younger flock (YF_{33}) demonstrated better incubation results than the older flock (OF_{49}). The egg fertilization rate was 95.24 and 94.22%, respectively, chick hatchability as the percentage of the total of incubated eggs was 86.51 and 84.89%, respectively, and chick hatchability as the percentage of the total of fertilized eggs was 90.83 and 90.09%, respectively. Embryo mortality rate was 8.73 and 9.17% (YF_{33}), and 9.33 and 9.91% (OF_{49}). Regardless of the parent flock age, eggs that hatched female chicks had lower values of observed morphometric traits than those that hatched male chicks, except for the egg shape index (77.49–77.47%, respectively) which was higher by 0.02% in eggs which hatched female chicks, but this difference was not statistically significant ($P > 0.05$). Contrary to effects of the chick sex, the parent flock age had considerably larger effect on observed morphometric traits, as all morphometric indicators of eggs and hatched chicks of both sexes in the older flock (OF_{59}) had statistically significantly higher values ($P < 0.001$) than in the younger flock (YF_{33}). The only exception is the relative share of the chicken in the egg mass where the measured difference (−0.03%) was not statistically significant ($P > 0.05$). The phenotype correlation coefficients determined (r_p) between the egg mass before the incubation period and the egg shape index were statistically significant ($P < 0.01$; $P < 0001$), except between the egg mass of eggs which hatched female chicks and the egg shape index in the young flock (YF_{33}), whereby the calculated coefficient ($r_{xy} = 0.107$) was not statistically confirmed ($P > 0.05$). Furthermore, the egg mass and hatched chicken mass of both sexes increased with the age of the parent flock, and statistically significant absolute phenotype correlation ($P < 0.001$) was determined between these two indicators.

Keywords: morphometric measurements, sex, eggs, chicks

1. Introduction

Keeping, breeding, and management of the parent flock during the eggs for fertilization production period, as well as the incubation of eggs for hatching, are very specific, specialized, and complex stages of the production process. Apart from the genotype (breed, hybrid), some nongenetic factors (age of the parent flock, keeping system and diet, morphometric egg traits, egg storage period, incubation conditions) play important role in the management of the one-day chicken production of the heavy hybrid (meat) and the light type hybrid (eggs for consumption), resulting from the specifics of the poultry reproduction.

The main difference between the reproduction of the birds—poultry—and the mammals is the fact that birds do not give birth to live offspring as mammals do. Instead, the new organism develops outside of the female's uterus, in the egg. The poultry breeds by means of eggs which must be fertilized in order to produce the young—offspring. Well fertilized egg for hatching is an embryo "package" with all the necessary nutrients which facilitate its development until it's hatched and for another few days after the hatching [1].

Effects of the nongenetic factors, primarily effects of age on egg incubation values, but also effects of the egg mass, length of the storage period, mechanical and structural egg characteristics on the effective production of the one-day chicken, were extensively studied, particularly in the case of the broiler parents (heavy hen hybrids). Pure breeds and parent flocks of various light line hybrid hens were studied on a much smaller scale in this context.

Studies of Skewea et al. [1–14] were conducted, among other reasons, to demonstrate effects of age of different broiler parent flocks of various hybrids on incubation values of eggs (fertilization rate, hatchability, and the quality of one-day old chicken), as well as their phenotype correlation. The general observations would be that the egg mass and the newly hatched chicken mass increases with the age of the broiler parents; the fertilization and hatchability rate reaches its peak in the middle of the production cycle; relative share of the chick in the egg mass increases slightly with the age; and while the egg shape index in various stages of the production cycle demonstrated substantial variability. The absolute phenotype correlation between the egg mass before incubation and the one-day old chicks was determined in most of the cases, regardless of the chicken sex.

Narushin and Romanov [5, 15] point out that the egg shape index has significant effects on incubation indicators, and thus that eggs of abnormal shapes should not be used for incubation because they prevent the normal embryo development which results in increased embryo mortality during the incubation period.

This observation was confirmed by Mitrović et al. [16] who studied mechanical (physical) and incubation values of eggs of two different hen breeds (Naked Neck hen and Sombor kaporka) kept in semi-intensive keeping systems. In their study, the egg shape index was 71.01 and 72.04%, respectively, and the relative share of the chicken in the egg mas was 73.61 and 76.68%, respectively. Abanikannda and Leigh [17] report higher egg shape index in eggs which hatched male chicks (Anak and Marshall hybrids) than in eggs which hatched female chicks (75.25 and 74.53%; 76.27 and 76.00%). In case of the Ross hybrid, the egg shape index measured was 76.09% (male chicks) and 76.41% (female chicks).

In general, the literature survey shows that most of the authors studied effects of the nongenetic factors on egg incubation values and the production of one-day chicken of both sexes coming from broiler parents. This is understandable since both sexes of the heavy hybrids are used for the meat production. However, in case of light hybrids egg incubation, male chicks are generally destroyed, while female chicks are used for the breeding and the commercial egg production.

However, Lichovníková et al. [18, 19] show in their studies that the male chicks of ISA Brown and Hy-Line can be successfully used in organic production of the quality broiler meat. In line with this, studies [20–22] point out that for humane reasons and from the animal welfare aspect, the fattening of the male chicks of light line hybrids in intensive broiler meat production is not economically justified, but in extensive and semiextensive conditions and with a prolonged fattening period, it is possible to produce quality meat with a higher protein share and a lower abdominal fat percentage.

Apart from the abovementioned authors [16, 17], research studying effects of nongenetic factors on the quality of hatching eggs and on incubation results and the quality of hatched chicks of male and female one-day chicks of the parent flocks of different breeds and light line hybrids was conducted by [5, 23–28], as well as [29, 30]. Similar research, but in case of the wild/nondomesticated birds (sparrow-*Passer*, falcon-*Falco*, northern lapwing-*Vanellus vanellus*), was conducted by [31–33].

For these reasons, the main purpose of this study was to investigate the effects of light ISA Brown hybrid parent flock's age on the morphometric egg traits (egg mass, egg length and width, egg shape index, egg volume, absolute and relative egg mass loss until day 18 of incubation, hatched female and male chicks mass, chicken share in the egg mass) and the incubation values of eggs for hatching (fertilization rate and chicken hatchability), as well as determination of the phenotype correlation between more important among the observed traits.

The ultimate aim of this research was to attempt to establish the sex of the future offspring on the basis of the morphometric measurements of the mentioned pre-incubation characteristics of eggs, which would potentially allow to incubate eggs selectively in the future, that is, only those eggs which will hatch female chicks—potential laying hens producing eggs for consummation.

2. Materials and methods

The experimental part of this study—incubation of the light line ISA Brown hybrid parent flock eggs—was conducted in the incubation station of the private poultry farm "Jugokoka," in Belgrade, Serbia.

Eggs originated from the flock bred on the parent farm in an installation which housed 5000 laying hens and 620 roosters during the production phase. With the purpose to determine morphometric egg traits and incubation egg values as well as the sex of the hatched chicks, eggs were collected from both the younger flock (YF) aged 33 weeks (YF_{33}) and the older flock (OF) when the flock was 49 weeks old (OF_{49}).

In both cases, incubated eggs were 4 or 5 days old. A total of 252 eggs were randomly collected for the first incubation (YF), and 225 for the second incubation (OF)—a total of 477 eggs. All eggs were laid in incubators of equal capacities and produced by the same manufacturer. The number and percentage of fertilized eggs were recorded at each round of laying of eggs into the incubator, as well as the number and percentage of hatched chicks out of the number of incubated and fertilized eggs, the number and percentage of eggs with dead embryos, and the progression of the mass loss until day 18 of the incubation period. This means that morphometric indicators of all eggs were individually measured before these were laid into the incubator. Egg mass, length and width of each egg were measured, marked on the shell with pencil, and each egg was disinfected in formaldehyde vapors. Upon completion of the incubation period, individual mass of the newly hatched chicks, both female and male, was measured (different sexes of this breed have different feather color).

Before the eggs were laid into the incubator, following measurements were made: egg shape index and egg volume, absolute and relative egg mass loss until day 18 of incubation, and the percentage of the chicken in the egg mass (relative share of the chicken in the egg mass).

Egg shape index (ESI) was calculated according to the following formula: ESI = EW (egg width)/EL (egg length) × 100. Particular attention was given to those eggs which hatched live and healthy chicken.

Egg volume was calculated according to the formula [34]: $V = (\pi/6) \times L \times W^2$ where V = egg volume; W = egg width, L = egg length; π = constant, 3.1416.

When moved to the incubator hatching tray (day 18), eggs were individually placed in separate compartments in order to be sure which chicken was hatched from which egg. Based on the egg mass and the newly hatched chicken mass, relative share of the chicken in the egg mass was calculated, that is, chick percentage (CP) in the egg mass, according to the following formula: –CP = [(chick mass – CM)/(egg mass – EM) × 100].

Information obtained in this way was entered into a database designed by the statistical software IBM SPSS statistics Version 22 (2013). Basic data processing was conducted by applying the standard variation statistical methods (descriptive statistics): arithmetic mean ($\bar{x}$), [...] arithmetic mean error ($S_{\bar{x}}$), standard deviation (S), and variation coefficient (VC).

Difference in significance between observed morphometric traits was tested by application of the corresponding variance analysis (two-level factorial experiment—2 flock ages × 2 sexes) with uneven numbers of repetition per treatment—classes, including interaction.

Following is the mathematical model of the variance analysis:

$Y_{ijk} = \mu + FA_i + ESC + (FA \times ESC)_{ij} + e_{ijk}$,

where:

Y_{ijk}—observed morphometric value in i-flock age, of the j-egg/chick sex, and k-repetition;

μ—general mean;

FA_i—effect of i-flock age (YF_{33} and OF_{49});

ESC_j—effect j-egg sex-chick during incubation period;

$(FA \times ESC)_{ij}$—effect of interaction between i-flock age and j-chick sex;

e_{ijk}—accidental error.

Based on the variance analysis and the results of the $F_{exp.}$ values, all significant and very significant differences were graded by Tukey test. Additionally, phenotype correlation coefficients (r_p) between observed morphometric traits, primarily between the egg mass and egg shape index and other morphometric indicators, with stress on the chick's sex, were calculated according to the relevant formula.

3. Results

For the initial experiment, a total of 477 eggs for hatching were incubated in two different time periods. When the parent flocks were 33 weeks old (YF_{33}) and 49 weeks old (OF_{49}), 252 eggs taken from the younger flock (YF_{33}) and 225 eggs from the older flock (OF_{49}) were incubated.

Regardless of the age and sex of the hatched chicks, total number of fertilized eggs was 452 (94.76%), nonfertilized 25 (5.24%), and there were 43 eggs with dead embryo (9.01% out of the number of eggs incubated; 9.51% out of the number of eggs fertilized), while the total number of hatched chicks was 409 (85.74% out of incubated eggs; 90.49% out of fertilized eggs).

In general, we can conclude that the younger flock (YF_{33}) had higher incubation values than the older flock (OF_{49}) as the fertilization rate was 95.24 and 94.22%,

Indicators	$\bar{x}$	*n*	$S_{\bar{x}}$	*S*	VC
Young parent flock (YF33)					
Female chicks					
Egg mass before incubation (g)	58.06	118	0.28	3.09	5.32
Egg length (cm)	5.48	118	0.01	0.13	2.37
Egg width (cm)	4.29	118	0.01	0.11	2.56
Egg shape index (%)	78.18	118	0.10	1.13	1.44
Egg volume (cm^3)	52.82	118	0.34	3.71	7.03
Egg mass loss until day 18 of incubation (g)	6.58	118	0.04	0.48	7.29
Egg mass loss until day 18 of incubation (%)	11.34	118	0.06	0.65	5.73
One day chick mass (g)	38.80	118	0.21	2.28	5.88
Relative chick share in the egg mass (%)	66.84	118	0.14	1.49	2.23
Male chicks					
Egg mass before incubation (g)	58.83	100	0.35	3.48	5.91
Egg length (cm)	5.51	100	0.01	0.13	2.36
Egg width (cm)	4.30	100	0.01	0.11	2.56
Egg shape index (%)	78.15	100	0.09	0.94	1.20
Egg volume (cm^3)	53.51	100	0.38	3.83	7.16
Egg mass loss until day 18 of incubation (g)	6.63	100	0.05	0.53	7.94
Egg mass loss until day 18 of incubation (%)	11.28	100	0.07	0.73	6.47
One day chick mass (g)	39.57	100	0.26	2.61	6.59
Relative chick share in the egg mass (%)	67.26	100	0.18	1.84	2.74
Both sexes					
Egg mass before incubation (g)	58.41	218	0.22	3.29	5.63
Egg length (cm)	5.49	218	0.01	0.13	2.37
Egg width (cm)	4.29	218	0.01	0.11	2.56
Egg shape index (%)	78.16	218	0.07	1.05	1.34
Egg volume (cm^3)	53.14	218	0.26	3.78	7.11
Egg mass loss until day 18 of incubation (g)	6.60	218	0.03	0.50	7.58
Egg mass loss until day 18 of incubation (%)	11.31	218	0.05	0.69	6.10
One day chick mass (g)	39.15	218	0.17	2.46	6.28
Relative chick share in the egg mass (%)	67.03	218	0.11	1.67	2.49
Old parent flock (OF49)					
Female chicks					
Egg mass before incubation (g)	62.92	101	0.21	2.12	3.37
Egg length (cm)	5.71	101	0.01	0.09	1.58
Egg width (cm)	4.38	101	0.01	0.07	1.60
Egg shape index (%)	76.69	101	0.03	0.33	0.43
Egg volume (cm^3)	57.35	101	0.28	2.77	4.83
Egg mass loss until day 18 of incubation (g)	7.23	101	0.04	0.42	5.81
Egg mass loss until day 18 of incubation (%)	11.48	101	0.04	0.42	3.66
One day chick mass (g)	42.14	101	0.16	1.62	3.84
Relative chick share in the egg mass (%)	66.96	101	0.06	0.61	0.91
Male chicks					
Egg mass before incubation (g)	63.90	90	0.31	2.98	4.66
Egg length (cm)	5.73	90	0.01	0.10	1.74
Egg width (cm)	4.40	90	0.01	0.09	2.04
Egg shape index (%)	76,72	90	0.03	0.25	0.33
Egg volume (cm^3)	58.05	90	0.35	3.30	5.69
Egg mass loss until day 18 of incubation (g)	7.47	90	0.05	0.53	7.09
Egg mass loss until day 18 of incubation (%)	11.68	90	0.04	0.38	3.25
One day chick mass (g)	42.93	90	0.23	2.22	5.17
Relative chick share in the egg mass (%)	67,16	90	0,06	0,53	0,79

Indicators	$\bar{x}$	n	$S\bar{x}$	S	VC
Both sexes					
Egg mass before incubation (g)	63.38	191	0.19	2.60	4.10
Egg length (cm)	5.72	191	0.01	0.10	1.75
Egg width (cm)	4.39	191	0.01	0.08	1.82
Egg shape index (%)	76.70	191	0.02	0.30	0.39
Egg volume (cm^3)	57.68	191	0.22	3.03	5.28
Egg mass loss until day 18 of incubation (g)	7.34	191	0.03	0.48	6.54
Egg mass loss until day 18 of incubation (%)	11.57	191	0.03	0.41	3.54
One day chick mass (g)	42.51	191	0.14	1.96	4.61
Relative chick share in the egg mass (%)	67.06	191	0.04	0.58	0.86
Young and old parent flock(YF33 and OF49)					
Female chicks					
Egg mass before incubation (g)	60.30	219	0.24	3.62	6.00
Egg length (cm)	5.59	219	0.01	0.16	2.86
Egg width (cm)	4.33	219	0.01	0.10	2.31
Egg shape index (%)	77.49	219	0.08	1.13	1.46
Egg volume (cm^3)	54.91	219	0.27	4.01	7.30
Egg mass loss until day 18 of incubation (g)	6.88	219	0.04	0.55	7.99
Egg mass loss until day 18 of incubation (%)	11.41	219	0.04	0.56	4.91
One day chick mass (g)	40.34	219	0.18	2.60	6.44
Relative chick share in the egg mass (%)	66.90	219	0.08	1.17	1.75
Male chicks					
Egg mass before incubation (g)	61.23	190	0.30	4.12	6.73
Egg length (cm)	5.61	190	0.02	0.16	2.85
Egg width (cm)	4.35	190	0.01	0.11	2.53
Egg shape index (%)	77.47	190	0.07	1.00	1.29
Egg volume (cm^3)	55.66	190	0.31	4.24	7.62
Egg mass loss until day 18 of incubation (g)	7.03	190	0.05	0.67	9.53
Egg mass loss until day 18 of incubation (%)	11.47	190	0.04	0.62	5.40
One day chick mass (g)	41.16	190	0.21	2.95	7.17
Relative chick share in the egg mass (%)	67.21	190	0.10	1.38	2.05
Both sexes					
Egg mass before incubation (g)	60.73	409	0.19	3.88	6.39
Egg length (cm)	5.60	409	0.01	0.16	2.86
Egg width (cm)	4.34	409	0.01	0.11	2.53
Egg shape index (%)	77.48	409	0.05	1.07	1.38
Egg volume (cm^3)	55.26	409	0.20	4.13	7.48
Egg mass loss until day 18 of incubation (g)	6.95	409	0.03	0.62	8.92
Egg mass loss until day 18 of incubation (%)	11.44	409	0.03	0.59	5.15
One day chick mass (g)	40.72	409	0.14	2.80	6.88
Relative chick share in the egg mass (%)	67.04	409	0.06	1.28	1.91

Table 1.
Mean values and variability of morphometric traits of eggs (chicks) which hatched chicks.

respectively, chick hatchability percentage out of the number of incubated (laid) eggs was 86.51 and 84.89%, respectively, and 90.83 and 90.09% out of the number of fertilized eggs. Embryo mortality was 8.73 and 9.17% (YF_{33}), while in the older flock (OF_{49}), it was 9.33 and 9.91%.

Detailed measurements of morphometric indicators of eggs which hatched female and male chicks (the most important egg category) were made. Their mean values and absolute and relative variability measures are given in **Table 1**, and the difference significance is given in **Table 2**.

In the younger flock (YF_{33}), average mass of eggs which hatched female chicks was 58.06 g, and 58.83 g was the mass of eggs which hatched male chicks (**Table 1**). Difference determined (−0.77 g) was not statistically confirmed

(P > 0.05). Correspondingly, in the older flock (OF_{49}), the average mass of eggs which hatched female chicks was smaller by 0.98 g (**Tables 1** and **2**). The difference was not statistically significant (P > 0.05). In YF_{33}, egg shape index which hatched female chicks was 78.18%, while of those which hatched male chicks, it was 78.15%, and the difference of 0.03 was not statistically significant (P > 0.05). In contrast to YF_{33}, in OF_{49} eggs which hatched male chicks had higher egg shape index (76.69–76.72%), but the difference (−0.03%) was not significant (P > 0.05). Egg volume, both in YF_{33} and OF_{49}, was greater in case of eggs which hatched male chicks (YF_{33} = 52.82and 3.51 cm^3; OF_{49} = 57.35 and 58.05 cm^3), differences calculated (−0.69 cm^3 and −0.70 cm^3) were not statistically confirmed (P > 0.05).

Indicators (traits)	$X_{ž}$–M_{xs}	Difference	Significance	$t_{exp.}$
Young parent flock (YF33) - sexes				
Egg mass before incubation (g)	58.06–58.83	−0.77	NS	1.738
Egg length (cm)	5.48–5.51	−0.03	NS	0.754
Egg width (cm)	4.29–4.30	−0.01	NS	0.697
Egg shape index (%)	78.18–78.15	0.03	NS	0.212
Egg volume (cm^3)	52.82–53.51	−0.69	NS	1.348
Egg mass loss until day 18 of incubation (g)	6.58–6.63	−0.05	NS	0.785
Egg mass loss until day 18 of incubation (%)	11.34–11.28	0.06	NS	0.757
One day chick mass (g)	38.80–39.57	−0.77	*	2.359
Relative chick share in the egg mass (%)	66.84–67.26	−0.42	NS	1.879
Old parent flock (OF49) - sexes				
Egg mass before incubation (g)	62.92–63.90	−0.98	NS	1.878
Egg length (cm)	5.71–5.73	−0.02	NS	1.085
Egg width (cm)	4.38–4.40	−0.02	NS	1.303
Egg shape index (%)	76.69–76.72	−0.03	NS	0.496
Egg volume (cm^3)	57.35–58.05	−0.70	NS	1.593
Egg mass loss until day 18 of incubation (g)	7.23–7.47	−0.24	*	2.547
Egg mass loss until day 18 of incubation (%)	11.48–11.68	−0.20	*	2.484
One day chick mass (g)	42.14–42.93	−0.79	*	1.984
Relative chick share in the egg mass (%)	66.98–67.16	−0.18	NS	1.543
Young and old parent flock YF33 and OF49) - sexes				
Egg mass before incubation (g)	60.30–61.23	−0.93	*	2.430
Egg length (cm)	5.59–5.61	−0.02	NS	1.261
Egg width (cm)	4.33–4.35	−0.02	NS	1.926
Egg shape index (%)	77.49–77.47	0.02	NS	0.188
Egg volume (cm^3)	54.91–55.66	−0.75	NS	1.837
Egg mass loss until day 18 of incubation (g)	6.88–7.03	−0.15	*	2.486
Egg mass loss until day 18 of incubation (%)	11.41–11.45	−0.06	NS	1.002
One day chick mass (g)	40.34–41.16	−0.82	**	2.988
Relative chick share in the egg mass (%)	66.90–67.21	−0.31	*	2.458
Parent flock age, young – old (YF33 and OF49)				
Egg mass before incubation (g)	58.41–63.38	−4.97	***	16.784
Egg length (cm)	5.49–5.72	−0.23	***	19.842
Egg width (cm)	4.29–4.39	−0.10	***	10.385
Egg shape index (%)	78.16–76.70	1.46	***	18.562
Egg volume (cm^3)	53.14–57.68	−4.54	***	13.264
Egg mass loss until day 18 of incubation (g)	6.60–7.34	−0.74	***	15.196
Egg mass loss until day 18 of incubation (%)	11.31–11.57	−0.26	***	4.551
One day chick mass (g)	39.15–42.51	−3.36	***	15.131
Relative chick share in the egg mass (%)	67.03–67.06	−0.03	NS	0.228

** means p<0.05; ** means p<0.01; *** means p<0.001.*

Table 2.
Difference significance of mean values of some morphometric traits of eggs which hatched chicks (YF_{33} and OF_{49}).

Eggs—chick sex	Flock age	n	r_{xy}	Correlation significance
Eggs which hatched female chicks	YF_{33}	118	0.261**	Weak
	OF_{49}	101	0.430***	Medium
Eggs which hatched male chicks	YF_{33}	100	0.107^{NS}	Very weak
	OF_{49}	90	0.498***	Medium
Total eggs which hatched	YF_{33}	218	0.188**	Very weak
either sex chick	OF_{49}	191	0.442***	Medium
Eggs which hatched female chicks	YF_{33} + OF_{49}	219	−0.286***	Weak
Eggs which hatched male chicks	YF_{33} + OF_{49}	190	−0.353***	Weak
Total eggs which hatched either sex chick	YF_{33} + OF_{49}	409	−0.314***	Weak

*** means p<0.01; *** means p<0.001.*

Table 3.
Phenotype correlation coefficients between egg mass of both flocks (YF_{33} and OF_{49}) and egg shape index.

In both age groups, female chicks had smaller mass (YF_{33} = 38.80–39.57 g; OF_{49} = 42.14–42.93 g), and determined differences were statistically relevant: $P < 0.05$ and $P < 0.01$, respectively. Relative share of the chick in the egg mass in both flocks was also greater in case of male chicks, but these differences were not statistically significant ($P > 0.05$). Regardless of the parent flock age, eggs which hatched female chicks had lower values of the observed morphometric traits, except for the egg shape index (77.49 – 77.47%), which was higher by 0.02% in eggs which hatched female chicks, but the difference determined was not statistically significant ($P > 0.05$).

Contrary to the chick's sex, parent flock's age had more significant effects on observed morphometric traits of both sexes (**Tables 1** and **2**). All observed morphometric indicators of both eggs and newly hatched chicks of both sexes were statistically considerably higher ($P < 0.001$) in OF_{49} than in the younger flock YF_{33}, except the relative chicken share in the egg mass, whereby the relative chicken share was also greater in OF_{49}; however, the obtained difference (−0.03%) was not statistically significant ($P > 0.05$). Egg shape index was higher in case of YF_{33} (78.16 – 76.70%), and the determined difference of 1.46% was statistically significant ($P < 0.001$).

Among all observed morphometric traits, the phenotype correlation (r_p) was calculated between the egg mass before the incubation period and the egg shape index as the most relevant indicator for the purpose of this study. Calculated values for both flocks and both sexes are given in **Table 3**.

There was a very weak, weak, and medium phenotype correlation between the egg mass before incubation and the egg shape index, and the correlation coefficients were statistically confirmed ($P < 0.01$; $P < 0.001$), except between the egg mass of those eggs which hatched female chicks and the egg shaped index in the younger flock YF_{33}, whereby the calculated coefficient (r_{xy} = 0.107) was not statistically confirmed ($P > 0.05$). Apart from the mentioned strength of the phenotype correlation, total correlation on the global level was found between the egg mass before incubation and the mass of the newly hatched chicks, that is, the correlation coefficient values were above 0.900.

4. Discussion

It is an established fact that the egg laying intensity grows with the age of the parent flock, that is, the rate of the egg fertilization and the chick hatchability grows up to a certain flock age, after which point it decreases. Most of the morphometric indicators (egg mass and newly hatched chicken mass in the first place) normally grow throughout the whole production cycle, which was confirmed by our study and the studies of numerous other authors who treated this matter. These fluctuations are more prominent in case of eggs which hatched male chicks.

Abudabos [9] in two broiler parent flocks (Cobb and Ross) and in parent flocks of different ages (26, 44, 32, and 36 weeks of age) report that the egg mass, relative egg mass loss until day 19 of incubation, newly hatched chicken mass, and the share (percentage) of the chick in the egg mass grow with the age in both parent flocks, while the percentage of fertilized eggs and chick hatchability decreases.

Embryo mortality is higher in case of older broiler parents, particularly in case of Cobb hybrid of 44 weeks of age, when the embryo mortality rate recorded was over 12%. Comparably, Ulmer-Franco et al. [8] report that the age of broiler parents of the Cobb 500 hybrids and egg mass have influence on incubation values of eggs for fertilization. With flock's growing age—from week 29 to week 59, egg mass increased from 53.8 to 71.3 g, relative egg mass loss until day 18 of incubation decreased from 12.8 to 11.9%, egg fertilization rate increased from 76.7 to 94.4%, while the chick hatchability out of the number of fertilized eggs decreased from 88.0 to 87.0%. Total embryo mortality was 10.6% (29 weeks old flock) and 9.4% (59 weeks old flock). Also in the Cobb 500 parent flock, 500 [1, 11] report statistically significant ($P < 0.001$) effect of age on the egg mass and newly hatched chicken mass increase, while the largest number of fertilized eggs was observed in the middle of the production cycle (97.05%), followed by the beginning of the cycle (96.09%), and the smallest was recorded in the final stage of the production cycle (93%). The chick hatchability out of the number of incubated and the number of fertilized eggs was also at its peak in the middle of the production cycle and amounted to 81.14 and 83.94%, respectively. Alsobayel et al. [12] report that the genotype (Arbor Acres, Cobb, and Ross) and the age of the broiler parents (30–35, 40–45, 50–55 weeks) have effects on the average egg mass and one-day chicken mass and also on the share of the chicken in the egg mass.

Depending on the type of hybrid (genotype), egg mass fluctuated between 63.7 and 64.7 g, one-day-old chicken mass between 44.5 and 45.4 g, and the share of chicken in the egg mass varied from 68.1 to 69.7%. In all three hybrids, the growth of egg mass (59.3–68.9 g) and chick mass (41.4–48.4 g) simultaneously with the growth of the flock's age was recorded, while the share of the chick in the egg mass was largest in the last weeks (69.7%) and smallest in the middle of the production cycle (687%). Similar chicken share in the egg mass in 10 different broiler hybrid parents—breeds (between 66.9 and 70.4%) is reported by [7]. Furthermore, loss of egg mass up to day 18 of incubation varied from 12.1 to 13.7%.

Pinchasov [3] makes a slightly different conclusion compared with previously mentioned researchers, which is that effects of laying hens age are less significant for the chick mass than the egg mass which has a considerably greater effect on the one-day-old chick mass. In line with these conclusions, [5] point out that physical characteristics of eggs (egg mass, egg shell thickness and porosity, egg shape index) play an important role in the embryo development process and hatchability success rate. Effects of egg mass on the newly hatched chicken mass, embryo mortality, and percentage of the chicken in the egg mass (progression of the egg mass loss during incubation period) are reported by [6] in case of the hybrid Ross SL 2000 parent flock, raised from 35 to 49 weeks of age. Light eggs (average mass 54.59 g) hatched chicken of 38.11 g of average mass, medium eggs (58.89 g) hatched the chicks of 40.74 g, and heavy eggs (63.10 g) hatched the chicks of 43.18 g of average mass. Loss due to embryo mortality was greater in light (7.83%) and heavy eggs (7.90%), in comparison with eggs of medium weight (6.67%). Furthermore, heavy (large) eggs had the greatest egg mass loss during incubation period. i.e., the smallest eggs had the largest share of the chick mass in the egg mass (69.81%), followed by the eggs of medium weight (69.17%), while the largest eggs had the smallest share (68.43%). Wilson [4, 6, 10] report the relative share of the chick mass in the egg mass fluctuating from 67 to 70%, from 62 to 76%, and from 68.43 to 69.81%, respectively. Similar conclusions related to effects of age and the egg weight group on incubation values in the Hubbard Classic broiler parents were made by [13, 14]. In the light egg weight group (average mass 63.09 g), average mass of newly hatched male chicks was 43.33 g; in the medium weight group (68.85 g), it was 48.40 g; and in the heavy weight group (74.81 g), it was 52.36 g. In case of female chicken, the average mass was 43.29, 48.24, and 52.38 g, respectively. The relative share of the chicken mass in the egg mass was 69.97, 70.22, and 70.38% (male chicks), and 69.84, 70.12, and 70.17% (female chicks).

In an interesting research, [5, 15] indicate considerable effects of the egg shape index on incubation indicators and point out that eggs of anomalous shape should not be used for incubation because they prevent normal embryo development which results in a higher embryo mortality rate during incubation period. This conclusion was confirmed by [16] who studied mechanical (physical) and incubation values of eggs of two hen breeds (Naked Neck and Sombor kaporka) raised in semi-intensive keeping system. They report egg shape index of 71.01 and 72.04%, respectively, for the two breeds and 73.61 and 76.68% for the relative share of the chicken in the egg mass. Abanikannda and Leigh [17] report higher egg shape index of eggs which hatched male chicks (Anak and Marshall hybrids) than in eggs which hatched female chicks (75.25 and 74.53%; 76.27 and 76.00%). Regarding the eggs coming from the Ross hybrid, egg shape index was 76.09% (male chicks) and 76.41% (female chicks).

Based on presented and discussed results of cited authors, we can conclude that most of them studied or reported effects of nongenetic factors on incubation values and morphometric egg indicators and production of one-day-old chicken of both sexes of different genotypes (hybrids) of broiler parents. This is understandable because hybrids of both sexes are used for the broiler meat production. However, when it comes to light hybrids egg incubation, male chicks are killed, while females are used for breeding and raising, i.e., for the commercial production of eggs for consummation.

Kocevski et al. [25, 26] studied effects of age on the mass, strength, fertility, and hatchability of eggs in two light line hybrids (ISA Brown and DeKalb White) on production of eggs for consummation and for fertilization.

Egg mass was considerably affected ($P \leq 0.05$) by age, but not by the line (genotype), although eggs of the (commercial) ISA Brown line were somewhat heavier than those of the DeKalb White line. The heaviest egg mass was that of older poultry, while eggs laid by younger hens had smaller mass. In younger commercial laying hens and parent [...] flocks laying hens produced eggs with stronger shells than those laid by older hens. Average chicken hatchability during the 6 months period of egg production was 70.50% (Isa Brown) and 73.64% (DeCalb). Zita et al. [29] report egg mass to grow with hen's age in three genotypes (ISA Brown, Hisex Brown, and Moravia BSL), while the egg shape index decreased.

Average egg mass of the ISA Brown laying hens was 54.00 g at the beginning of the production cycle, 62.78 g in the middle, and 63.42 g at the end of the production cycle (54–60 weeks of age), while the egg shape index decreased from 78.52 to 76.64%, and 75.09%. Duman et al. [30] are of opinion that the standard egg shape index of commercial laying hens fluctuates between 72.2 and 75.9% (average value –74.3%), of the pointy shape between 68.0 and 71.9% (average 71.0%), while the egg shape index of round eggs ranged from 76.1 to 82.3% (average 78.8%).

Furthermore, the authors make a conclusion that the egg shape index affects certain consummation egg qualities and report certain phenotype correlation between these qualities.

Results obtained in our research partly correspond with conclusions of previous authors who primarily studied different genotypes of broiler parents. Research most similar to ours was conducted by [27] with one of the aims to determine chick's potential sex before egg incubation (Super Nick White Layer) using morphometric measurements of eggs (weight, length, width, shape index, volume). Their conclusion is that length, width, shape index, and volume before incubation period have certain effect on the future chick sex, especially egg shape index and egg width and length.

In case of 54 weeks old laying hens, egg shape index was 71.1% (male), 75.5% (female), and 75.3% (both sexes), while the average egg mass was 60.8 g (male), 60.9 g (female), and 60.8 g (both sexes). Two-hundred forty-four (244) chicks of both sexes were hatched out of 300 incubated eggs (the hatchability rate was 81.33%). Terčič and Pestotnik [28] recorded egg incubation values of the hybrid Prelux-G parent younger (24 weeks) and older (65 weeks) flocks. Older parent flock produced heavier eggs (66.63–56.77 g), than the younger one, had higher relative egg mass loss up to week 18 of incubation (11.26–11.14%), male and female chicks were heavier, embryo mortality was higher (13.52–11.36%), while the chick hatchability out of the number of incubated and fertilized eggs was at a lower level. Male-female chick rate was 1.09 in the younger flock and 1.16 in the older flock. In both flocks, male chicks were slightly heavier than the female. In general, our results are to a large extent compatible with results obtained by [27, 28] and even [24]. Narushin [24] reports an average egg shape index of 75.20% (69.70–80.10%) and egg volume of 60.19 cm^3 (52.00–70.40 cm^3) of eggs laid by the 65-week-old Hy-Line Brown hens, regardless of the chick sex.

These results agree with those of [23] who reports the egg length of 5.90 cm, width 4.40 cm, and egg shape index around 75.0%. Additionally, [31–33] analyzed morphometric indicators of the wild birds' eggs with a purpose to determine possible bird sex before the incubation period and reached similar but questionable results.

Certain authors report phenotype correlation between the parent flock's age in hens and even other types of poultry, laying intensity, hatching egg mass, newly hatched offspring mass, and the relative share of the chick in the egg mass. Skewea et al. [2] point out that the egg mass increases with the age of the parent flock in domestic birds and also that eggs of different sizes (mass) have different physical (external) and chemical (internal) characteristics which affect the hatchability percentage out of the number of fertilized eggs and the quality of hatched chicks. Explicitly, one-day-old chick mass is tightly related with the preincubation egg mass, that is, there is a strong correlation between them. Furthermore, heavier chicks have smaller yolk, while lighter chicks have a larger yolk (food reserve), which enables them to survive a longer period of time before they can obtain exogenous food source.

In line with these results, Suarez et al. [35] report total correlation between the age and the egg mass of Arbor Acres broiler parents of 29–57 weeks of age and medium phenotype correlation between the age and chick mass, and the determined correlation coefficient was statistically significant at $P < 0.001$. Farooq et al. [36] report statistically significant ($P < 0.05$) correlation coefficient ($r_p = 0.496$) between the egg mass and one-day chick mass of the pure breed Rhode Island Red. By calculating the phenotype correlation coefficients between the egg mass, chick mass, and the hatchability percentage out of the number of incubated and fertilized eggs, these authors have also confirmed statistically significant correlation ($P < 0.05$), which is to a large extent in line with our results.

In both ethical and economic contexts, as an addition to this discussion, we are including here the studies of [18, 19], who used male chicks of the ISA Brown and Hy-Line hybrids in organic production of broiler meat, where the fattening period lasted 49 and 90 days, and 51 days, and obtained quality organic broiler meat. Gerken et al. [20–22] point out that from the humane and welfare aspect, it is not economically justified to use male chicks of all light line hybrids for fattening in intensive broiler meat production, but in extensive and semiextensive conditions, with a prolonged fattening period, it is possible to produce good-quality meat with higher protein content and lower percent of abdominal fat. Finally, Weissmann et al. [22] report greater average mass of newly hatched male chicks, than of female chicks by 0.5, 2.1, and 0.02 g in three parent flocks of the light type Lohmann—Germany.

5. Conclusion

Based on the research undertaken with an aim to establish effects of the light line hybrid ISA Brown parent flock's age on incubation values of eggs (fertilization and chick hatchability) and morphometric indicators (pre incubation egg mass, length, width, and egg shape index, absolute and relative egg mass loss until day 18 of incubation, hatched chicken mass and relative share of the chick in the egg mass) and their possible effects on the sex of the hatched chicks, two conclusion can be made. First, in comparison with the older flock (OF_{49}), the younger flock (YF_{33}) demonstrated better incubation values of eggs, and all its morphometric indicators were lower, with the exception of the egg shape index which was statistically significantly greater ($P < 0.001$) by 1.46% in the YF_{33}. Second conclusion is that the eggs which hatched female chicks (YF_{33} + OF_{49}) had lower values of observed morphometric traits than those with male chicks, except the egg shape index.

Author details

Milena Milojević*, Živan Jokić and Sreten Mitrović
Faculty of Agriculture, University in Belgrade, Belgrade, Serbia

*Address all correspondence to: milojevic.milena23@gmail.com

References

[1] Mitrović S, Radojčić Dimitrijević M, Perić L, Stanišić G, Pandurević T. Influence of the Cobb 500 hybrid parent age and egg storage period on incubation parameters. Biotechnology in Animal Husbandry. 2017;**33**(4):409-423

[2] Skewea PA, Wilson HR, Mather FB. Correlation among egg weight, chick weight, and yolk sac weight in Bobwhite quail (*Colinus virginianus*). Florida Scientist. 1988;**51**:159-162

[3] Pinchasov Y. Relationship between the welght of hatching eggs and subsequent early performance of broiler chicks. British Poultry Science. 1991;**32**:109-115

[4] Wilson HR. Interrelationships of egg size, chick size, post hatching growth and hatchability. World's Poultry Science Journal. 1991;**47**:5-20

[5] Narushin VG, Romanov MN. Egg physical characteristics and hatchability. World's Poultry Science Journal. 2002;**58**:297-303

[6] Miclea V, Zăhan M. Eggs weight influence on the incubation of light hen breeds eggs. Buletin USAMV-CN. 2006;**63**:107-110

[7] Wolanski NJ, Renema RA, Robinson FE, Carney VL, Fancher BI. Relationship Among Egg Characteristics, Chick Measurements, and Early Growth Traits in Ten Broiler Breeder Strains. Poultry Science. 2007;**86**:1784-1792

[8] Ulmer-Franco AM, Fasenko GM, O'Dea Christopher EE. Hatching egg characteristics, chick quality, and broiler performance at breeder flock ages and from 3 egg weights1. Poultry Science. 2010;**89**(12):2735-2742

[9] Abudabos A. The effect of broiler breder strain and parent flock age on hatchability and fertile hatchability. International Journal of Poultry Science. 2010;**9**(3):231-235

[10] Traldi AB, Menten JFM, Silva CS, Rizzo PV, Pereira PWZ, Santarosa J. What determines hatchling weight: Breeder age or incubated egg weight? Brazilian Journal of Poultry Science. 2011;**13**(4):283-285

[11] Mitrović S, Pandurević T, Stanišić G, Đekić V, Đermanović V, Jež G. The effect of the broiler parents age and the period of egg storage in incubation indicators. In: Third International Scientific Symposium Agrosym Jahorina 2012. 2012. pp. 559-565

[12] Alsobayel AA, Almarshade MA, Albadry MA. Effect of breed, age and storage period on egg weight, egg weight loss and chick weight of commercial broiler breeders raised in Saudi Arabia. Journal of the Society of Agricultural Science. 2013;**12**:53-57

[13] Iqbal J, Knan SH, Mukhtar N, Ahmed T, Pasha RA. Effects of egg size (weight) and age on hatching performance and chick quality of broiler breeder. Journal of Applied Animal Research. 2016;**44**(1):54-64

[14] Iqbal J, Mukhtar N, Ur Rehman Z, Hassan Khan S, Ahmed T, Sefdar Anjum M, et al. Effect of egg weight on the egg quality, chick quality, and broiler performance at the later stages of production (week 60) in broiler breeders. The Journal of Applied Poultry Research. 2017;**26**:183-191

[15] Aşcı E, Durmuş I. Effect of egg shape index on hatching characteristics in hens. Turkish Journal of Agriculture–Food Science and Tehnology. 2015;**3**(7):583-587

[16] Mitrović S, Djermanović V, Djekić V, Milojević M, Simić D. Comparative

studies on the reproductive and productive traits of new hampshire and sombor crested chicken breeds reared in semi-extensive production system. In: Proceedings of the International Symposium on Animal Science 2014, Belgrade-Zemun. 2014. pp. 61-67

[17] Abanikannda OTE, Leigh AO. Statistical determination of chick sex from pre-hatch egg measurements. International Journal of Livestock Production. 2015;**6**(7):87-90

[18] Lichovníková M, Jandásek J, Jůzl M, Dračková E. The meat quality of layer males from free range in comparasion with fast growing chickens. Czech Journal of Animal Science. 2009;**54**:490-497

[19] Choo YK, Oh ST, Lee KW, Kang CW, Kim HW, Kim CJ, et al. The growth performance, carcass characteristics, and meat quality of egg-type male growing chicken and white-mini broiler in comparison with commercial broiler (Ross 308). Korean Journal for Food Science of Animal Resources. 2014;**34**(5):622-629

[20] Gerken M, Jaenecke D, Kreuzer M, Martin DG. Growth, behavior and carcass characteristics of egg-type cockerels compared to male broilers. World's Poultry Science Journal. 2003;**59**(1):46-49

[21] Koening M, Hahn G, Damme M, Schmutz M. Utilization of laying-type cockerels as coquelets: Influence of genotype and diet characteristics on growth performance and carcass composition. Archiv fur Geflugelkunde. 2012;**76**(3):197-202

[22] Weissmann A, Förster A, Gottschalk J, Reitemeir S, Krautwald-Junghanns ME, Preisinger R, et al. *In ovo*-gender identification in laying hen hybrids: Effects on hatching and production performance. European Poultry Science. 2014;**78**:1-12

[23] Grashorn M. Geflügelzucht. In: Scholtyssek S, editor. Geflügel. Stuttgart: Ulmer Verlag; 1987. pp. 176-215

[24] Narushin VG. Egg geometry calculation using the measurements of length and breadth. Poultry Science. 2005;**84**:482-484

[25] Kocevski D, Nikolova N, Kuzelov A. The Influence of strain and age on some egg quality parameters of commercial laying hens. Biotechnology in Animal Husbandry. 2011;**27**(4):1649-1658

[26] Nikolova N, Kocevski D, Kuzelov A. Influence of genotype on eggshell strength and the hatchability of laying parent stock flock. Biotechnology in Animal Husbandry. 2011;**27**(4):1659-1666

[27] Yilmaz-Dikmen B, Dikmen S. A morphometric method of sexing white layer eggs. Brazilian Journals of Poultry Science. 2013;**15**(3):169-286

[28] Terčić D, Pestotnik M. Effects of flock age, prestorage heating of eggs, egg position during storage and storage duration on hatchability parameters in layer parent stock. Acta Agriculturae Slovenica. 2016;**5**:132-142

[29] Zita L, Tůmová E, Štolc L. Effects of genotype, age and their interaction on egg quality in brown-egg laying hens. Acta Veterinaria Brno. 2009;**78**:85-91

[30] Duman M, Şekeroğlu A, Yıldırım A, Eleroğlu H, Camcı Ö. Relation between egg shape index and egg quality characteristics. European Poultry Science. 2016;**80**:1/9-9/9

[31] Cordero PJ, Griffith SC, Aparicio JM, Parkin DT. Sexual dimorphism in house sparrow eggs. Behavioral Ecology Sociobiology. 2000;**48**:353-357

[32] Burnham W, Sandfort C, Belthoff JR. Peregrine falcon eggs: Egg

size, hatchling sex, and clutch sex ratios. Condor. 2003;**105**:327-335

[33] Lislev T, Byrkjedal I, Borge T, Saetre GP. Egg size in relation to sex of embryo, brood sex ratios and laying sequence in northern lapwings Vanellus vanellus. Journal Zoological Society of London. 2005;**267**:81-87

[34] Hoyt DF. Practical methods for estimating volume and fresh weight of bird eggs. Auk. 1979;**96**:73-77

[35] Suarez ME, Wilson HR, Mather FB, Wilcox CJ, Mcpherson BN. Effect of strain and age of the broiler breeder female on incubation time and chick weight. Poultry Science. 1997;**76**:1029-1036

[36] Farooq M, Durrani FR, Aleem M, Chand N, Muquarrb AK. Egg traits and hatching performance of Desi, Fayumi and Rhode Island Red xhicken. Pakistan Journal of Biological Sciences. 2001;**4**:909-911